Tissue Elasticity Imaging
Volume 1: Theory and Methods

Tissue Elasticity Imaging

Volume 1: Theory and Methods

Edited by

S. Kaisar Alam
Imagine Consulting LLC
Dayton, NJ, United States

The Center for Computational Biomedicine Imaging
and Modeling (CBIM)
Rutgers University
Piscataway, NJ, United States

Brian S. Garra
Division of Imaging, Diagnostics, and Software Reliability
Office of Science and Engineering Laboratories, Center for Devices
and Radiological Health, FDA, Silver Spring, MD, United States

ELSEVIER

Elsevier
Radarweg 29, PO Box 211, 1000 AE Amsterdam, Netherlands
The Boulevard, Langford Lane, Kidlington, Oxford OX5 1GB, United Kingdom
50 Hampshire Street, 5th Floor, Cambridge, MA 02139, United States

Notices
Knowledge and best practice in this field are constantly changing. As new research and
experience broaden our understanding, changes in research methods, professional
practices, or medical treatment may become necessary.

Practitioners and researchers must always rely on their own experience and knowledge in
evaluating and using any information, methods, compounds, or experiments described
herein. In using such information or methods they should be mindful of their own safety
and the safety of others, including parties for whom they have a professional
responsibility.

To the fullest extent of the law, neither the Publisher nor the authors, contributors, or
editors, assume any liability for any injury and/or damage to persons or property as a
matter of products liability, negligence or otherwise, or from any use or operation of any
methods, products, instructions, or ideas contained in the material herein.

Library of Congress Cataloging-in-Publication Data
A catalog record for this book is available from the Library of Congress

British Library Cataloguing-in-Publication Data
A catalogue record for this book is available from the British Library

ISBN: 978-0-12-809661-1

For information on all Elsevier publications visit our website at
https://www.elsevier.com/books-and-journals

Publisher: Susan Dennis
Acquisition Editor: Anita Koch
Editorial Project Manager: Lindsay Lawrence
Production Project Manager: Paul Prasad Chandramohan
Cover Designer: Matthew Limbert

Typeset by TNQ Technologies

Contents

Contributors..xi
About the editors...xiii
Foreword..xv
Preface xvii
Acknowledgments...xxi

CHAPTER 1   An early history of elasticity imaging......................1
Robert M. Lerner
1. Overview and personal observations from a radiologist..............1
2. Early history of tissue elasticity determination 3
   2.1 Palpation...3
   2.2 Oestreicher and von Gierke (1950s)..............................3
   2.3 Early tissue motion studies (1970 to mid-1980s)..............4
3. The early era of imaging tissue stiffness (late 1980s to
   mid-1990s)..6
   3.1 Vibration amplitude sonoelastography (sonoelasticity)........6
   3.2 Compression elastography...7
4. Quantitative tissue stiffness determination and imaging
   (1990 to present)..7
   4.1 Quantitative tissue stiffness determination.....................7
   4.2 Further expansion of tissue elasticity imaging
       (1994 to present) 8
   4.3 Microscopic tissue elasticity imaging............................10
5. Conclusion/discussion..10
Acknowledgments..11
References...12

CHAPTER 2   The governing theory of elasticity imaging..............17
Salavat R. Aglyamov
1. Introduction..17
2. Displacement and strain..17
3. Forces, stress, and equilibrium equation 21
4. Stress-strain relation for an elastic material and elastic
   constants...22
5. Equilibrium equations 24
6. Cylindrical and spherical coordinate systems.........................25
7. Basic solutions...26
   7.1 Uniaxial deformation...26

7.2 Uniaxial deformation of the tissue with spherical
   inclusion 28
8. Dynamic deformation..30
9. Examples of dynamic problems 32
   9.1 Plane wave propagation 32
   9.2 Motion of solid sphere under dynamic load...................33
10. Viscoelastic models 34
   10.1 Viscoelastic tissue response..............................34
   10.2 Viscoelastic response to harmonic excitation...............38
   10.3 Generalized viscoelastic models..........................38
11. Dynamic deformation of a viscoelastic medium...................39
12. Summary 42
References..42

CHAPTER 3   Vibration sonoelastography 45
      Kevin J. Parker
1. Early results..45
2. Theory 46
3. Vibration phase gradient sonoelastography 50
4. Crawling waves 52
5. Clinical results..54
6. Reverberant shear wave fields..................................55
7. Conclusion..56
References..56

CHAPTER 4   Introduction to quasi-static elastography 61
      Paul E. Barbone, Assad A. Oberai and Timothy J. Hall
1. Introduction and background.....................................61
2. Deformation application and measurement 62
   2.1 Motion tracking algorithms................................63
   2.2 Motion tracking performance and error....................65
   2.3 Tracking large deformations 66
3. Interpretation of the measured deformation...........................67
   3.1 The quasi-static approximation............................67
   3.2 Strain..68
   3.3 The elastic approximation 69
   3.4 The one-dimensional assumption.............................70
   3.5 Strain image interpretation 71
4. Beyond linear elastic imaging: biomechanical imaging.............73
   4.1 Nonlinear elastic imaging................................73
   4.2 Poroelastic imaging.....................................74
5. Summary..75
Acknowledgments..76
References..76

CHAPTER 5   Acoustic radiation force and shear wave elastography techniques **85**
Arsenii V. Telichko, Carl D. Herickhoff and Jeremy J. Dahl

1. Introduction 85
2. Physical basis for acoustic radiation force from ultrasonography 86
    2.1 The acoustic radiation force..86
3. Acoustic radiation force imaging techniques...........................90
    3.1 Acoustic radiation force applied to tissues.....................90
    3.2 Acoustic radiation force impulse imaging 92
    3.3 Preliminary applications of acoustic radiation force impulse imaging...97
    3.4 Vibroacoustography...98
    3.5 Preliminary applications of vibroacoustography 101
    3.6 Harmonic motion imaging..102
    3.7 Viscoelastic response imaging and model-based aproaches 102
4. Shear wave elastography techniques...................................104
    4.1 Shear wave generation by radiation force.....................104
    4.2 Shear wave imaging techniques.................................105
    4.3 Shear wave tracking methods....................................114
References..119

CHAPTER 6   Magnetic resonance elastography **129**
Bogdan Dzyubak and Kevin J. Glaser

1. Introduction 129
2. Acquisition 131
    2.1 Generating and delivering mechanical waves................131
    2.2 Imaging the waves with magnetic resonance imaging......132
3. Inversions 134
    3.1 Overview of inversions and processing........................134
    3.2 Magnetic resonance elastographic outputs....................136
4. Applications..138
    4.1 Liver 138
    4.2 Brain 141
    4.3 Tumors 142
    4.4 Other organs 144
5. Artifacts and quality control...145
6. Summary and conclusions 148
References..149

CHAPTER 7 Reconstructive elastography.............................**155**
Marvin M. Doyley
1. Introduction 155
2. Solving the forward elasticity problem............................ 156
3. Solving the inverse elastography problem......................... 156
  3.1 Quasi-static elastography.. 156
  3.2 Dynamic elastography based on local frequency
      estimation.. 157
  3.3 Transient elastography based on arrival time
      estimation.. 158
4. Advanced reconstruction methods 158
  4.1 Viscoelasticity.. 160
  4.2 Nonlinearity... 161
5. Discussion 161
References.. 162

**CHAPTER 8 Lateral and shear strain imaging for ultrasound
            elastography**..**167**
Tomy Varghese
1. Introduction 167
2. Classification of lateral and shear strain estimation methods.... 169
3. One-dimensional deformation tracking and estimation............ 169
  3.1 Incompressibility assumption 169
  3.2 Weighted interpolation and recorrelation approach.......... 170
  3.3 Perpendicular insonification using dual transducers......... 170
  3.4 Altered point spread functions and synthetic aperture
      system 171
  3.5 Angular beam-steered data acquisition approach 171
4. Two-dimensional deformation tracking and estimation............ 173
  4.1 Subsample displacement estimation 173
  4.2 Direct axial and lateral displacement estimation using
      two-dimensional kernels... 173
  4.3 Coupled axial and lateral displacement estimation
      using two-dimensional kernels 174
  4.4 Two-dimensional processing of angular beam-steered
      data acquisitions... 174
  4.5 Model-based elastography 174
  4.6 Minimization and regularization.................................. 174
  4.7 Plane wave compounding and directional
      beamforming.. 175

5. Clinical applications of lateral and shear strain estimation 175

    5.1 Radial and circumferential strain 178

6. Conclusion 178

References 179

CHAPTER 9   Optical elastography on the microscale 185

Philip Wijesinghe, Brendan F. Kennedy and
David D. Sampson

1. Introduction 185

    1.1 Brief history of optical elastography 186

    1.2 Optical elastography: a matter of scale 188

2. Optical coherence elastography 190

    2.1 Optical coherence tomography 190

    2.2 Measuring displacement in optical coherence
        elastography 194

    2.3 Quasi-static optical coherence elastographic methods 196

    2.4 Dynamic optical coherence elastographic methods 200

    2.5 Probe-based optical coherence elastography 203

    2.6 Computational inverse methods in optical coherence
        elastography 205

3. Brillouin microscopy 208

4. Other techniques 212

5. Outlook 214

References 215

Index 231

Contributors

Salavat R. Aglyamov
Department of Mechanical Engineering, University of Houston, Houston, TX, United States

Paul E. Barbone
Department of Mechanical Engineering, Boston University, Boston, MA, United States

Jeremy J. Dahl
Department of Radiology, Stanford University, Stanford, CA, United States

Marvin M. Doyley
Department of Electrical and Computer Engineering, University of Rochester, Rochester, NY, United States

Bogdan Dzyubak
Department of Medical Physics, Mayo Clinic, Rochester, MN, United States

Kevin J. Glaser
Medical Physics, Mayo Clinic, Rochester, MN, United States

Timothy J. Hall
Department of Medical Physics, University of Wisconsin, Madison, WI, United States

Carl D. Herickhoff
Department of Radiology, Stanford University, Stanford, CA, United States

Brendan F. Kennedy
BRITElab, Harry Perkins Institute of Medical Research, QEII Medical Centre, Nedlands, WA, Australia; Department of Electrical, Electronic and Computer Engineering, School of Engineering, The University of Western Australia, Perth, WA, Australia

Robert M. Lerner
Department of Clinical Imaging, University of Rochester, Rochester, NY, United States; Department of Diagnostic Imaging, Rochester General Hospital, Rochester Regional Health, Rochester, NY, United States

Assad A. Oberai
Department of Aerospace and Mechanical Engineering, University of Southern California, Los Angeles, CA, United States

Kevin J. Parker
William F. May Professor of Engineering, Professor of Electrical and Computer Engineering, of Biomedical Engineering, and of Imaging Sciences (Radiology), University of Rochester, Rochester, NY, United States; Dean Emeritus, School of Engineering & Applied Sciences, University of Rochester, Rochester, NY, United States

David D. Sampson
Optical+Biomedical Engineering Laboratory, Department of Electrical, Electronic and Computer Engineering, The University of Western Australia, Perth, WA, Australia; University of Surrey, Surrey, United Kingdom

Arsenii V. Telichko
Department of Radiology, Stanford University, Stanford, CA, United States

Tomy Varghese
Department of Medical Physics University of Wisconsin School of Medicine and Public Health University of Wisconsin—Madison, Madison, WI, United States

Philip Wijesinghe
Optical+Biomedical Engineering Laboratory, Department of Electrical, Electronic and Computer Engineering, The University of Western Australia, Perth, WA, Australia; BRITElab, Harry Perkins Institute of Medical Research, QEII Medical Centre, Nedlands, WA, Australia

About the editors

S. Kaisar Alam, Ph.D.

President and Chief Engineer, Imagine Consulting LLC, Dayton, NJ, United States

Visiting Research Faculty, Center for Computational Biomedicine Imaging and Modeling (CBIM), Rutgers University, Piscataway, NJ, United States

Adjunct Faculty, Electrical & Computer Engineering, The College of New Jersey (TCNJ), Ewing, NJ, United States

Dr. S. Kaisar Alam received his B.Tech (Honors) from IIT, Kharagpur, India. Following a 3-year stint as a Lecturer at RUET, Bangladesh, he came to the University of Rochester, Rochester, New York, for graduate studies and received his M.S. and Ph.D. degrees in electrical engineering in 1991 and 1996, respectively. After spending 3 years (1995–1998) as a postdoctoral fellow at the University of Texas Health Science Center, Houston, Dr. Alam was a Principal Investigator at Riverside Research, New York, from 1998 to 2013, working on a variety of research topics in biomedical imaging. He was the Chief Research Officer at Improlabs Pte Ltd, an upcoming tech startup in Singapore until 2017. Then he founded his own consulting company for biomedical image analysis, signal processing, and medical imaging. He has also been involved in training and mentoring high school students. He has been a visiting research professor at CBIM, Rutgers University, Piscataway, New Jersey (since 2013), a visiting professor at IUT, Gazipur, Bangladesh (2010 and 2012), and an adjunct faculty at The College of New Jersey (TCNJ), Ewing, New Jersey (since 2017).

Dr. Alam has been active in research for more than 30 years. His research interests include diagnostic and therapeutic applications of ultrasound and optics, and signal/image processing with applications to medical imaging. The areas of his most active research include elasticity imaging and quantitative ultrasound; he is among a few researchers with experience in both quasistatic and dynamic elasticity imaging. Dr. Alam has written over 40 papers in international journals and holds several patents. He is a coauthor of the textbook *Computational Health Informatics* (to be published late 2019 or early 2020 by *CRC Press*). He is a Fellow of AIUM, a Senior Member of IEEE, and a Member of Sigma Xi, AAPM, ASA, and SPIE. Dr. Alam has served in the AIUM Technical Standards Committee and the Ultrasound Coordinating Committee of the RSNA Quantitative Imaging Biomarker Alliance (QIBA). He is an Associate Editor of *Ultrasonics* (Elsevier) and *Ultrasonic Imaging* (Sage). Dr. Alam was a recipient of the prestigious Fulbright Scholar Award in 2011–2012.

Brian S. Garra, M.D.

Division of Imaging, Diagnostics, and Software Reliability, Office of Science and Engineering Laboratories, Center for Devices and Radiological Health, FDA, Silver Spring, MD, United States

Dr. Brian S. Garra completed his residency training at the University of Utah and spent 3 years as an Army radiologist in Germany before returning to Washington DC and the National Institutes of Health in the mid 1980s. After 4 years at the NIH, he joined the faculty of Georgetown University as Director of Ultrasound. In 1998, he left Georgetown to become Professor & Vice Chairman of Radiology at the University of Vermont/Fletcher Allen Healthcare. In 2009, Dr. Garra returned to the Washington DC area as Chief of Imaging Systems & Research in Radiology at the Washington DC Veterans Affairs Medical Center. In April 2010, he also joined the FDA as an Associate Director in the Division of Imaging and Applied Mathematics/OSEL. In 2018, he left the VA and currently splits his time between the FDA and private practice radiology in Florida.

Dr. Garra's clinical activities include spinal MRI and general ultrasound. His research interests include PACS, digital signal processing, and quantitative ultrasound including Doppler, ultrasound elastography, and photoacoustic tomography. He was chair of the FDA radiological Devices Panel from 1999 to 2002 and has been involved in the approval of several new technologies including high resolution breast ultrasound, the first digital mammographic system, the first computer-aided detection system for mammography, and the first computer-aided nodule detection system for chest radiographs as well as the ultrasound contrast agent albunex. He also led the team that developed the AIUM breast ultrasound accreditation program, and helped develop the ARDMS registry in breast ultrasound. He is currently also Vice Chairman of the Ultrasound Coordinating Committee of the RSNA Quantitative Imaging Biomarker Alliance (QIBA) and is the Principal Author of the forthcoming QIBA Ultrasound Shear Wave Speed Profile which will provide a standard approach to acquisition of shear wave speed data for research, clinical application, and regulatory testing.

Foreword

Given the heavy relatively successful use of manual palpation over the past few thousand years, the ultrasound community, and medicine in general, was very excited to understand and realize the possibility of measuring and imaging the stiffness of tissues. This included tissues too deep for manual palpation. Improving the spatial and quantitative fidelity of elasticity images was addressed aggressively. Also pursued were many extensions related to elastic properties, such as the anisotropy of elasticity, the complex elastic modulus (viscous and elastic components), and elasticity as a function of time under compression.

This two-volume book *Tissue Elasticity Imaging* extensively covers the principles, implementation, and applications of all these approaches to image the biomechanical properties of tissues. The achieved and future biomedical applications of these many capabilities are also well explained, as are important optical and magnetic resonance imaging techniques that followed, and that sometimes leaped ahead of the many ultrasound developments.

These rapid advances are brought to life for the reader of these books by physicians and other imaging scientists and engineers who made leading advances in each of the covered areas. I initially wished to list key lead authors with a summary of their contributions, but that would essentially be repeating most of the table of contents. The editors of these books, Drs. Brian Garra and S. Kaisar Alam, excelled in recruiting the many luminaries to author the various chapters, defining the topics, and editing the work for readability by the target audience of imaging scientists, engineers, entrepreneurs, clinicians, and operators of the systems. The work should serve as a definitive reference for those teaching and those writing shorter explanations for various groups. This is a much-needed work in the field. Luckily, it will not be the last, as advances are and will continue to be made.

Paul L. Carson, Ph.D.
University of Michigan
Ann Arbor, Michigan
United States
July 14, 2019

Preface

Since its modest beginning in the late 1980s to early 1990s, elastography has gained wide acceptance in many clinical applications, e.g., detection, diagnosis, and treatment monitoring. To assess the growth of elastography, we performed a PubMed search for "elastography." The total number of results was 4711 if we searched only the title. We have observed that some papers on elastography do not include "elastography" in the title but include it in the abstract. Accordingly, we also performed a title/abstract search for "elastography": the number of papers went up to 7912. To provide a perspective on the rapid growth, these numbers were 1 and 1, respectively, if we limited the search to only the year 1991. These numbers increased to 16 and 22 (title/abstract) in 2001, 265 and 399 (title/abstract) in 2011, and 729 and 1305 (title/abstract) in 2018. Clearly from these yearly numbers, the ascent of elastography has been rapid, especially during the last decade.

Physicians have known for a long time that tissue elasticity changes with (or due to) disease and routinely used palpations to aid in diagnostic evaluations. If the reader ever went to a physician with an abdominal complaint, the physician probably palpated the abdomen, including the liver. Hippocrates (a Greek physician who lived during Greece's Classical period and is widely regarded as the "father of medicine") wrote about abdominal swellings in *The Book of Prognostics*: "...Such swellings as are soft, free from pain, and yield to the finger...and are less dangerous than the others. ...then, as are painful, hard, and large, indicate danger of speedy death; but such as are soft, free of pain, and yield when pressed with the finger, are more chronic than these."

Manual palpation, however, is subjective and highly dependent on the physician expertise. The measurements are nonquantitative and not very useful for small or deep lesions. Several researchers explored the clinical use of tissue elasticity in the 1980s. Eventually, Robert Lerner and Kevin Parker published the first journal paper on dynamic elastography (vibration sonoelastography) in 1988. Jonathan Ophir introduced quasi-static elastography in 1991. Many other elastography variants have been invented since then, and a brief history describing many of them may be found in Chapter 1 of Volume 1. Elastography methods do not typically suffer from the limitations of manual palpation. Furthermore, quantitative elastography allows objective monitoring of change over time. Typically, medical imaging modalities measure and display parameters that vary only a few percent between normal and pathological tissues. In contrast, elastography modalities (especially the modalities that image a modulus) can exploit parameter ranges of up to six orders of magnitude! Elastography is probably the only modality with this (very large dynamic range) advantage.

Dr. Brian Garra and I have been involved with elastography since its early days. We discussed editing a reference book on elastography several times in the past. We felt a few of years ago that the time was finally right for us to put this book together. As an Associate Editor of the Elsevier Journal *Ultrasonics*, I knew our Publisher (at the time) Ysabel Ermers. We approached Ysabel, and she put us in touch with Elsevier's Acquisition Editor Dr. Anita Koch. With Anita's help, we finalized the plan for the book. The book was approved soon afterward. Brian and I wanted the book to be useful for introducing someone to elasticity imaging as well as a reference for someone more advanced in the art. Some of the specifics in the chapters of both volumes will become somewhat outdated within a short time. However, the basics and the general information will remain useful. The readers can search the Internet (e.g., Google, PubMed, etc.) and contact the authors in this book and other experts for guidance on the state of the art. The readers can also consult the companion website for this book at https://www.elsevier.com/books-and-journals/book-companion/978-0-12-809661-1.

There were many options with respect to the organization of the book. We decided to divide the book into two volumes. Volume 1 discusses theory and methods of elasticity imaging, and Volume 2 discusses clinical applications of elasticity imaging modalities. In Volume 1, Chapter 1 takes the readers through a brief history of elastography, starting with some discussion about preimaging days. Chapter 2 provides a unified view of the governing theory of elastography. (Individual chapters in Volume 1 have expanded on the theory for each modality, as needed.) Chapter 3 describes vibration sonoelastography, the first elasticity imaging method. It is followed by a detailed description of quasi-static elastography in Chapter 4. A thorough treatment of dynamic elastography techniques based on acoustic radiation force and shear wave is provided in Chapter 5. Chapter 6 describes magnetic resonance elastography. Inverse problems and modulus construction are briefly treated in Chapter 7. Chapter 8 describes lateral and shear strain imaging. The volume concludes with a detailed chapter on optical elastography (Chapter 9).

In Volume 2, nine chapters discuss several major clinical applications of elastography. This volume can also serve to introduce basic scientists to an array of clinical applications, their current challenges, and future prospects. Even after three decades of development, elastography is a rapidly expanding field. Given the ever-increasing number of labs, researchers, and commercial endeavors, we believe that such progress (in new methods and clinical applications) is likely to continue for many years.

We recruited leading researchers to write the chapters and would like to thank all the authors who contributed. In addition, we would like to thank the reviewers who provided helpful comments for all the chapters. Their service was crucial in ensuring the quality of the chapters. The names of the reviewers are indicated below in an alphabetical order to acknowledge their service.

S. Kaisar Alam
Dayton, New Jersey, USA
October 1, 2019

Chapter reviewers:
Volume 1: Theory and methods
Arun K. Thittai
Assad A. Oberai
David Bradway
EEW Van Houten
Guy Cloutier
James F. Greenleaf
Jean-Luc Gennisson
Kirill Larin
Mark Palmeri
Marvin M. Doyley
Matthew Urban
Michael Richards
Salavat Aglyamov
Thomas A. Krouskop
Tom Seidl
Tomek Czernuszewicz
Yogesh Kannan Mariappan

Acknowledgments

Editing this important reference book was much harder and at the same time, much more fulfilling than I could have ever imagined. First and foremost, I want to thank the Almighty. He gave me the power to pursue my dreams and this book. I could never have done this without my faith in Him. This book happened because He wished it to be.

I am ever grateful to my deceased parents who always encouraged me to pursue my dreams. Thank you my dear wife, daughter, and son for your constant patience and support, especially during difficult times. My younger brother and sister have been my source of strength since they were born. Their spouses and children have been a source of inspiration and joy for me. I have a large number of uncles, aunts, cousins, nephews, and nieces, who have always supported me. I am lucky to have all of you as my family.

I also want to thank many individuals whom I regard as mentors and friends. They include my childhood mentor Dr. Kazi Khairul Islam, my doctoral advisor Dr. Kevin J. Parker, my postdoc supervisor late Dr. Jonathan Ophir, my former supervisors Dr. Ernie Feleppa and late Dr. Fred Lizzi, and my coeditor Dr. Brian Garra. (Brian also provided the artwork used to design the cover.).

I am also indebted to many family members, friends, and colleagues, and it would be impossible to thank them all individually. I am lucky to have been your family, friend, and colleague. Thank you all!

Last but not the least, thanks to everyone in the Elsevier team. Special thanks to our Acquisition Editor (Dr. Anita Koch), Editorial Project Managers (Lindsay Lawrence, Jennifer Horigan, and Amy Clark), Project Manager (Paul Prasad Chandramohan), Cover Designer (Matthew Limbert), and many other individuals who worked behind the scenes to make this book a reality.

S. Kaisar Alam
Dayton, New Jersey, USA
October 1, 2019

An early history of elasticity imaging

Robert M. Lerner[1,2]

[1]*Department of Clinical Imaging, University of Rochester, Rochester, NY, United States;*
[2]*Department of Diagnostic Imaging, Rochester General Hospital, Rochester Regional Health, Rochester, NY, United States*

1. Overview and personal observations from a radiologist

Tissue elasticity imaging provides medical or biological images with pixels that qualitatively or quantitatively correspond to measures of tissue stiffness related to clinical palpation. Qualitative elasticity imaging refers to a region of interest that responds differently to a perturbing force than the adjacent tissue, whereas quantitative elasticity imaging refers to a region of interest where a measured value is assigned that is a fundamental mechanical property such as elasticity. It was developed to provide objective biomechanical information to complement conventional medical ultrasonographic imaging. Conventional medical ultrasonographic imaging (B-scan) is based on relative amplitudes of backscattered longitudinal waves (echogenicity), which show no direct correlation to organ and lesion stiffness [1,2]. To better understand tissue elasticity imaging's place in history, a brief description of medical ultrasonography and tissue stiffness considerations is necessary.

Medical ultrasonographic imaging equipment initially mimicked underwater sonar using the speed of sound in water for designing ultrasonographic equipment as an echolocation system presuming biological tissue behaved much like water. When considering a relevant biological tissue as composed mainly of water, the optimum parameters for medical imaging required low megahertz frequencies for adequate depth of penetration of longitudinal waves into deep tissues, with minimal attenuation and with best image detail (smallest wavelength). Initially, bistable images of anatomy were produced from specular reflected echoes originating from organ boundaries and interfaces based on longitudinal wave (bulk modulus) impedance mismatches. These images were able to identify fluid, gas, bone, or calcium but could not differentiate specific organs by their echogenicity. By quantifying the strengths of the weak backscattered echo amplitudes from the small scattering centers in tissue, grayscale (B-scan or brightness mode) images were produced that depicted patient anatomy in a range of contrast detail with different echogenicity patterns. Although some organs in the normal state had characteristic relative echogenicity when compared with other organs, no elastic or viscous mechanical tissue properties could be gleaned from the images, as echogenicity is a complex

Tissue Elasticity Imaging. https://doi.org/10.1016/B978-0-12-809661-1.00001-7

interaction of the basic speckle pattern of the ultrasonographic instrument and the distribution and strength of the scattering centers in the tissue/organ [3,4]. The images of relative tissue echogenicity based on bulk impedance (longitudinal wave propagation) mismatches have been extraordinarily useful in medicine, depicting normal and abnormal anatomy, organ size, focal lesions, abnormal masses, fluid collections, relative tissue motion, and subjective compliance of tissues to applied transducer pressure. Because shear waves do not propagate through water and are rapidly attenuated in biological tissues at megahertz frequencies [5], they were not considered useful for medical ultrasonographic.

Recognition that subjective tissue stiffness by palpation was not correlated to echogenicity became the motivation for developing a method to image stiffness after a pilot study suggested tissue stiffness as detected by clinical palpation was a better predictor of prostate cancer than echogenicity [1,2,6].

"If it's not hard, it's not cancer" was a quote by Charles Huggins (Nobel laureate for the discovery of the hormonal treatment of prostate cancer) that was subsequently related to the author by Harry Fischer, MD, a noted X-ray contrast media researcher and former chairman of the Radiology Department at the University of Rochester.

Clearly, there was more to tissue stiffness (hardness) than was being depicted by changes in echogenicity on conventional ultrasonographic equipment. This led to an early project to image objective tissue stiffness with ultrasonography, which would be independent of echogenicity. Complementary studies to measure the elastic properties of prostate tissue in vitro as compared to pathology were subsequently reported [1,7—10].

My initial studies (circa 1982 in collaboration with Professor Robert Waag of the University of Rochester) started with graded compression of a stack of two sponges, one stiffer than the other, with detection of the radio frequency (RF) ultrasound signals from regions in each sponge subjected to increasing degrees of compression. Offline computer processing of the data to detect the correlation length of the sponge scattering elements from each sponge showed that for a certain degree of compression, the stiff sponge maintained its correlation length, whereas, at the same compression level, the softer sponge could no longer show a correlation length. Although the concept showed promise in distinguishing a hard from soft sponge, the computational time was long and seemed too complex at that time to consider for real-time medical imaging and that approach was abandoned.

The realization that there was a need for a method of imaging tissue stiffness in "real time" for practical radiologic imaging led to experiments where tissue-mimicking phantoms were subjected to low-frequency vibrations or mechanical thrusts (low audible range, 50—200 Hz). The vibrations propagated into the phantoms from below with an above range-gated Doppler ultrasound transducer detecting the peak local velocities as a function of distance through the specimen. By translating the transducer and vibration source across the specimen in increments that matched the Doppler gate, a two-dimensional image was created of relative stiffness [2,11] (also see Chapter 3 of this volume). Related work (see Section 2) to characterize tissue stiffness was also underway by other researchers, although not with the stated goal of producing images of tissue stiffness.

2. Early history of tissue elasticity determination

2.1 Palpation

Palpation is the examination of structures by touching the surface over an area of concern with the goal of identifying and characterizing the deeper tissues. It has been an important physician skill for the detection of underlying anatomic and pathologic conditions for thousands of years [12].

Palpation can give an indication of organ size and also may detect internal organ abnormalities such as overall stiffness (which may relate to fibrosis, scarring, tumor, or inflammation) or focal abnormalities such as tumors or nodules. Abnormalities detected outside organs include tumors, fluid collections (abscesses, hematomas, cysts, seromas), and bone abnormalities. Vascular pulsations may also be detected by putting a finger over an artery for pulse rate and rhythm determination and estimation of blood pressure. Two examples of medical tests that may be viewed as quantitative palpation (initially performed by clinical palpation but with poor reproducibility and accuracy) are blood pressure measurements and ocular tonometry, which detect intravascular blood pressure and intraocular pressure by measuring a response to an applied pressure or force, respectively. For example, in blood pressure measurements, systole is determined as the lowest pressure that allows a pulse sound to be detected by a stethoscope or Doppler ultrasonography as the pressure in the cuff is reduced from a high enough pressure to obstruct the pulse or flow. The intraocular pressure is determined by measuring a deformation of the cornea to a known pressure pulse (puff of air) and comparing to a standard.

A variety of early attempts to investigate tissue stiffness in a relative or quantitative manner that would correlate to palpation were explored for several decades by researchers using instrumentation to objectively monitor strain (change in dimension per unit of initial dimension after stress is applied) or motion imparted to tissues using various perturbing stresses (force fields per unit area). See Chapters 2 and 4 of this volume for details of how stress and strain precisely relate to elasticity.

2.2 Oestreicher and von Gierke (1950s)

von Gierke et al. used a strobe light and camera to image surface wave propagation patterns over the human thigh, produced by a piston source in contact with the skin at 64 Hz. The patterns were recorded at a distance from the focal harmonic force applied to the skin. Surface wavelength and wave speed were determined [13], which could be related to material properties of an ideal semi-infinite medium [14].

They did not relate the surface wave speed to shear waves (slow waves) or longitudinal waves (fast waves) but it is apparent (from the wave speeds recorded as approximately several meters per second) that they were observing effects of shear wave disturbances in the tissue and the longitudinal waves were not detected. This was the first quantitative experimental observation of surface wave propagation in humans. Surface waves are recognized to be associated with speeds near shear waves [14].

This work followed a very elegant mathematic treatment of an oscillating sphere in a viscoelastic medium by Oestreicher [15]. The work was largely unexploited until its application as a foundation for tissue elasticity imaging was recognized and applied to enhance the current understanding of the basic science of tissue elasticity imaging [16,17] (also see Chapters 2, 3, and 5 of this volume).

2.3 Early tissue motion studies (1970 to mid-1980s)

Although not explicitly stated, the implication for these studies was the presumption that tissue motion resulting from applied force fields could ultimately allow for the determination of objective relative (qualitative) or absolute (quantitative) tissue stiffness (and other mechanical properties, e.g., viscosity) when appropriate stress-strain physical models were applied. The applied force fields were from a variety of sources such as transmitted cardiac pulsations and controllable external forces from mechanical pistons, acoustic horns, speakers, or puffs of air. Internal tissue motion was detected by ultrasonography for these early studies, although surface wave motion had been explored by photographic techniques also.

2.3.1 Instrument-enhanced palpation

An objective relative assessment of tissue response to external force was attempted in the early stages of medical ultrasonography by clinical investigators who palpated tissues during scanning with static B-scanners, interleaving global images with selected areas where tissue motion was depicted by M-mode and later with real-time B-mode, providing a more global assessment of tissue response to simultaneous palpation or compression by the ultrasound transducer [18–23]. These efforts to glean more stiffness information than was readily available from a static image using commercial instrumentation are to be applauded and served to challenge researchers to conceive of more reproducible and objective techniques. This technique is still useful in ultrasound practice for the detection of sliding motion of organs or tumors with respect to other structures.

2.3.2 Tissue stimulation by cardiac pulsations or natural sources

Wilson and Robinson presented an RF M-mode ultrasound signal processing technique to measure small displacements of liver tissue caused by the radial expansion of arteries within the liver from cardiac pulsations. They were able to calculate the velocity of tissue motion from the trajectory of a constant phase point and integrate the velocity over time to estimate displacement [24].

Dickinson and Hill used the correlation coefficient between successive A-scan lines to measure the amplitude and frequency of tissue motion. They defined a correlation parameter to characterize the changes of the interrogated region between the successive A-scans. For small displacements, they assumed the decorrelation was proportional to displacement [25]. Tristam et al. [26] further developed the technique to investigate the responses of normal and cancerous liver to cardiac pulsation. De Jong et al. [27] also used a modified correlation technique to measure tissue motion.

Fetal lung elasticity was investigated as an important parameter of fetal lung maturity by Birnholz and Farrell. They tried to qualitatively determine the stiffness of fetal lungs by evaluating the local compression of the lung adjacent to the heart compared to the more distant lung, which would compress relatively less depending on the lung stiffness and distance from the heart [28]. Adler et al. developed more quantitative estimates by applying correlation techniques to digitized M-mode images and estimated a parameter that characterizes the range of transmitted cardiac motion in fetal lungs. The parameter is a measure of the temporally and spatially averaged systolic to diastolic deformation per unit epicardial displacement [29].

Holen et al. [30] observed a characteristic Bessel-band Doppler spectrum when using Doppler ultrasonography to examine unusually oscillating heart valves. Taylor [31] showed that the expression for the Doppler spectrum of a scattered Doppler signal from a vibrating target is similar to that of a pure-tone frequency modulation process under certain conditions.

Cox and Rogers studied the Doppler ultrasound response of fish auditory organs to low-frequency sound. The vibration amplitude of the hearing organ was determined by comparing the ratio of the carrier and the first side band of the Doppler spectrum [32].

These experimental techniques and mathematic solutions all made contributions to relative and absolute tissue strain and velocity measurements, but without absolute measures of stress, strain, and boundary conditions of the target tissues, as well as appropriate mathematic models, intrinsic material elasticity values could not be quantitated independent of the experimental setup.

2.3.3 Tissue stimulation by externally controllable sources

Eisensher et al. [33] used M-mode ultrasonography to monitor the frequency content of tissue motion induced in breast and liver tissues by a 1.5-Hz vibration source. They found that the quasi-static compression response from benign lesions was characteristically sinusoidal, whereas that from malignant tumors tended to be more flat, i.e., more nonlinear.

Sato et al. [34] investigated nonlinear interactions between ultrasound and lower frequency pump waves in tissues at the time when Parker and Lerner [11] and Krouskop and Levinson [35] were using linear methods to investigate the propagation of vibrations inside tissues.

Krouskop et al. reported one of the first attempts at a quantitative measurement of tissue elasticity using gated pulse Doppler to detect tissue motion subjected to an external vibration. The set of equations relating tissue properties and movements reduces to simple forms under assumptions of isotropy and incompressibility. Determining tissue elasticity then reduces to measuring peak tissue displacements and gradients. They suggested possible absolute tissue stiffness could be determined in a very small region, i.e., 0.5×0.5 mm, within a homogeneous medium [35].

Except for the work of Krouskop, the tissue elasticity data was qualitative and lacking sufficient detail for the determination of a fundamental tissue property independent of the experimental setup that could translate to an absolute measure of tissue stiffness (elasticity, i.e., bulk modulus or Young's modulus of elasticity; neglecting density variations in biological soft tissues, which are very small compared with elasticity variations). Although, in general, absolute stiffness measurements could not be obtained, relative measurements could prove very valuable to detect focal lesions in tissue if the data were reliable over an extended area allowing comparison of the target tissue to the adjacent tissue.

3. The early era of imaging tissue stiffness (late 1980s to mid-1990s)

3.1 Vibration amplitude sonoelastography (sonoelasticity)

To our knowledge, the first published image of relative stiffness was a crude grayscale map proportional to pulse-Doppler-detected vibration motion in a tissue-mimicking phantom containing a hard inclusion, which was subjected to an external mechanical stimulus [2,11]. This technique was called "sonoelasticity." After initial proof of concept in phantoms and in vitro animal studies, real-time color Doppler vibrational images were demonstrated in animal and human tissues [1,36] (also see Chapter 3 of this volume).

Real-time modified color Doppler observation of tissue vibration amplitude images during deliberate variation of the frequencies of the external vibration source in the range of 50–200 Hz resulted in variable centimeter-sized modal patterns corresponding to wave speeds of 1–3 m/s, similar to published values of shear wave speed in tissues [13,36]. The modal patterns were very sensitive to the frequency changes and became more complex at higher frequencies. Hard inclusions in the phantoms disturbed the modal patterns. This supported the hypothesis that the observed sonoelasticity modes sensitive to tissue stiffness were related to shear rather than longitudinal waves. This recognition that tissue stiffness correlated more closely with shear wave propagation than with longitudinal wave propagation was likely the motivation for several research laboratories to direct their tissue characterization efforts toward shear waves [37] (also see Chapter 5 of this volume).

A real-time vibration amplitude (qualitative) imaging (sonoelasticity) study of in vitro prostate specimens showed better sensitivity and predictive value for cancer detection than conventional B-scan alone [8]. The cancerous regions in the specimens showed less relative motion than the adjacent normal tissue. A mathematic model for vibration amplitude sonoelastography was completed, showing that the shear wave elastographic contrast was orders of magnitude greater than the contrast based on echogenicity [38–40].

Later, the principles of vibration sonoelastography were also applied using vocal fremitus as the external vibration source with the patients' own voice (as in humming a vowel sound) in conjunction with Doppler display of vibration. However, the limited control over amplitude, frequency, and the complications of the effect of varying echogenicity on the Doppler display all contributed to a variable and patient-dependent response using vocal fremitus [41].

3.2 Compression elastography

Ophir et al. introduced compression elastography as an imaging method to display relative stiffness based on local tissue strain changes induced by a "modest" (2%) compression applied to a B-scan real-time imaging transducer. B-scan RF information from the backscattered ultrasonography before and after compression was used to calculate local strain by correlation analysis. Stiffer tissues would undergo less strain than soft tissues under the same applied stress. Images, in principle, would be simple to interpret but required the application of a uniform stress to the surface and intervening tissues superficial and deep to the target, which is not easily accomplished in practice. Within these limitations, qualitative (relative) stiffness images were obtained [42].

This concept was initially introduced in some commercial medical ultrasonographic equipment and created a platform for early clinical studies to advance the field of relative stiffness imaging of tissues. This concept gained enough success to be currently available on nearly all commercial clinical ultrasonographic systems.

This topic is reviewed in more detail in Chapter 4 of this volume.

4. Quantitative tissue stiffness determination and imaging (1990 to present)

4.1 Quantitative tissue stiffness determination

Yamakoshi and Sato et al. developed a vibration phase gradient approach that maps the amplitude and phase of the low-frequency shear wave propagation inside tissues, which can be used to derive the elastic and viscous characteristics of the tissue [37]. The rate of change of phase could yield a quantitative estimate of tissue stiffness (see Chapter 3 of this volume).

Transient elastography, a method for the measurement of the shear wave elastic modulus or shear wave speed in liver, was the first application of an elastographic method developed for a specific application that met with widespread clinical acceptance. The method, using an external pistonlike mechanical stimulation applied to a patient's skin, transmitted shear wave pulses into the liver where ultrasonographic monitoring of the resulting liver tissue motion along an axial path yielded a measure of shear wave speed [43]. The technique was incorporated in an instrument called FibroScan, which has had clinical success in staging the degree of liver fibrosis

[44,45]. FibroScan determines shear wave speed based on shear wave tissue motion detected with ultrasound waves along the axial beam of the ultrasound transducer. Intuitively, shear waves propagate at right angles to longitudinal waves; however, longitudinal shear wave propagation along the axial path, a nonintuitive result, is predicted by the analysis of Oestreicher's work [15,17,46].

The rapid acceptance of shear wave speed (in meters per second) or elasticity (in kilopascals) as a clinical parameter for assessing the degree of liver fibrosis was likely the stimulus for the flurry of activity that followed with the goal of generating images of absolute tissue stiffness in terms of tissue elasticity or shear wave speed. Shear modulus G is related to shear wave speed v_s by the expression

$$G = 3 \cdot \rho \cdot v_s^2, \qquad\qquad 1.1$$

where ρ is the density.

Quantitative imaging of the elastic properties of tissues involves perturbation of the tissue and measurement of the tissue response over space and time, with details depending on whether it is based on compression elastography, mechanical imaging, or shear wave imaging. Knowledge of the surface geometry of the tissue under examination and the mechanical stimulation to the tissue as a function of space and time is required to mathematically process the data using inverse methods when applied to an appropriate model [47–51] (also see Chapter 7 of this volume).

4.2 Further expansion of tissue elasticity imaging (1994 to present)

Work at the University of Paris under Professor Mathias Fink demonstrated that a transient shear wave tracking approach could produce clinically useful estimates of shear wave speed [43]. This concept was successfully commercialized into an instrument called FibroScan. The recognition that FibroScan could quantify the fundamental tissue property stiffness (i.e., shear wave speed [in meters per second] or elasticity [in kilopascals]; see Eq. (1.1)) led to a rapid expansion of developmental work leading to research instruments and eventually clinical trials for other medical applications. The ensuing instruments could obtain a localized shear wave speed in a region of interest defined on a B-scan image. Liver cross-sectional images of quantitative shear wave speed in tissues had been obtained earlier using magnetic resonance imaging (MRI) and an external low-frequency source of shear waves. However, due to the limited patient access to MRI, magnetic resonance elastography had not yet achieved widespread application despite its elegant capabilities [52].

Subsequently, cross-sectional images of shear wave speed in tissues were obtained using other ultrasonographic techniques and eventually optical methods. These major developments are the subjects of later chapters in this volume and include vibration sonoelastography, quasi-static elastography, acoustic radiation force impulse (ARFI) imaging, shear wave imaging, optical computed tomographic elastography, and MRI elastography. MRI elastography, in principle, could detect three-dimensional motion, thus permitting a nonisotropic material's stress tensor to be measured.

Longitudinal ultrasound pulses focused in tissues result in sound absorption leading to acoustic momentum transfer that produces tissue motion and are a source for localized shear waves (tissue motion). As the shear wave speed in tissues is approximately 1000 times slower than the longitudinal wave speed, propagation of the focal tissue shear displacements or motion can be detected or imaged using the same transducer that applied the radiation force [53–56]. Thus propagation of the focal tissue displacements and velocities could be tracked to quantitate shear wave speed. Some of these concepts have been used commercially (ARFI, shear wave elasticity imaging, and supersonic imaging) and avoid the, sometimes cumbersome, externally applied low-frequency shear wave source.

Use of acoustic radiation force as a means of generating and propagating shear waves into deep tissues using focused ultrasound pulses from conventional imaging transducers has been a major contribution for ultrasound-based imaging of cross-sectional tissue stiffness [57]. This technique can be traced to contributions by several investigators. Nightingale [58] had used acoustic radiation force to produce streaming motion of liquids for the characterization of cysts in breast tissue. Using this concept, Nightingale and Trahey [59] reported a clinical study to differentiate cysts from solid lesions. Subsequently, they applied it to perturb breast tissue for motion analysis. They realized that radiation force itself could be used to create an image (ARFI) [56,60].

Sarvazyan [54] applied acoustic radiation force to produce localized tissue motion and to take advantage of the resulting shear waves, which he detected propagating perpendicular to the longitudinal axially directed ultrasound beam. An advantage of this technique is that the stimulated tissue is small in size and the shear wave propagation is limited in range so that boundary conditions do not complicate the shear wave propagation [61].

Sarvazyan also developed an approach called mechanical imaging in which he applied an array of surface mechanical stimulators and detectors of the responding stress patterns at the surface producing data that could be processed using inversion models to provide quantitative tissue constants [62].

Additional elasticity determination refinements and imaging techniques that have had success in the laboratory are shear wave dispersion [63], single tracking location methods that suppress speckle noise in shear wave velocity estimation [64], and laboratory and clinical trials using crawling wave [65], single tracking location [66], X-ray and computed tomographic elastography [67], and photoacoustic elastography [68].

Some of these topics are subjects of later chapters of this volume. Because of the emerging interest in elastography and given its clinical importance, Kevin Parker and colleagues at the Rochester Center for Biomedical Ultrasound sponsored a special workshop in Washington, DC in June 1994, inviting Drs. Ophir, Levinson, Bamber, and others from the international community to share research insights. This may have been the first dedicated workshop on elastography.

Later in the 1990s, Professors Ophir and Parker would agree on the need for a dedicated conference where experts from multiple disciplines of biomechanics, cellular biology, imaging sciences, radiology, and biophysics could have extended discussions about the rapidly expanding world of elastography. Urged on by Dr. S. Kaisar Alam, who had worked with both Ophir and Parker, the International Tissue Elasticity Conference was launched with the first conference in October of 2002 in Niagara Falls, Canada. This conference flourished with now over 13 meetings held and is currently chaired by Professor Jeff Bamber.

4.3 Microscopic tissue elasticity imaging

Extension of elasticity imaging to the microscopic domain is occurring that portends to rapidly expand knowledge at the intra- and extracellular levels. Optical elastography has been done using optical computed tomography (OCT, see Chapter 9 of the volume) and other methods such as Michelson laser vibrometry [69]. The mechanical excitation for OCT can use external mechanical stimulation or ARFI in addition to optical absorption, leading to thermally generated acoustic waves.

These techniques use light scattered from semitransparent tissues to generate images depicted in gray scale similar to ultrasound echo images, but with much finer resolution for tracking mechanically excited tissue motion and production of microscopic (subcellular level) tissue elasticity imaging.

Photoacoustic elastography can provide high-resolution elasticity images. In one implementation, the compression is applied manually and the resulting strain map is estimated from photoacoustic signals [68]. Photoacoustic imaging uses light pulses focused within a tissue to generate heat depending on the optical absorption coefficient of the specific tissue, which causes local thermal expansion to create sharp tissue motion that results in acoustic waves. The range of optical absorption coefficients may be very wide, permitting a new basis for contrast and tissue characterization related to the photochemical absorption in addition to tissue stiffness.

Tissue elasticity imaging on the microscopic level will likely open new areas of understanding of the fibrotic response of tissues at the extracellular matrix and intracellular levels. Stem cell differentiation has been shown to be sensitive to the stiffness of its milieu, which may trigger stem cells to differentiate into cancer cells or fibrocytes that impacts liver cirrhosis, idiopathic pulmonary fibrosis, and systemic sclerosis [70,71]. Photoacoustic absorption may target melanin or heme with and without oxygen to provide superior contrast to their neighboring tissues. Perhaps special stains with specific photoabsorption frequencies and intracellular organelle affinities could lead to new applications.

5. Conclusion/discussion

Imaging and understanding tissue elasticity at the macroscopic tissue organizational level (100 micrometers to millimeters) and microscopic intracellular and extracellular matrix levels portends to significantly impact our understanding of biology and diseases. Future experiments and theories to understand real (lossless, strain in phase with stress) and imaginary (strain out of phase with stress) components of shear wave speed in tissues and simulated tissue phantoms as a function of relative amounts of a two-phase system composed of fluid and a supporting solid matrix simulating fibrous tissue, cell membrane, or intracellular organelles may be rewarding. Mathematic methods, some labeled as "inverse problems" (see Chapter 7 of this volume), continue to be developed that allow for the realities of experimental designs such as finite tissue boundaries, realistic stimulating force profiles in time and space, and anisotropy of the material under evaluation to obtain absolute biomechanical properties from elastography data [51]. These methods have been applied to study additional mechanical tissue properties that may have clinical relevance, such as shear wave dispersion, anisotropy, porosity, and nonlinearity [51].

Optical/photoacoustic elastography may allow cellular biologists to target intracellular organelles using special stains that can be tuned to optical wavelengths so that optical modulation and motion detection could produce elastographic images of intracellular structures and evaluate their surrounding stiffness.

It is also expected that elastography techniques will be extended to more imaging platforms of different modalities, and with specialized approaches tuned to different applications and pathologic conditions, including those affecting the musculoskeletal and cardiovascular systems, along with large organs and the brain. More complex measures of tissue anisotropy and viscosity will add to the existing elastographic assessments.

Tissue elasticity imaging is a great example of how researchers working with clinical colleagues advanced medical imaging from a qualitative clinical pattern recognition system to a diagnostic quantitative imaging system. Fundamental parameters such as shear wave speed or bulk shear wave elasticity as objective measures of liver fibrosis could allow clinicians to treat and follow-up patients with liver disease, thus avoiding some biopsies and directing biopsies to productive sites. A recent Google interrogation of elastography produced 930,000 references. Elasticity imaging is currently being extended to other organs (breast, thyroid, brain, muscle, kidney, skin, cervix, and placenta, to name a few) for improved clinical management and possible reduced numbers of biopsies. Multimodality development has allowed tissue elasticity imaging concepts to extend beyond ultrasonography to magnetic resonance and optical techniques, where new insights into fundamental understanding of diseases and biology are sure to be forthcoming.

Acknowledgments

The helpful perspectives from Drs. Jeffrey Bamber and Kevin Parker are greatly appreciated in addition to valuable editorial assistance from Drs. S. Kaisar Alam and Brian Garra. The expert assistance by Linda Weidman in the preparation of this chapter was essential.

References

[1] R.M. Lerner, S.R. Huang, K.J. Parker, "Sonoelasticity" images derived from ultrasound signals in mechanically vibrated tissues, Ultrasound Med. Biol. 16 (3) (1990) 231−239.

[2] R.M. Lerner, K.J. Parker, J. Holen, R. Gramiak, R.C. Waag, Sonoelasticity: medical elasticity images derived from ultrasound signals in mechanically vibrated targets, Acoust. Imaging 16 (1988) 317−327.

[3] J.C. Bamber, G.R. ter Haar, in: C.R. Hill (Ed.), Physical Principles of Medical Ultrasonics, John Wiley & Sons, New York-Chichester-Brisbane-Toronto, 1986.

[4] C.B. Burckhardt, Speckle in ultrasound b-mode scans, IEEE Trans. Son. Ultrason. 25 (1) (1978) 1−6.

[5] L.A. Frizzell, E.L. Carstensen, J.F. Dyro, Shear properties of mammalian-tissues at low megahertz frequencies, J. Acoust. Soc. Am. 60 (6) (1976) 1409−1411.

[6] R.M. Lerner, Linear array transrectal ultrasound, evaluation of the prostate: a review of 85 cases, in: Annual Meeting, Northeastern Section, American Urological Association., Buffalo, NY, 1984.

[7] K.J. Parker, S.R. Huang, R.A. Musulin, R.M. Lerner, Tissue response to mechanical vibrations for "sonoelasticity imaging", Ultrasound Med. Biol. 16 (3) (1990) 241−246.

[8] D.J. Rubens, M.A. Hadley, S.K. Alam, L. Gao, R.D. Mayer, et al., Sonoelasticity imaging of prostate cancer: in vitro results, Radiology 195 (2) (1995) 379−383.

[9] L.S. Taylor, B.C. Porter, G. Nadasdy, P.A. di Sant'Agnese, D. Pasternack, et al., Three-dimensional registration of prostate images from histology and ultrasound, Ultrasound Med. Biol. 30 (2) (2004) 161−168.

[10] T.A. Krouskop, T.M. Wheeler, F. Kallel, B.S. Garra, T. Hall, Elastic moduli of breast and prostate tissues under compression, Ultrason. Imaging 20 (4) (1998) 260−274.

[11] R.M. Lerner, K.J. Parker, Sonoelasticity images derived from ultrasound signals in mechanically vibrated targets, in: Seventh European Communities Workshop, Nijmegen, The Netherlands, 1987.

[12] D. Berger, A brief history of medical diagnosis and the birth of the clinical laboratory. Part 1–Ancient times through the 19th century, Med. Lab. Obs. 31 (7) (1999), 28−30, 32, 34−40.

[13] H.E. von Gierke, H.L. Oestreicher, E.K. Franke, H.O. Parrack, W.W. Wittern, Physics of vibrations in living tissues, J. Appl. Physiol. 4 (12) (1952) 886−900.

[14] K.F. Graff, Wave motion in elastic solids, in: Oxford Engineering Science Series, Clarendon Press, Oxford, 1975 (Chapter 6).

[15] H.L. Oestreicher, Field and impedance of an oscillating sphere in a viscoelastic medium with an application to biophysics, J. Acoust. Soc. Am. 23 (6) (1951) 704−714.

[16] E.L. Carstensen, K.J. Parker, R.M. Lerner, Elastography in the management of liver disease, Ultrasound Med. Biol. 34 (10) (2008) 1535−1546.

[17] E.L. Carstensen, K.J. Parker, Oestreicher and elastography, J. Acoust. Soc. Am. 138 (4) (2015) 2317−2325.

[18] E. Ueno, E. Tohno, S. Soeda, Y. Asaoka, K. Itoh, et al., Dynamic tests in real-time breast echography, Ultrasound Med. Biol. 14 (1988) 53−57.

[19] C.R. Hill, A. Kratochwil, European Federation of Societies for Ultrasound in Medicine and Biology., in: Medical Ultrasonic Images : Formation, Display, Recording, and Perception: Based on Papers Presented at a European Symposium Held in Brussels,

January 31-February 1, 1981. International congress series, vol. 541, Amsterdam Excerpta Medica, 1981, pp. 292–298.

[20] C.M. Gros, G. Dale, B. Gairand, Breast echography: criteria of malignancy and results, in: Recent Advances in Ultrasound Diagnosis, Excerpta Medica, Amsterdam, 1978.

[21] P.B. Guyer, K.C. Dewbury, D. Warwick, J. Smallwood, I. Taylor, Direct contact B-scan ultrasound in the diagnosis of solid breast masses, Clin. Radiol. 37 (5) (1986) 451–458.

[22] M. Tristam, D.C. Barbosa, D.O. Cosgrove, J.C. Bamber, C.R. Hill, Application of Fourier analysis to clinical study of patterns of tissue movement, Ultrasound Med. Biol. 14 (8) (1988) 695–707.

[23] J.C. Bamber, L. De Gonzalez, D.O. Cosgrove, P. Simmons, J. Davey, et al., Quantitative evaluation of real-time ultrasound features of the breast, Ultrasound Med. Biol. 14 (Suppl. 1) (1988) 81–87.

[24] L.S. Wilson, D.E. Robinson, Ultrasonic measurement of small displacements and deformations of tissue, Ultrason. Imaging 4 (1) (1982) 71–82.

[25] R.J. Dickinson, C.R. Hill, Measurement of soft tissue motion using correlation between A-scans, Ultrasound Med. Biol. 8 (3) (1982) 263–271.

[26] M. Tristam, D.C. Barbosa, D.O. Cosgrove, D.K. Nassiri, J.C. Bamber, et al., Ultrasonic study of in vivo kinetic characteristics of human tissues, Ultrasound Med. Biol. 12 (12) (1986) 927–937.

[27] P.G.M. De Jong, T. Arts, A.P.G. Hoeks, R.S. Reneman, Determination of tissue motion velocity by correlation interpolation of pulsed ultrasonic echo signals, Ultrason. Imaging 12 (2) (1990) 84–98.

[28] J.C. Birnholz, E.E. Farrell, Fetal lung development: compressibility as a measure of maturity, Radiology 157 (2) (1985) 495–498.

[29] R.S. Adler, J.M. Rubin, P.H. Bland, P.L. Carson, Quantitative tissue motion analysis of digitized M-mode images: gestational differences of fetal lung, Ultrasound Med. Biol. 16 (6) (1990) 561–569.

[30] J. Holen, R.C. Waag, R. Gramiak, Representations of rapidly oscillating structures on the doppler display, Ultrasound Med. Biol. 11 (2) (1985) 267–272.

[31] K.J. Taylor, Absolute measurement of acoustic particle velocity, J. Acoust. Soc. Am. 59 (3) (1976) 691–694.

[32] M. Cox, P.H. Rogers, Automated noninvasive motion measurement of auditory organs in fish using ultrasound, J. Vib. Acoust. 109 (1) (1987) 55–59.

[33] A. Eisensher, E. Schweg-Toffer, G. Pelletier, G. Jacquemard, La palpation echographique rythmee-echosismographie, J. Radiol. 64 (1983) 225–261.

[34] T. Sato, A. Fukusima, N. Ichida, H. Ishikawa, H. Miwa, et al., Nonlinear parameter tomography system using counterpropagating probe and pump waves, Ultrason. Imaging 7 (1) (1985) 49–59.

[35] T.A. Krouskop, D.R. Dougherty, F.S. Vinson, A pulsed Doppler ultrasonic system for making noninvasive measurements of the mechanical properties of soft tissue, J. Rehabil. Res. Dev. 24 (2) (1987) 1–8.

[36] K.J. Parker, R.M. Lerner, Sonoelasticity of organs: shear waves ring a bell, J. Ultrasound Med. 11 (8) (1992) 387–392.

[37] Y. Yamakoshi, J. Sato, T. Sato, Ultrasonic imaging of internal vibration of soft tissue under forced vibration, IEEE Trans. Ultrason. Ferroelectr. Freq. Control 37 (2) (1990) 45–53.

[38] L. Gao, K.J. Parker, S.K. Alam, D. Rubens, R.M. Lerner, Theory and application of sonoelasticity imaging, Int. J. Imaging Syst. Technol. 8 (1) (1997) 104–109.

[39] K.J. Parker, D. Fu, S.M. Graceswki, F. Yeung, S.F. Levinson, Vibration sonoelastography and the detectability of lesions, Ultrasound Med. Biol. 24 (9) (1998) 1437−1447.

[40] R.M. Cramblitt, K.J. Parker, Generation of non-Rayleigh speckle distributions using marked regularity models, IEEE Trans. Ultrason. Ferroelectr. Freq. Control 46 (4) (1999) 867−874.

[41] C. Sohn, A. Baudendistel, Differential diagnosis of mammary tumors with vocal fremitus in sonography: preliminary report, Ultrasound Obstet. Gynecol. 6 (3) (1995) 205−207.

[42] J. Ophir, I. Cespedes, H. Ponnekanti, Y. Yazdi, X. Li, Elastography: a quantitative method for imaging the elasticity of biological tissues, Ultrason. Imaging 13 (2) (1991) 111−134.

[43] L. Sandrin, M. Tanter, J.L. Gennisson, S. Catheline, M. Fink, Shear elasticity probe for soft tissues with 1-D transient elastography, IEEE Trans. Ultrason. Ferroelectr. Freq. Control 49 (4) (2002) 436−446.

[44] N. Ganne-Carrie, M. Ziol, V. de Ledinghen, C. Douvin, P. Marcellin, et al., Accuracy of liver stiffness measurement for the diagnosis of cirrhosis in patients with chronic liver diseases, Hepatology 44 (6) (2006) 1511−1517.

[45] J. Boursier, J. Vergniol, A. Sawadogo, T. Dakka, S. Michalak, et al., The combination of a blood test and Fibroscan improves the non-invasive diagnosis of liver fibrosis, Liver Int. 29 (10) (2009) 1507−1515.

[46] E.L. Carstensen, K.J. Parker, Physical models of tissue in shear fields, Ultrasound Med. Biol. 40 (4) (2014) 655−674.

[47] C. Sumi, A. Suzuki, K. Nakayama, Estimation of shear modulus distribution in soft tissue from strain distribution, IEEE Trans. Biomed. Eng. 42 (2) (1995) 193−202.

[48] F. Kallel, M. Bertrand, Tissue elasticity reconstruction using linear perturbation method, IEEE Trans. Med. Imaging 15 (3) (1996) 299−313.

[49] M.M. Doyley, P.M. Meaney, J.C. Bamber, Evaluation of an iterative reconstruction method for quantitative elastography, Phys. Med. Biol. 45 (6) (2000) 1521−1540.

[50] A. Samani, J. Bishop, D.B. Plewes, A constrained modulus reconstruction technique for breast cancer assessment, IEEE Trans. Med. Imaging 20 (9) (2001) 877−885.

[51] M.M. Doyley, Model-based elastography: a survey of approaches to the inverse elasticity problem, Phys. Med. Biol. 57 (3) (2012) R35−R73.

[52] R. Muthupillai, D.J. Lomas, P.J. Rossman, J.F. Greenleaf, A. Manduca, et al., Magnetic-resonance elastography by direct visualization of propagating acoustic strain waves, Science 269 (5232) (1995) 1854−1857.

[53] T. Sugimoto, S. Ueha, K. Itoh, Tissue hardness measurement using the radiation force of focused ultrasound, in: Ultrasonics Symposium, 1990. Proceedings., IEEE 1990, 1990.

[54] A.P. Sarvazyan, O.V. Rudenko, S.D. Swanson, J.B. Fowlkes, S.Y. Emelianov, Shear wave elasticity imaging: a new ultrasonic technology of medical diagnostics, Ultrasound Med. Biol. 24 (9) (1998) 1419−1435.

[55] J. Bercoff, M. Tanter, M. Fink, Supersonic shear imaging: a new technique for soft tissue elasticity mapping, IEEE Trans. Ultrason. Ferroelectr. Freq. Control 51 (4) (2004) 396−409.

[56] K.R. Nightingale, M.L. Palmeri, R.W. Nightingale, G.E. Trahey, On the feasibility of remote palpation using acoustic radiation force, J. Acoust. Soc. Am. 110 (1) (2001) 625−634.

[57] S. Catheline, F. Wu, M. Fink, A solution to diffraction biases in sonoelasticity: the acoustic impulse technique, J. Acoust. Soc. Am. 105 (5) (1999) 2941−2950.

[58] K.R. Nightingale, P.J. Kornguth, W.F. Walker, B.A. McDermott, G.E. Trahey, A novel ultrasonic technique for differentiating cysts from solid lesions: preliminary results in the breast, Ultrasound Med. Biol. 21 (6) (1995) 745–751.

[59] K.R. Nightingale, P.J. Kornguth, G.E. Trahey, The use of acoustic streaming in breast lesion diagnosis: a clinical study, Ultrasound Med. Biol. 25 (1) (1999) 75–87.

[60] K.R. Nightingale, R.W. Nightingale, M.L. Palmeri, G.E. Trahey, A finite element model of remote palpation of breast lesions using radiation force: factors affecting tissue displacement, Ultrason. Imaging 22 (1) (2000) 35–54.

[61] A.P. Sarvazyan, Method and Device for Shear Wave Elasticity Imaging, Patent number 5,606,971. Issued by U.S.P.a.T. Office, March 4, 1997.

[62] A. Sarvazyan, Mechanical imaging: a new technology for medical diagnostics, Int. J. Med. Inform. 49 (2) (1998) 195–216.

[63] C.T. Barry, C. Hazard, Z. Hah, G. Cheng, A. Partin, et al., Shear wave dispersion in lean versus steatotic rat livers, J. Ultrasound Med. 34 (6) (2015) 1123–1129.

[64] E.C. Elegbe, S.A. McAleavey, Single tracking location methods suppress speckle noise in shear wave velocity estimation, Ultrason. Imaging 35 (2) (2013) 109–125.

[65] Z. Wu, K. Hoyt, D.J. Rubens, K.J. Parker, Sonoelastographic imaging of interference patterns for estimation of shear velocity distribution in biomaterials, J. Acoust. Soc. Am. 120 (1) (2006) 535–545.

[66] J.H. Langdon, E. Elegbe, S.A. McAleavey, Single tracking location acoustic radiation force impulse viscoelasticity estimation (STL-VE): a method for measuring tissue viscoelastic parameters, IEEE Trans. Ultrason. Ferroelectr. Freq. Control 62 (7) (2015) 1225–1244.

[67] G.W. Kim, B.H. Han, M.H. Cho, S.Y. Lee, X-ray elastography: a feasibility study, in: Annual International Conference of the IEEE Engineering in Medicine and Biology Society. 2009, 2009.

[68] P. Hai, J. Yao, G. Li, C. Li, L.V. Wang, Photoacoustic elastography, Opt. Lett. 41 (4) (2016) 725–728.

[69] J. Li, C.-H. Liu, A. Schill, M. Singh, Y.V. Kistenev, et al., A comparison study of optical coherence elastography and laser Michelson vibrometry, in: SPIE BIOS, SPIE, 2016.

[70] T.R. Cox, J.T. Erler, Remodeling and homeostasis of the extracellular matrix: implications for fibrotic diseases and cancer, Dis. Model. Mech. 4 (2) (2011) 165–178.

[71] A.J. Engler, S. Sen, H.L. Sweeney, D.E. Discher, Matrix elasticity directs stem cell lineage specification, Cell 126 (4) (2006) 677–689.

The governing theory of elasticity imaging

2

Salavat R. Aglyamov

Department of Mechanical Engineering, University of Houston, Houston, TX, United States

1. Introduction

Elasticity imaging is a general term for a set of diagnostic techniques to evaluate the mechanical properties of soft tissues remotely and noninvasively. Biophysical basis of elasticity imaging is the fact that cellular changes in the tissues leading to pathologic conditions often result in macroscopic changes in the mechanical properties of tissues. From this point of view, elasticity imaging can be considered as a combination of biomechanics, i.e., mechanics applied to the fields of biology and medicine, and biomedical imaging. Typically, all elasticity imaging techniques follow a similar approach based on (1) the application of a mechanical excitation, which can be externally applied, internally generated, or naturally occurring; (2) the measurement of tissue mechanical response (i.e., displacements, velocities, strains, etc.) using various imaging modalities where a multitude of tissue motion estimation methods chosen based on the type of mechanical excitation used offer a broad variety of elastography techniques; and (3) the interpretation of the measured mechanical response to assess the mechanical properties of tissue. The final step, often referred to as reconstruction, is based on the principles of continuum mechanics and inverse problem solution. To fully describe the biomechanics of soft biological tissues and, therefore, to be able to reconstruct mechanical properties of the tissue based on the solution of inverse problems given by the data obtained using imaging techniques, sophisticated mechanical models with a large number of parameters are required. In most practical cases, however, a comprehensive mechanical characterization of the tissue is not necessary, and mechanical models can be significantly simplified. This chapter briefly introduces the theoretic basics of governing principles used in elasticity imaging with the focus on the response of the soft tissues to mechanical loads. For a more complete description of the principles of soft tissue biomechanics, the reader is referred to the literature[1-4,4a,5-6].

2. Displacement and strain

Consider the initial stage of elastography procedure in which a mechanical load is applied to soft biological tissues. Under the action of the mechanical load, soft

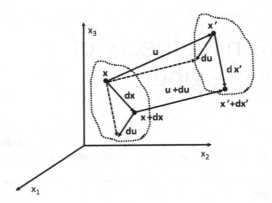

FIGURE 2.1

Displacement and deformation of the tissue.

tissues change shape and/or volume. This process, called deformation, can be described mathematically as follows. The location of every particle in tissues can be defined by a position vector $x = (x_1, x_2, x_3)$ in a three-dimensional rectangular Cartesian frame of reference. If the tissue is deformed, every particle moves to a new position x', as shown in Fig. 2.1. The vector u is called the displacement vector of the particle:

$$u = x' - x \qquad (2.1)$$

If the displacement vector is known for every point in the tissue, deformation of tissue can be entirely described by the displacement field. Both vectors $u(x)$ and $x'(x)$ are the functions of the coordinates x of the point before deformation. We assume that the function $x'(x)$ is continuous and has a unique inverse such that both the function and its inverse are one-to-one functions.

If the body is deformed, the distance between its points changes. Therefore the change in the distance between points can be considered as a measure of deformation. For two close points with position vectors x and $x + dx$ the square of the distance between the points is $dl^2 = dx_1^2 + dx_2^2 + dx_3^2$ before deformation and $dl'^2 = dx_1'^2 + dx_2'^2 + dx_3'^2$ after deformation, where dx and dx' are the vectors connecting these points before and after deformation, respectively (see Fig. 2.1). The difference between displacement vectors at the points x and $x + dx$ is $du = dx' - dx$. Using Einstein summation rule, the components of the vector du can be expressed through the components of the vector dx:

$$du_n = \frac{\partial u_n}{\partial x_k} dx_k \qquad (2.2)$$

Therefore the difference between the squares of the length elements can be expressed as

$$dl'^2 - dl^2 = 2\varepsilon_{nm} dx_n dx_m, \qquad (2.3)$$

$$dl'^2 - dl^2 = 2\varepsilon'_{nm} dx'_n dx'_m,$$

where ε_{nm} and ε'_{nm} are called the finite strain tensors and refer to the original and current coordinate systems, respectively:

$$\varepsilon_{nm} = \frac{1}{2} \left(\frac{\partial u_n}{\partial x_m} + \frac{\partial u_m}{\partial x_n} + \frac{\partial u_k}{\partial x_n} \frac{\partial u_k}{\partial x_m} \right) \qquad (2.4)$$

$$\varepsilon'_{nm} = \frac{1}{2} \left(\frac{\partial u_n}{\partial x'_m} + \frac{\partial u_m}{\partial x'_n} - \frac{\partial u_k}{\partial x'_n} \frac{\partial u_k}{\partial x'_m} \right)$$

The tensor ε_{nm} in Eq. (2.4) is called the Green's or Lagrangian finite strain tensor, while ε'_{nm} is often referred to as Eulerian finite strain tensor. As a tissue is deformed, strain tensor defines the change in the element of length, and, by definition, both tensors are symmetric. In a rigid-body motion, where the distance between all points in a tissue is kept constant, the components of the strain tensor are zero as follows from Eq. (2.3). The finite strain tensor is introduced without any assumption regarding the level of deformation. It means that Eq. (2.4) can be used for the analysis of the geometrically nonlinear behavior of tissues, where the deformation is large. However, in most problems of elastography, the deformations are so small that the squares of the first partial derivatives of the components of displacement vector are negligible compared with the first-order terms. For small strain theory, or geometrically linear case, there is no difference between Lagrangian and Eulerian tensors, and the last terms in Eq. (2.4) can be neglected. The finite strain tensors in Eq. (2.4) are reduced to an infinitesimal strain tensor, which will be referred to as the strain tensor:

$$\varepsilon_{nm} = \frac{1}{2} \left(\frac{\partial u_n}{\partial x_m} + \frac{\partial u_m}{\partial x_n} \right) \qquad (2.5)$$

To visualize a geometric meaning of the components of the strain tensor in Eq. (2.5), we consider several simple examples of tissue deformation presented in Fig. 2.2. Consider an infinitesimal tissue volume in the form of a cube whose edges are parallel to the coordinate axes. The first invariant of the strain tensor (Eq. 2.5) is

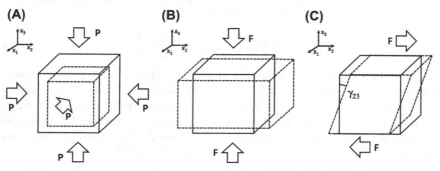

FIGURE 2.2

(A) Hydrostatic compression. (B) Uniaxial compression. (C) Shear deformation.

equal to the sum of diagonal components and has an important physical interpretation as a dilatation or relative volume change. The volume of this region before deformation dV is the product of the elements of length along coordinate axes: $dx_1dx_2dx_3$, while after deformation the new volume is $dV' = dx'_1dx'_2dx'_3$. For small strain theory, neglecting higher order terms, the change in volume per unit of original volume is the sum of diagonal components of the linear strain tensor or the divergence of the displacement vector:

$$\frac{dV' - dV}{dV} = \Theta = \varepsilon_{11} + \varepsilon_{22} + \varepsilon_{33} = \nabla \cdot \boldsymbol{u}$$

$$= \frac{\partial u_1}{\partial x_1} + \frac{\partial u_2}{\partial x_2} + \frac{\partial u_3}{\partial x_3}$$

(2.6)

For example, if a tissue is under hydrostatic compression, i.e., normal forces are applied uniformly to all surfaces of the tissue, the only way for the body to deform is through a change in volume, as shown in Fig. 2.2A, and the sum of diagonal components of the strain tensor completely characterizes the tissue deformation.

The individual diagonal components of the strain tensor represent normal strains in coordinate directions, i.e., the change in length per unit of original length. For example, if a tissue is under uniaxial compression or tension (see Fig. 2.2B), only the diagonal components of the strain tensor $\varepsilon_{11}, \varepsilon_{22}, \varepsilon_{33}$ are not zero, and they reflect the relative changes in the sample length along the coordinates x_1, x_2, x_3, respectively:

$$\frac{dx'_n - dx_n}{dx_n} = \varepsilon_{nn} = \frac{\partial u_n}{\partial x_n}$$

(2.7)

Here, no summation over the index n is assumed. As follows from the definition, strain is assumed to be negative under compression and positive under tension. In Fig. 2.2B, where the compressive force is applied in the x_3 direction the component ε_{33} is negative and the components ε_{11} and ε_{22} are positive.

Off-diagonal terms of the strain tensor ε_{nm} are called shear strains and represent one-half of the small angle change γ_{nm} between original lines along x_n and x_m axes at right angle to one another before deformation:

$$\gamma_{nm} \approx \sin(\gamma_{nm}) = 2\varepsilon_{nm} = \frac{\partial u_n}{\partial x_m} + \frac{\partial u_m}{\partial x_n}$$

(2.8)

A pure shear deformation is shown in Fig. 2.2C, where a shear force is applied sideways on the sample. In Fig. 2.2C, two initially orthogonal lines along x_2 and x_3 directions have an angle 90-γ_{23} after deformation. In a pure shear deformation, there is no change in volume, only the shape of the body changes. Because no changes in the length along coordinates x_i are produced, all the diagonal terms of the strain tensor are zero. In the example presented in Fig. 2.2C, only components ε_{23} and ε_{32} in the strain tensor are nonzero.

3. Forces, stress, and equilibrium equation

There are two types of forces we consider here: body forces and surface forces. Body forces act on all material particles in the tissue, whereas surface forces act on surface elements whether it is the bounding surface or internal surface. Examples of body forces are gravity, inertial, and electromagnetic forces. Contact forces between bodies or pressure are types of surface forces. For example, all forces shown in Fig. 2.2 are surface forces.

When external forces are applied to a tissue, the internal forces arise to maintain the tissue's original shape. Such forces are higher for stiffer tissues; therefore, deformation of stiffer tissues requires more significant effort. The intensity of the internal forces characterizes the magnitude of the mechanical response and is called stress. Stress is usually defined as a force per unit cross-sectional area. Although the concept of stress is similar to that of pressure, and stress has the same dimension as pressure, i.e., pascals (Pa), stress takes into account force and area orientations. Therefore for every point in the tissue, stress is defined by a tensor called the stress tensor $\boldsymbol{\sigma}$.

Consider the area of an infinitesimal element in a surface ds, with ds having an orientation given by the outward unit normal vector $\boldsymbol{n}$, as shown in Fig. 2.3. If the force vector $\boldsymbol{df}$ is applied on the surface element ds, we can define a traction vector $\boldsymbol{\tau}$ as follows:

$$\tau_n = \frac{df_n}{ds}.\qquad(2.9)$$

The traction vector $\boldsymbol{\tau}$ is different for different orientations of the area ds. Because area orientation is characterized by the unit outward normal vector $\boldsymbol{n}$, we have

$$\tau_n = n_m \sigma_{mn},\qquad(2.10)$$

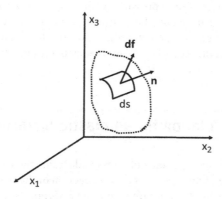

FIGURE 2.3

The concept of stress.

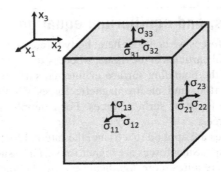

FIGURE 2.4

Notations for the components of stress tensor.

In Eq. (2.10), the index m denotes the orientation of the area and the index n denotes the orientation of the force vector. In the linear theory, the area ds may be considered in either the reference or the current coordinate systems with only negligible difference, and the stress tensor is symmetric $\sigma_{nm} = \sigma_{mn}$. Notations of the stress tensor components are illustrated in Fig. 2.4 for an infinitesimal cubic volume with sides parallel to coordinate axes. For example, for the cube surface orthogonal to the x_3 axis, the component of stress acting normal to the surface in x_3 direction is σ_{33}, while σ_{31} and σ_{32} are the components in the x_1 and x_2 directions, respectively. Therefore similar to strain, diagonal components of the stress tensor are called normal stress, while off-diagonal components are called shear stress, and represent the stress acting in the plane of the area. It follows from Eq. (2.10) that a stress tensor component is positive if both the outward normal vector and the traction vector point in the same (positive or negative) direction. For instance, all stress components presented in Fig. 2.4 are positive, while for hidden surfaces, the positive directions for stress are opposite. Following this rule, if the normal stress produces a tension, it is positive and it is negative for compression. Consider, as an example, a hydrostatic compression (Fig. 2.2A) in which the uniform pressure P acts on the surface of the tissue. Because force is always orthogonal to the tissue surface, all off-diagonal components of the stress tensor are zero, whereas the diagonal components are negative: $\sigma_{11} = \sigma_{22} = \sigma_{33} = -P$.

4. Stress-strain relation for an elastic material and elastic constants

The relation between the components of stress and strain tensors defines mechanical properties of tissues, i.e., how significant is the reaction of the tissue (stress) to deformation (strain). In general, this relation could be very complex, but here we assume

Hooke's law in which tissue is considered as an elastic body and the stress tensor is linearly proportional to the strain tensor:

$$\sigma_{nm} = C_{nmkl}\varepsilon_{kl}, \tag{2.11}$$

where C_{nmkl} is a tensor of elastic constants, which is independent of the strain and stress. Usually, Hooke's law is considered as an appropriate model for biological tissue if deformation is small. The relation (2.11) is also called physically linear stress-strain relation to underline that this type of linearity reflects physical properties of the tissue and is different from geometric linearity, which is characteristic of deformation, but not tissue property. In practice, in most cases, when the strain tensor components are less than 0.1, the linear stress-strain relation is considered as a good approximation.

The elastic properties of a material described by Eq. (2.11) depend on the direction and these types of materials are called anisotropic. In general, a tensor of elastic constants C_{nmkl} has 81 components, but due to the symmetry of both the stress and strain tensors, only 21 components are independent. Anisotropic material is a good approximation for many biological tissues that are made from fibers, such as muscle tissues. However, it is difficult to solve an inverse problem with so many parameters; therefore, very often, an isotropic material is considered, in which all directions are equal:

$$\sigma_{nm} = \lambda\Theta\delta_{nm} + 2\mu\varepsilon_{nm}, \tag{2.12}$$

where Θ is the first invariant of the strain tensor, as defined in Eq. (2.6); δ_{nm} is a Kronecker symbol; and λ and μ are called the Lame constants. The second Lame constant μ is also called the shear elastic modulus. Therefore, the elastic properties of a linear isotropic medium can be characterized by two independent parameters. Other elastic constants that are often used to describe tissue elasticity are Young's modulus E, Poisson's ratio ν, and bulk modulus K. They are connected with the Lame constants as

$$\lambda = \frac{E\nu}{(1+\nu)(1-2\nu)}, \quad \mu = \frac{E}{2(1+\nu)}, \quad K = \lambda + \frac{2}{3}\mu. \tag{2.13}$$

The bulk modulus, K, accounts for the resistance of the tissue to changing volume and relates the external pressure to the cubical dilatation, $K\Theta = -P$, for a uniform hydrostatic state of stress (see Fig. 2.2A). The shear elastic modulus, μ, quantifies a deformation without change in volume, where only the shape of the body changes (see Fig. 2.2C). In Eq. (2.13), Young's modulus, the shear modulus, and the bulk modulus are all positive; therefore, the Poisson's ratio for isotropic materials is strictly bounded between -1 and 0.5, but for most materials, it is in the range between 0.0 and 0.5. By definition, Poisson's ratio is the ratio of relative contraction to relative expansion in the uniaxial compression (see Fig. 2.2B), such that in uniaxial compression $\nu = -\varepsilon_{33}/\varepsilon_{11} = -\varepsilon_{33}/\varepsilon_{22}$. Therefore, Poisson's ratio also characterizes a material's compressibility and is 0.5 for incompressible materials. Because water is the major constituent of soft biological tissues, tissues are characterized by a low degree of compressibility. This means that the bulk modulus is orders of magnitude larger than the shear modulus (i.e., K, $\lambda \gg \mu$, and $\nu \approx 0.5$). Thus tissues are generally assumed to be nearly incompressible materials, i.e., volume change is minimal for any type of load. For incompressible materials the divergence of the displacement vector is zero:

$$\Theta = \nabla \cdot \boldsymbol{u} = 0. \tag{2.14}$$

Under these conditions, the stress-strain relation (2.12) can be reduced to

$$\sigma_{nm} = p\delta_{nm} + 2\mu\varepsilon_{nm}, \tag{2.15}$$

where p is an internal pressure defined as

$$p = \lim_{\substack{\lambda \to \infty \\ \Theta \to 0}} \lambda\Theta. \tag{2.16}$$

A consequence of the low compressibility of soft tissues is that Young's modulus is about three times its shear modulus (i.e., $E = 3\mu$). It allows the elastic behavior of many soft tissues to be characterized by only one elastic constant, i.e., the Young's or shear modulus, and this constant can be used to quantify tissue elasticity regardless of the type of deformation.

5. Equilibrium equations

In the equilibrium state the internal stresses for each volume should be in balance with all body forces, including inertia forces. Therefore, for every volume V and surface S the sum of the surface and body forces is zero:

$$\int_S \tau_n ds + \int_V F_n dV = \int_S n_m \sigma_{mn} ds + \int_V F_n dV = 0, \tag{2.17}$$

where F_n are the components of the body force per unit volume $\boldsymbol{F}$. Converting the surface integral to a volume integral using the divergence theorem of Gauss, and taking into account the symmetry of the stress tensor, Eq. (2.17) can be written as

$$\int_V \frac{\partial \sigma_{nm}}{\partial x_m} dV + \int_V F_n dV = 0. \tag{2.18}$$

As Eq. (2.18) should be satisfied for arbitrary volume V, we can write in vector notation as

$$\nabla \cdot \boldsymbol{\sigma} + \boldsymbol{F} = 0. \tag{2.19}$$

Eq. (2.19) is called the equilibrium equation, which appears in expanded form in the rectangular Cartesian coordinate system as

$$\frac{\partial \sigma_{11}}{dx_1} + \frac{\partial \sigma_{12}}{\partial x_2} + \frac{\partial \sigma_{13}}{\partial x_3} + F_1 = 0,$$

$$\frac{\partial \sigma_{21}}{dx_1} + \frac{\partial \sigma_{22}}{\partial x_2} + \frac{\partial \sigma_{23}}{\partial x_3} + F_2 = 0, \tag{2.20}$$

$$\frac{\partial \sigma_{31}}{dx_1} + \frac{\partial \sigma_{32}}{\partial x_2} + \frac{\partial \sigma_{33}}{\partial x_3} + F_3 = 0.$$

Eq. (2.19) must be satisfied at all internal points of the tissue. In combination with Hooke's law in Eq. (2.12) and the strain-displacement relation (Eq. 2.5), we have the set of governing equations to characterize deformation of the linear isotropic tissue. In addition, prescribed conditions on displacement or stress components must be satisfied on the parts of the boundary of the tissue B_u and B_σ, respectively:

$$u_n = u_n^0 \text{ on } B_u, \tag{2.21}$$

$$\sigma_{nm} n_m = \tau_n^0 \text{ on } B_\sigma.$$

where n_m are the components of the unit outward normal to the boundary of the tissue and u_n^0 and τ_n^0 are the components of the displacement and traction vectors, respectively, of the boundary.

Sometimes it is convenient to have the equilibrium equations expressed in terms of displacement vector. For a homogeneous tissue, in which the elastic moduli do not depend on spatial coordinates, combining Eqs. (2.5), (2.12), and (2.19) gives

$$(\lambda + \mu)\nabla\nabla \cdot \boldsymbol{u} + \mu\nabla^2\boldsymbol{u} + \boldsymbol{F} = 0, \tag{2.22}$$

which are independent of the coordinate system and are called the Navier-Cauchy equations. Taking into account Eqs. (2.14) and (2.15) for incompressible case, Eq. (2.19) has the form

$$\nabla p + \mu\nabla^2\boldsymbol{u} + \boldsymbol{F} = 0, \tag{2.23}$$

$$\nabla \cdot \boldsymbol{u} = 0.$$

6. Cylindrical and spherical coordinate systems

Because of the geometry of tissues, sometimes it is convenient to use the components of the strain tensor in other coordinate systems than the rectangular Cartesian system. For example, if objects such as vessel or eye are considered, cylindrical and spherical coordinate systems can be more appropriate to use. For reference, we give here the corresponding representation of the strain tensor in Eq. (2.5) and the equilibrium Eq. (2.19) in cylindrical (r, θ, z) and spherical (r, φ, ϑ) coordinate systems [2].

In the cylindrical coordinate system the components of the strain tensor have the form

$$\varepsilon_{rr} = \frac{\partial u_r}{\partial r}, \quad \varepsilon_{\theta\theta} = \frac{1}{r}\frac{\partial u_\theta}{\partial \theta} + \frac{u_r}{r}, \quad \varepsilon_{zz} = \frac{\partial u_z}{\partial z}, \quad \varepsilon_{rz} = \frac{1}{2}\left(\frac{\partial u_r}{\partial z} + \frac{\partial u_z}{\partial r}\right), \tag{2.24}$$

$$\varepsilon_{r\theta} = \frac{1}{2}\left(\frac{1}{r}\frac{\partial u_r}{\partial \theta} + \frac{\partial u_\theta}{\partial r} - \frac{u_\theta}{r}\right), \quad \varepsilon_{\theta z} = \frac{1}{2}\left(\frac{\partial u_\theta}{\partial z} + \frac{1}{r}\frac{\partial u_z}{\partial \theta}\right)$$

Equilibrium equations:

$$\frac{\partial \sigma_{rr}}{dr} + \frac{1}{r}\frac{\partial \sigma_{r\theta}}{\partial \theta} + \frac{\partial \sigma_{rz}}{\partial z} + \frac{\sigma_{rr} - \sigma_{\theta\theta}}{r} + F_r = 0,$$

$$\frac{\partial \sigma_{r\theta}}{dr} + \frac{1}{r}\frac{\partial \sigma_{\theta\theta}}{\partial \theta} + \frac{\partial \sigma_{\theta z}}{\partial z} + \frac{2}{r}\sigma_{r\theta} + F_\theta = 0, \tag{2.25}$$

$$\frac{\partial \sigma_{rz}}{dr} + \frac{1}{r}\frac{\partial \sigma_{\theta z}}{\partial \theta} + \frac{\partial \sigma_{zz}}{\partial z} + \frac{1}{r}\sigma_{rz} + F_z = 0.$$

In the spherical coordinate system, the components of the strain tensor have the form

$$\varepsilon_{rr} = \frac{\partial u_r}{\partial r}, \quad \varepsilon_{\varphi\varphi} = \frac{1}{r\sin\vartheta}\frac{\partial u_\varphi}{\partial \varphi} + \frac{u_r}{r} + \text{ctg}\vartheta\frac{u_\vartheta}{r}, \quad \varepsilon_{\vartheta\vartheta} = \frac{1}{r}\frac{\partial u_\vartheta}{\partial \vartheta} + \frac{u_r}{r}, \quad \varepsilon_{r\varphi}$$

$$= \frac{1}{2}\left(\frac{1}{r\sin\vartheta}\frac{\partial u_r}{\partial \varphi} - \frac{u_\varphi}{r}\right),$$

$$\varepsilon_{r\varphi} = \frac{1}{2}\left(\frac{1}{r\sin\vartheta}\frac{\partial u_r}{\partial \varphi} - \frac{u_\varphi}{r} + \frac{\partial u_\varphi}{\partial r}\right), \quad \varepsilon_{r\vartheta} = \frac{1}{2}\left(\frac{1}{r}\frac{\partial u_r}{\partial \vartheta} - \frac{u_\vartheta}{r} + \frac{\partial u_\vartheta}{\partial r}\right), \tag{2.26}$$

$$\varepsilon_{\varphi\vartheta} = \frac{1}{2}\left(\frac{1}{r}\frac{\partial u_\varphi}{\partial \vartheta} - \frac{u_\varphi}{r}\text{ctg}\vartheta + \frac{1}{r\sin\vartheta}\frac{\partial u_\vartheta}{\partial \varphi}\right).$$

Equilibrium equations:

$$\frac{\partial \sigma_{rr}}{dr} + \frac{1}{r\sin\vartheta}\frac{\partial \sigma_{r\varphi}}{\partial \varphi} + \frac{1}{r}\frac{\partial \sigma_{r\vartheta}}{\partial \vartheta} + \frac{2\sigma_{rr} - \sigma_{\varphi\varphi} - \sigma_{\vartheta\vartheta} + \sigma_{r\vartheta}\text{ctg}\vartheta}{r} + F_r = 0,$$

$$\frac{\partial \sigma_{r\varphi}}{dr} + \frac{1}{r\sin\vartheta}\frac{\partial \sigma_{\varphi\varphi}}{\partial \varphi} + \frac{1}{r}\frac{\partial \sigma_{\varphi\vartheta}}{\partial \vartheta} + \frac{3\sigma_{r\varphi} + 2\sigma_{\varphi\vartheta}\text{ctg}\vartheta}{r} + F_\varphi = 0, \tag{2.27}$$

$$\frac{\partial \sigma_{r\vartheta}}{dr} + \frac{1}{r\sin\vartheta}\frac{\partial \sigma_{\varphi\vartheta}}{\partial \varphi} + \frac{1}{r}\frac{\partial \sigma_{\vartheta\vartheta}}{\partial \vartheta} + \frac{3\sigma_{r\vartheta} + (\sigma_{\vartheta\vartheta} - \sigma_{\varphi\varphi})\text{ctg}\vartheta}{r} + F_\vartheta = 0.$$

7. Basic solutions

7.1 Uniaxial deformation

Perhaps the most common test of elastic properties for tissues and tissuelike materials is the uniaxial compression/tension test schematically shown in Fig. 2.2B. By subjecting a sample of tissue to a controlled compressive or tensile deformation along a single axis, the applied load and changes in dimensions can be measured to obtain a stress-strain profile of the sample. Uniaxial tests are actively used in elastography to measure Young's modulus of the phantoms and soft tissues because of the simplicity of the experiment and the straightforward interpretation of the results. Consider a prismatic bar with dimensions a_1, a_2, a_3, such that $-a_1/2 \le x_1 \le a_1/2$, $-a_2/2 \le x_2 \le a_2/2$, and $0 \le x_3 \le a_3$, uniformly deformed in x_3 direction with strain ε_0. We assume that the deformation is homogeneous, i.e., the components of the strain tensor are independent of the coordinates. For example, if the uniaxial test is performed using two plates on the top and bottom of the sample, this assumption means that there is no friction between the plates and the sample. The displacement in x_3 direction is fixed on the bottom ($x_3 = 0$) and it is $a_3\varepsilon_0$ on the top of the sample ($x_3 = a_3$). No stress is applied on the surface of the sample except on the top and bottom; therefore, taking into account notations for the stress components (see Fig. 2.4), we can write down the mixed boundary condition for this uniaxial problem:

$$\sigma_{11} = \sigma_{12} = \sigma_{13} = 0, \quad \text{on } x_1 = -a_1/2, a_1/2,$$

$$\sigma_{22} = \sigma_{21} = \sigma_{23} = 0, \quad \text{on } x_2 = -a_2/2, a_2/2, \quad (2.28)$$

$$u_3 = \sigma_{31} = \sigma_{32} = 0, \quad \text{on } x_3 = 0,$$

$$u_3 = a_3\varepsilon_0, \quad \sigma_{31} = \sigma_{32} = 0, \quad \text{on } x_3 = a_3.$$

Because we have a homogeneous deformation here, any linear functions of the coordinates satisfy Eq. (2.22), and after taking into account the boundary conditions in Eq. (2.28), we have a solution in the form $u_1 = -\nu\, \varepsilon_0 x_1$, $u_2 = -\nu\, \varepsilon_0 x_2$, and $u_3 = \varepsilon_0 x_3$. As described earlier, only the diagonal components of the strain tensor are nonzero, whereas for the stress tensor, only the axial stress σ_{33} is nonzero. If the applied force F_0 and displacement on the sample surface are controlled in the experiment, the Young's modulus of the sample can be obtained using the simple relation:

$$\sigma_0 = E\varepsilon_0, \quad (2.29)$$

where $\sigma_0 = \sigma_{33} = F_0/S_0$ is the uniform stress in the sample and S_0 is a surface area of the sample, and $S_0 = a_1 a_2$ for prismatic bar. Note here that both stress σ_0 and strain ε_0 are negative in the compression test and positive in the tensile test.

The relation (2.29) could be applied not only to a prismatic bar but also for other shapes of the samples, including the cylindrical samples, as can be shown using (Eqs. 2.24 and 2.25). The boundary conditions in Eq. (2.28) are not always easy to satisfy in the experiment, and in the compression test, for example, the influence of the friction factor should be as small as possible to avoid an overestimation of the Young's modulus.

7.2 Uniaxial deformation of the tissue with spherical inclusion

Another example of the tissue response to external load is uniform deformation of the homogeneous tissue having inclusion inside, which has different elastic properties in comparison with surrounding tissue. An analytic solution for compressible medium has been derived by Goodier for spherical and cylindrical inclusions [7]. Similar solutions for an incompressible medium can be readily derived from Goodier's results as the value of Poisson's ratio v approaches 0.5 [8]. This problem is a simple approximation of the typical situation in static elastography in which a compressive force is applied to the surface of the tissue that is continuously imaged with ultrasonography or other imaging technology (see Chapter 4 for details). The induced axial strain in the direction of the applied deformation is inversely proportional to the Young's modulus at that point and, therefore, an axial strain image (or "elastogram") reflects the distribution of the elasticity within the interrogated tissue. Therefore, for example, a stiff lesion in the breast tissue can be visualized.

Consider a uniaxial deformation of incompressible tissue with a small spherical inclusion of radius R, as shown in Fig. 2.5A, such that R is much smaller than the characteristic size of the tissue. Similar to the previous example, the tissue uniformly deformed in x_3 direction with strain ε_0. In this case, we can assume that the medium is unlimited and compressive stress is applied in infinity, where $|x_3| \to \infty$. The problem is formulated in a spherical system of coordinates (r, φ, ϑ), and the origin of the coordinate system is placed at the center of the inclusion. The polar axis of both systems of coordinates is along an applied deformation, that is, an angle φ is between a radius vector and the deformation direction (see Fig. 2.5A). It is assumed that the surrounding tissue and inclusion have Young's moduli of E_0 and E_1, respectively. Because the problem is axisymmetric, the φ-component of the displacement vector $(u_r, u_\varphi, u_\vartheta)$ is zero, and dependence on the angle φ is not considered. For both the

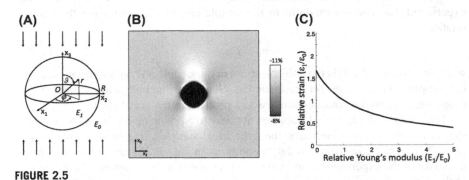

FIGURE 2.5

A) Uniform deformation of the spherical inclusion in an unlimited homogeneous medium. (B) Spatial distribution of the axial strain ε_{33} in the central plane ($x_1 = 0$) for a hard inclusion $E_1/E_0 = 2$ and 10% of the external compressive deformation ($\varepsilon_{33} = -0.1$). (C) Relative axial strain ε_{33} inside an inclusion as a function of the relative Young's modulus of inclusion.

surrounding tissue and the inclusion, Eq. (2.23) should be satisfied, in which no body forces are assumed. On the surface of the inclusion $r = R$, there are the conditions of continuity for displacement and stress u_r, u_ϑ, σ_{rr} and $\sigma_{r\vartheta}$. Taking into account Eq. (2.26), the solution for displacement vector in spherical coordinates (u_r, u_φ, u_ϑ) can be obtained as

$$u_r = \frac{1}{4} U(r)(1 + 3\cos(2\vartheta)), \qquad u_\varphi = 0,$$

$$u_\vartheta = -\frac{1}{4}\left(2U(r) + r\frac{\partial U(r)}{\partial r}\right)\sin(2\vartheta),$$

$$U(r) = \begin{cases} A\varepsilon_0 r, & r \leq R \\ \varepsilon_0 r\left(1 + 5B\left(\dfrac{R}{r}\right)^3 - 3B\left(\dfrac{R}{r}\right)^5\right), & r \geq R \end{cases}, \qquad (2.30)$$

$$A = \frac{5}{3+2k}, \quad B = \frac{1-k}{3+2k}, \quad k = \frac{E_1}{E_0}.$$

Here, we assume that the solution is limited at the center of the system of coordinates and must match the strain applied at infinity: $\lim\limits_{r\to\infty} U(r)/r = \varepsilon_0$. After transformation of the vector components to the rectangular coordinate system the solution has the form

$$u_1 = -\frac{x_1}{2r}\left(U(r) - \frac{x_3^2}{r^2}V(r)\right), \quad u_2 = -\frac{x_2}{2r}\left(U(r) - \frac{x_3^2}{r^2}V(r)\right), \qquad (2.31)$$

$$u_3 = \frac{x_3}{2r}\left(2U(r) - \frac{(x_1^2 + x_2^2)}{r^2}V(r)\right),$$

$$V(r) = U(r) - r\frac{\partial U(r)}{\partial r}, \quad r = \sqrt{x_1^2 + x_2^2 + x_3^2}.$$

In a homogeneous medium $k = 1$, this solution represents the uniaxial deformation of the incompressible tissue, considered previously $u_1 = -\varepsilon_0 x_1/2$, $u_2 = -\varepsilon_0 x_2/2$, and $u_3 = \varepsilon_0 x_3$. For an inhomogeneous medium $k \neq 1$, the solution in (Eqs. 2.30 and 2.31) approaches the uniaxial deformation far away from the inclusion, where $r \to \infty$. In Fig. 2.5B, the axial strain ε_{33} image for the stiff inclusion ($k = 2$) embedded in a homogeneous background is shown for the central plane ($x_1 = 0$). The central dark region indicates a region of inclusion with low deformation and high stiffness. Inside the inclusion the axial strain is a constant $\varepsilon_{33} = \varepsilon_1 = A\varepsilon_0$, such that parameter A defines the strain contrast between the inclusion and the surrounding tissue. Based on this analysis, the normalized axial strain $\varepsilon_1/\varepsilon_0$ in the direction of the applied deformation within the inclusion is shown in Fig. 2.5C as functions of the relative Young's modulus of the inclusion E_1/E_0. Fig. 2.5C demonstrates that the dependence of the strain contrast on elasticity contrast is nonlinear, and the axial strain distribution can be considered only as a rough approximation of the elasticity distribution in the tissue. In model problem, using the functional relation presented in Fig. 2.5C, the unknown relative Young's modulus of an inclusion can be evaluated based on the measured strain distribution and this could be considered as an example of the simplest inverse problem solution. In more complex geometry, however, such approach can lead to significant errors in estimating the Young's modulus distribution.

In two considered examples of static deformation, it is assumed that deformations vary slowly with respect to the measurement duration. This assumption is not always valid in the experiments, when the inertia terms and viscoelastic behavior cannot be neglected. For example, in the uniaxial test, the strain rate can be a significant factor in the measured stress level for viscous materials. Dynamic deformation and viscoelastic models are considered in the following.

8. Dynamic deformation

Static methods usually rely on slow tissue deformation where the inertia forces in the equilibrium equations can be neglected and no elastic waves propagate in the tissue. In dynamic elastography methods, the tissue is perturbed by impulsive loading or harmonic vibrations using mechanical devices or acoustic radiation force and the tissue motion is measured as a function of time. In this case, inertia terms, i.e., the product of acceleration and the mass per unit volume of the tissue, should be included in the governing equations to take into account the effect of elastic wave propagation. Eq. (2.19) can be written as

$$\nabla \cdot \boldsymbol{\sigma} + \boldsymbol{F} = \rho \frac{\partial^2 \boldsymbol{u}}{\partial t^2}, \tag{2.32}$$

where ρ is the tissue density. For a homogeneous tissue, Eqs. (2.22) and (2.23) are

$$(\lambda + \mu)\nabla\nabla \cdot \boldsymbol{u} + \mu\nabla^2 \boldsymbol{u} = \rho \frac{\partial^2 \boldsymbol{u}}{\partial t^2}, \tag{2.33}$$

for compressible and

$$\nabla p + \mu\nabla^2 \boldsymbol{u} = \rho \frac{\partial^2 \boldsymbol{u}}{\partial t^2} \tag{2.34}$$

$$\nabla \cdot \boldsymbol{u} = 0,$$

for incompressible cases. Also, initial conditions should be defined to solve Eqs. (2.32)–(2.34).

Shear wave elastography is based on the measurement of the speed of shear wave propagation in tissues (see Chapter 5) and, generally speaking, does not require knowledge about stress to evaluate absolute values of the elastic parameters [9].

Applying the Helmholtz decomposition to the displacement vector $u = \nabla\Phi + \nabla \times \Psi$ permits us to describe the displacement vector in terms of its rotation-free and divergence-free components [2]. Combining this representation with Eq. (2.33) and taking into account the vector relations $\nabla \times \nabla\Phi = 0$ and $\nabla \cdot \nabla \times \Psi = 0$ result in the equation for the scalar compressional wave potential Φ and vector shear wave potential Ψ:

$$\nabla\left[(\lambda + 2\mu)\nabla^2\Phi - \rho\frac{\partial^2\Phi}{\partial t^2}\right] + \nabla \times \left[\mu\nabla^2\Psi - \rho\frac{\partial^2\Psi}{\partial t^2}\right] = 0. \qquad (2.35)$$

One of the possible solutions of Eq. (2.35) can be obtained by setting both bracketed terms to zero, which results in two independent wave equations:

$$\nabla^2\Phi - \frac{1}{c_l^2}\frac{\partial^2\Phi}{\partial t^2} = 0, \qquad \nabla^2\Psi - \frac{1}{c_t^2}\frac{\partial^2\Psi}{\partial t^2} = 0, \qquad (2.36)$$

where

$$c_l = \sqrt{\frac{\lambda + 2\mu}{\rho}}, \qquad c_t = \sqrt{\frac{\mu}{\rho}}. \qquad (2.37)$$

Eq. (2.36) represents two types of waves propagating with velocities c_l and c_t, respectively: the compressional wave, or P-wave, and the shear wave, or S-wave. The propagation of compressional wave requires changes in volume and pressure. Therefore the speed of compressional wave c_l is also called the speed of sound. In terms of bulk modulus the equation for speed of sound can be rewritten as $c_l = \sqrt{(K + 4\mu/3)/\rho}$. In incompressible medium, the speed of sound goes to infinity, whereas in most soft biological tissues, it is close to the speed of sound in water: 1500 m/s.

In contrast to compressional wave, the shear wave propagates without change in volume and has a finite speed in incompressible medium. As discussed before, for soft tissues the bulk modulus is orders of magnitude larger than the shear modulus and as a result, shear wave speed is much smaller than the speed of sound, i.e., $c_l \gg c_t$. The speed of shear wave c_t plays a critical role in elastography because it is directly connected with shear elastic modulus. After measuring tissue motion and estimating the speed of shear wave in tissues, the shear modulus can be directly calculated based on Eq. (2.37). However, it is important to note that this equation is based on significant oversimplifications. Boundary conditions, tissue heterogeneity, and complex mechanical behaviors of tissues, such as viscoelasticity or anisotropy, can lead to more complex dependences between shear wave speed and tissue elasticity. Although compressional and shear wave motions cannot be always decoupled in the way described earlier, representation (2.36) allows to perform the analyses of the dynamic problems for many practical applications.

It is often useful to consider the dynamic problems of elastography in spectral domain assuming harmonic tissue excitation, where a tissue is subjected to periodic oscillations. Such an approach is often used, for example, in sonoelastography or magnetic resonance elastography, see Chapters 3 and 6, respectively. Also, this is one of the simplest ways to experimentally determinate the viscoelastic properties of tissues. On the other hand, any dynamic load can be presented as a superposition of many infinitely long wave trains of different frequencies using the Fourier transform, and tissue response to the dynamic load can be written as a combination of spectral components. Thus the problem of the dynamic tissue response can be transformed into one of determining the set of Fourier coefficients.

Consider the problem of the dynamic deformation of medium in the frequency domain assuming harmonically applied force with angular frequency ω. We assume that the external force, displacement vector, and the strain and stress tensor components are specified as being harmonic functions of time according to

$$u_n = \tilde{u}_n e^{i\omega t}, \quad \sigma_{nm} = \tilde{\sigma}_{nm} e^{i\omega t}, \quad \varepsilon_{nm} = \tilde{\varepsilon}_{nm} e^{i\omega t}, \quad \Theta = \tilde{\Theta} e^{i\omega t}, F_n = \tilde{F}_n e^{i\omega t}, \quad (2.38)$$

where $\tilde{\sigma}_{nm}$, $\tilde{\varepsilon}_{nm}$, $\tilde{u}_n$, $\tilde{F}_n$, and $\tilde{\Theta}$ are the amplitudes of stress and strain tensor components, displacement components, components of the body force, and dilatation, respectively, as defined in Eq. (2.6). Using the motion Eq. (2.32), the wave equations for steady-state harmonic motion is

$$\frac{\partial \tilde{\sigma}_{nm}}{\partial x_m} + \rho \omega^2 \tilde{u}_n + \tilde{F}_n = 0. \quad (2.39)$$

In general, for harmonic motion, the velocity of elastic wave, i.e., the velocity at which the phase of the wave propagates in space, depends on the frequency and is called the phase velocity.

9. Examples of dynamic problems
9.1 Plane wave propagation

Consider the propagation of a plane elastic wave in an unlimited, isotropic, and homogeneous medium, in which the displacements are the function of only one space coordinate, x_1, for example, and time. As there is no dependence on coordinates x_2 and x_3, Eq. (2.36) results in three independent equations:

$$\frac{\partial^2 u_1}{\partial x_1^2} - \frac{1}{c_l^2} \frac{\partial^2 u_1}{\partial t^2} = 0, \quad \frac{\partial^2 u_2}{\partial x_1^2} - \frac{1}{c_t^2} \frac{\partial^2 u_2}{\partial t^2} = 0, \quad \frac{\partial^2 u_3}{\partial x_1^2} - \frac{1}{c_t^2} \frac{\partial^2 u_3}{\partial t^2} = 0. \quad (2.40)$$

Eq. (2.40) is the simple one-dimensional wave equation for waves traveling in x_1 direction. For the compressional wave the particles are moving parallel to the direction of wave propagation. This type of wave is often called a longitudinal wave and c_l is called longitudinal wave speed. For the shear waves, the directions of the particle motion is orthogonal to x_1 direction and therefore shear waves are also called transverse waves. The general solution of Eq. (2.40) is the sum of two arbitrary functions: $f_1(t - x_1/c) + f_2(t + x_1/c)$, i.e., the sum of a wave traveling in positive and negative directions with velocity $c = c_l, c_t$.

For the plane harmonic shear wave, where displacements have the form in Eq. (2.38), and the wave propagates in the positive x_1 direction, the displacement in x_3 direction has the form

$$u_3 = \tilde{u}_3 e^{i\omega t} = A e^{i\omega\left(t - \frac{x_1}{c_t^*}\right)}, \tag{2.41}$$

where A is a constant and c_t^* is a phase velocity of shear wave. For the plane elastic wave the phase speed does not depend on frequency and $c_t^* = c_t$.

9.2 Motion of solid sphere under dynamic load

Consider a solid sphere of radius R embedded in an unlimited incompressible elastic medium [10]. Assume also that an external, time-dependent force F^{ext}, such as acoustic radiation force or magnetic force, is applied to the sphere. The motion in the incompressible medium is described by Eq. (2.34), while the sphere is not deformed, and the boundary conditions in terms of displacement for Eq. (2.34) are defined at the sphere surface. Elastic waves generated by the sphere motion attenuate far away from the sphere. It can be shown that the displacement of the sphere U in the direction of force can be described as [10]

$$U + \frac{R}{c_t}\frac{\partial U}{\partial t} + \frac{1}{9}(1 + 2\beta)\frac{R^2}{c_t^2}\frac{\partial^2 U}{\partial t^2} = \frac{F^{ext}}{6\pi\mu R}, \tag{2.42}$$

where the parameter β is the solid sphere density normalized by medium density. We consider that the external force applied to a solid sphere is impulsive, i.e.,

$$F^{ext} = \begin{cases} F_0, & 0 \le t \le t_0 \\ 0, & t > t_0 \end{cases}, \tag{2.43}$$

where t_0 is the duration of the force pulse and F_0 is amplitude of the force. The solution of Eq. (2.42) has the form

$$U(t) = \begin{cases} \dfrac{F_0}{6\pi\mu R}\left(1 + \dfrac{\xi_2}{\xi_1 - \xi_2}e^{\frac{c_t}{R}\xi_1 t} - \dfrac{\xi_1}{\xi_1 - \xi_2}e^{\frac{c_t}{R}\xi_2 t}\right), & 0 \le t \le t_0 \\[4mm] \dfrac{F_0}{6\pi\mu R}\left(\left(1 - e^{-\frac{c_t}{R}\xi_1 t_0}\right)\dfrac{\xi_2}{\xi_1 - \xi_2}e^{\frac{c_t}{R}\xi_1 t} - \left(1 - e^{-\frac{c_t}{R}\xi_2 t_0}\right)\dfrac{\xi_1}{\xi_1 - \xi_2}e^{\frac{c_t}{R}\xi_2 t}\right), \\[4mm] \quad t > t_0 \end{cases} \tag{2.44}$$

where $\xi_{1,2} = -\frac{3}{2}\cdot\frac{\sqrt{5-8\beta}\pm3}{1+2\beta}$ are the roots of an algebraic equation: $(1 + 2\beta)$ $\xi^2 + 9\xi + 9 = 0$. The solution (2.44) is a sum of two time-dependent exponential functions in which the real parts of the exponents are always negative. The examples

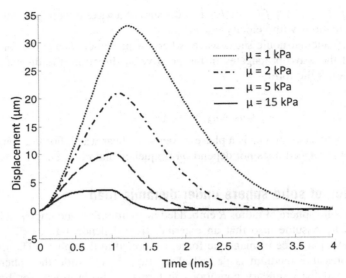

FIGURE 2.6

Displacement of the solid sphere in a medium with different shear elastic moduli for $R = 1$ mm, $\beta = 1$, $t_0 = 1$ ms, and $F_0 = 1$ mN.

of the sphere displacements as the functions of time for different shear moduli of a medium are presented in Fig. 2.6. For the infinitely long duration t_0 of the external force, the problem becomes static, the external force is compensated for by the elastic response of the medium, and the displacement of sphere is constant U_∞ as long as the external force is applied on the sphere:

$$U_\infty = \frac{F_0}{6\pi\mu R}. \tag{2.45}$$

10. Viscoelastic models

10.1 Viscoelastic tissue response

If biological tissue is considered as an elastic body, it is assumed that tissues have a capacity to store mechanical energy with no dissipation, i.e., the deformation is completely reversible. However, such assumption is correct only if deformation occurs with infinitesimal speed, where thermodynamic equilibrium is established at every moment of time. If the speed of deformation is finite, mechanical energy dissipates and the deformation process is irreversible [1,11]. Under isothermal conditions, energy dissipation is a result of the internal friction, or viscosity, and such mechanical behavior is called viscoelasticity. A viscoelastic body combines properties of a solid body to store mechanical energy and a viscous fluid, which does not store energy, but only dissipates it. The stress-strain relation of an elastic material does not depend on time, i.e., the material does not have a "memory," whereas the mechanical response of a viscoelastic material is not only determined by the current stress and strain but also defined by the history of deformation. Therefore, the viscoelastic stress-strain relation is a function of time and usually it is characterized through several typical viscoelastic responses to the time-dependent load. Most biological tissues demonstrate viscoelastic behavior including creep, relaxation, and hysteresis.

If the load is suddenly applied to the tissue, and is held constant thereafter, the tissue continues to deform and this phenomenon is called strain creep. If a tissue is suddenly deformed, and this deformation is maintained constant thereafter, the resulting stress decreases with time and this viscoelastic response is called stress relaxation. Here, the term "suddenly" means fast enough to induce a viscous response but not fast enough to induce elastic wave propagation in tissue, i.e., inertial terms are neglected for creep and relaxation phenomena. If a tissue is subject to cycles of loading and unloading and the response for each cycle is different, then it is said to display hysteresis. All these phenomena are found in most biological tissues and are used to characterize tissue viscoelasticity.

To describe viscoelastic stress-strain relation, many different models are used, but most of them are based on two one-dimensional rheologic models: the Maxwell model and the Kelvin-Voigt model (see Fig. 2.7A and B). These models can be represented as a combination of simple linear elements such as the linear spring (ideal elastic body obeying Hooke's law) and the linear dashpot (Newtonian viscous fluid obeying Newton's law).

Consider, for example, a state of a pure shear deformation shown in Fig. 2.2C, with nonzero stress and strain components σ_{23} and ε_{23}, respectively. As discussed previously, the ideal elastic body reacts instantly to the load, such that the stress is linearly proportional to strain at any moment of time t: $\sigma_{23}(t) = 2\mu\varepsilon_{23}(t)$. Stress in the viscous element is proportional to the strain rate: $\sigma_{23}(t) = 2\eta d\varepsilon_{23}(t)/dt$, where η is the shear viscosity coefficient.

Combining springs and dashpots, we can obtain different models of viscoelastic behavior. Viscoelastic behavior can be characterized through the nature of the tissue response to suddenly applied stress and strain, i.e., creep and relaxation responses. The unit step function is defined as

$$h(t) = \begin{cases} 0, & t < 0, \\ 1/2, & t = 0, \\ 1, & t > 0. \end{cases} \tag{2.46}$$

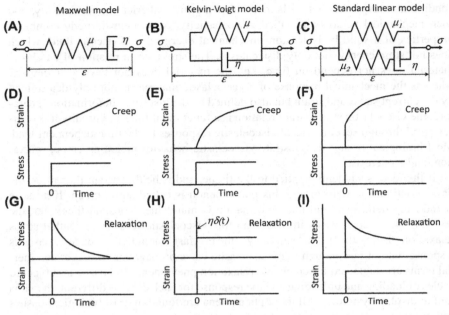

FIGURE 2.7

Rheologic models of viscoelastic body: (A) Maxwell model, (B) Kelvin-Voigt model, (C) Standard linear model, (D) Creep function of a Maxwell body, (E) Creep function of a Kelvin-Voigt model, (F) Creep function of a standard linear model, (G) Relaxation function of a Maxwell body, (H) Relaxation function of a Kelvin-Voigt body, and (I) Relaxation function of a standard linear model.

If stress or strain are step functions of time, i.e., $\sigma_{23} = \sigma_0 h(t)$ and $\varepsilon_{23} = \varepsilon_0 h(t)$, for creep and relaxation, respectively, we can define the creep function $J(t)$ and relaxation function $G(t)$ as

$$\varepsilon_{23} = J(t)\sigma_0, \tag{2.47}$$

$$\sigma_{23} = G(t)\varepsilon_0.$$

The resulting strain and stress are directly related to the creep and relaxation functions.

10.1.1 Maxwell model

The Maxwell model combines elastic and viscous elements arranged in series, as shown in Fig. 2.7A. If stress is applied to the Maxwell model, this stress produces strain in the spring, and the flow in the dashpot, such that total strain rate is

$$\frac{d\varepsilon_{23}}{dt} = \frac{1}{2\mu}\frac{d\sigma_{23}}{dt} + \frac{1}{2\eta}\sigma_{23}. \tag{2.48}$$

Solutions of Eq. (2.48) for creep and relaxation functions have the respective forms

$$J(t) = \frac{1}{2}\left(\frac{1}{\mu} + \frac{t}{\eta}\right)h(t), \tag{2.49}$$

$$G(t) = 2\mu e^{-(\mu/\eta)t}h(t).$$

Creep and relaxation functions for the Maxwell model are shown in Fig. 2.7D and G, respectively. As the creep response is unlimited for $t \to \infty$, the Maxwell body is referred to as the Maxwell fluid. The time constant $\tau_\sigma = \eta/\mu$ is called a relaxation time and it characterizes the rate of relaxation for the viscous element. A smaller relaxation time corresponds to a faster relaxation process.

10.1.2 Kelvin-Voigt model
For the Kelvin-Voigt model, elastic and viscous elements are connected in a parallel fashion, as shown in Fig. 2.7B. Both elements have the same strain, and the total stress is a sum of the stresses in both elements:

$$\sigma_{23} = 2\mu\varepsilon_{23} + 2\eta\frac{d\varepsilon_{23}}{dt}. \tag{2.50}$$

After the solution of Eq. (2.39), creep and relaxation functions are

$$J(t) = \frac{1}{2\mu}\left(1 - e^{-\left(\frac{\mu}{\eta}\right)t}\right)h(t), \tag{2.51}$$

$$G(t) = 2\eta\delta(t) + 2\mu h(t),$$

where $\delta(t)$ is a Dirac delta function. As shown in Fig. 2.7E, a sudden application of stress produces no immediate deflection because dashpot does not move immediately. The dashpot strain decreases exponentially and the creep function increases with time but with decreasing slope. As time becomes large, all the stress is carried by the spring and we have $\varepsilon_{23} = \sigma_0/2\mu$. For the relaxation function (see Fig. 2.7H) a sudden strain requires an infinite stress at the initial moment; however, after that moment, the relaxation function exhibits no stress relaxation. The time $\tau_\varepsilon = \eta/\mu$ is referred to as the retardation time and is analogous in meaning to the relaxation time: the time required for the rate of creep to approach nearly zero.

10.1.3 Standard linear body model
The Kelvin-Voigt and the Maxwell models are thus the simplest viscoelastic models. The Maxwell model corresponds to a viscoelastic fluid, whereas the Kelvin-Voigt represents a solid. Therefore the Kelvin-Voigt model is the most widely used viscoelastic model to describe soft tissue behavior under dynamic load. However, both models usually fail to represent real tissue behavior. For example, the Maxwell model demonstrates an unrealistic creep function, while the Kelvin-Voigt model shows no stress relaxation. To improve the representation of viscoelastic behavior,

more general models can be considered by combining a number of springs and dash-pots, increasing the number of viscoelastic parameters, and adding more exponential terms to the creep and relaxation functions. An example of such generalization is shown in Fig. 2.7C, F, and I, in which the standard linear body model is presented. This model is obtained by adding a spring in parallel to the Maxwell model and enables a realistic representation of both creep and relaxation processes. Two elastic parameters μ_1 and μ_2 are needed to represent the instantaneous and large-time elastic response, respectively.

10.2 Viscoelastic response to harmonic excitation

If a viscoelastic body is subjected to steady-state oscillatory conditions, stress and strain can be considered as harmonic functions of time:

$$\sigma_{23} = \widetilde{\sigma}_{23} e^{i\omega t}, \quad \varepsilon_{23} = \widetilde{\varepsilon}_{23} e^{i\omega t}, \tag{2.52}$$

where $\widetilde{\sigma}_{23}$ and $\widetilde{\varepsilon}_{23}$ are the amplitudes of shear stress and strain, respectively, and ω is the frequency of oscillations. Therefore, a differentiation with respect to time can be substituted by multiplication with $i\omega$, and the stress-strain relation can be written as

$$\sigma_{23} = 2\mu^*(i\omega)\varepsilon_{23}, \tag{2.53}$$

where $\mu^*(i\omega)$, the complex shear modulus, is a complex function of frequency. The real and imaginary parts of the complex modulus are referred to as the storage and loss moduli, respectively. Taking into account Eqs. (2.48) and (2.50), the complex shear moduli for the Maxwell and Kelvin-Voigt models have the forms

$$\mu^*(i\omega) = i\omega \left(\frac{i\omega}{\mu} + \frac{1}{\eta} \right)^{-1} \quad \text{for the Maxwell body}, \tag{2.54}$$

$$\mu^*(i\omega) = \mu + i\omega\eta \quad \text{for the Kelvin – Voigt body}.$$

10.3 Generalized viscoelastic models

Although we did consider creep and relaxation functions for shear deformation, the analogy between rheologic models and other states of uniform deformation can be established and the creep and relaxation functions for these states can be defined. Consideration of a more general three-dimensional viscoelastic problem requires a more general relationship between stress and strain tensors in the governing Eq. (2.19). The most general formulation of the linear viscoelastic stress-strain relation in the tensor form can be written as the following convolution integrals:

$$\sigma_{nm}(t) = \int_{-\infty}^{t} G_{nmkl}(t-\tau) \frac{d\varepsilon_{kl}(\tau)}{d\tau} d\tau, \tag{2.55}$$

$$\varepsilon_{nm}(t) = \int\limits_{-\infty}^{t} J_{nmkl}(t - \tau) \frac{d\sigma_{kl}(\tau)}{d\tau} d\tau.$$

Here, G_{nmkl} and J_{nmkl} are the generalized relaxation and creep functions, respectively, in the form of fourth-order tensors. The symmetry of the stress and strain tensors implies the symmetry of the relaxation and creep functions, similar to the tensor of elastic constants C_{nmkl}. Therefore, Eq. (2.55) is an analogue of the generalized Hooke's law in Eq. (2.11) for viscoelastic anisotropic materials. These representations are equivalent and may be obtained one from the other by reversing the roles of stress and strain. For isotropic materials, we can rewrite Eq. (2.55) as

$$\sigma_{nm} = \delta_{ij} \int\limits_{-\infty}^{t} \overline{\lambda}(t - \tau) \frac{\partial \Theta}{\partial \tau} d\tau + 2 \int\limits_{-\infty}^{t} \overline{\mu}(t - \tau) \frac{\partial \varepsilon_{nm}}{\partial \tau} d\tau. \tag{2.56}$$

The relaxation functions $\overline{\lambda}(t)$ and $\overline{\mu}(t)$ are analogous to the Lame constants in elasticity (see Eq. 2.12) and their specific form is defined by the chosen viscoelastic rheologic model. For the Kelvin-Voigt model of solids, Eq. (2.56) has the form

$$\sigma_{nm} = \lambda \Theta \delta_{nm} + 2\mu \varepsilon_{nm} + \zeta \frac{\partial \Theta}{\partial t} \delta_{nm} + 2\eta \frac{\partial \varepsilon_{nm}}{\partial t}, \tag{2.57}$$

where ζ is a viscosity coefficient corresponding to the Lame coefficient λ.

11. Dynamic deformation of a viscoelastic medium

For dynamic methods of elastography, viscous effects play an important role in tissue motion and could significantly influence the results of the measurements. Usually, if impulse or harmonic loads are applied, the motion is fast enough to produce elastic wave propagation. The previously derived viscoelastic stress-strain relations for quasi-static deformation apply in such cases, but additional, high-frequency effects of inertia terms should be taken into account, similar to the elastic case. Typical effects of viscosity to tissue motion include decrease of the amplitude and frequency of the vibrations, as well as higher attenuation of vibrations, especially high-frequency components. In a viscoelastic medium, elastic waves attenuate much faster than that in an elastic medium, and the phase speed of the wave grows with frequency. Although viscosity effects can confound elastographic results and lead to difficulties in interpreting elasticity measurements, the viscoelasticity used to describe tissue behavior opens new possibilities for medical diagnosis. Recent studies showed that viscosity can provide independent information about tissue conditions in addition to elasticity. For example, some pathologic changes in liver correlate with the changes in liver viscosity [12–14]. Therefore despite its name, elasticity imaging has not been exclusively considered as a means to measure elasticity, but rather it has been utilized as a way to estimate a set of different mechanical parameters that may be used.

Consider the harmonic loading of a viscoelastic medium. The relation between stress and strain components in Eq. (2.56) in the frequency domain has a form similar to that of Eq. (2.12):

$$\tilde{\sigma}_{nm} = \lambda^*(i\omega)\widetilde{\Theta}\delta_{nm} + 2\mu^*(i\omega)\tilde{\varepsilon}_{nm}, \tag{2.58}$$

except that the Lame moduli are replaced by the complex moduli $\lambda^*(i\omega)$ and $\mu^*(i\omega)$, respectively. These moduli are specified as functions of frequency for a particular viscoelastic model:

$$\lambda^*(i\omega) = \frac{i\omega\lambda\zeta}{\lambda + i\omega\zeta}, \quad \mu^*(i\omega) = \frac{i\omega\mu\eta}{\mu + i\omega\eta} \text{ for the Maxwell body,} \tag{2.59}$$

$$\lambda^*(i\omega) = \lambda + i\omega\zeta, \quad \mu^*(i\omega) = \mu + i\omega\eta \text{ for the Kelvin − Voigt body.}$$

For a homogeneous medium the equation of motion in the frequency domain is

$$(\lambda^* + \mu^*)\nabla\nabla \cdot \tilde{u} + \mu^*\nabla^2\tilde{u} + \rho\omega^2\tilde{u} = -\tilde{F}. \tag{2.60}$$

Usually, if elastic wave propagates in tissue, the different spectral components of the wave travel at different phase speeds and the wave changes its shape during propagation. This effect is called dispersion. The causes of dispersion could be propagation in the presence of boundaries, as it happens for guided waves in plates and other structural elements [15]. In an unlimited pure elastic medium, where no interaction with boundaries occurs, the shear wave propagates without dispersion and without changes in the shape. However, the presence of viscosity results in dispersion of shear waves in biological tissues.

Consider the propagation of a plane harmonic shear wave in viscoelastic and unlimited media. Similar to the last equation in (2.40), the wave propagates in x_1 direction, whereas the particles are oscillating in x_3 direction. For the amplitude of oscillation, the motion Eq. (2.39) reduces to the one-dimensional equation:

$$\frac{\partial^2\tilde{u}_3}{\partial x_1^2} + \frac{\rho\omega^2}{\mu^*(i\omega)}\tilde{u}_3 = 0. \tag{2.61}$$

Displacement in x_3 direction has the form

$$u_3 = \tilde{u}_3 e^{i\omega t} = Ae^{-\alpha_t^* x_1} e^{i\omega\left(t-\frac{x_1}{c_t^*}\right)}, \tag{2.62}$$

where A is a constant, c_t^* is the phase speed of a shear wave, and α_t^* is a wave attenuation. For a pure elastic medium the speed is defined by Eq. (2.37) and does not depend on frequency, and in addition, the wave propagates without any attenuation as a plane wave: $c_t^* = c_t$, $\alpha_t^* = 0$. In a viscoelastic medium, the speed and attenuation depend on the complex shear modulus as [11]

$$c_t^* = \frac{|\mu^*(i\omega)|}{\sqrt{\rho} Re\sqrt{\mu^*(i\omega)}} \quad , \tag{2.54}$$

$$\alpha_t^* = \frac{\sqrt{\rho}\omega Im\sqrt{\mu^*(i\omega)}}{|\mu^*(i\omega)|}.$$

For the Kelvin-Voigt model the speed and attenuation can be expressed through the shear moduli of elasticity and viscosity as

$$c_t^* = \sqrt{\frac{2(\mu^2 + \omega^2\eta^2)}{\rho\left(\mu + \sqrt{\mu^2 + \omega^2\eta^2}\right)}}, \tag{2.55}$$

$$\alpha_t^* = \sqrt{\frac{\rho\omega^2\left(\sqrt{\mu^2 + \omega^2\eta^2} - \mu\right)}{2(\mu^2 + \omega^2\eta^2)}},$$

Because attenuation grows with frequency, high-frequency components of the wave attenuate faster. Such representations are often used in shear wave elastography to estimate tissue viscosity based on the dispersion of the shear waves or attenuation with distance. Fig. 2.8 presents the examples of dispersion curves for the Kelvin-Voigt model, with various shear viscosity coefficients.

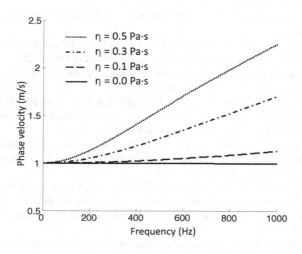

FIGURE 2.8

Shear wave dispersion in the Kelvin-Voigt viscoelastic medium for different values of the shear viscosity coefficient η, and $\mu = 1$ kPa, $\rho = 1000$ kg/m^3.

12. Summary

In this chapter the theoretic basics of governing principles used in elasticity imaging are briefly introduced. The chapter contains a description of both elastic and viscoelastic behaviors of soft tissues. Special attention is devoted to the mechanical response of soft tissues to various types of loads, which are used in elastography. Several examples of analytic solutions for static and dynamic problems are presented to demonstrate the tissue reaction to external loads. The basic ideas, equations, and formulas used in elasticity imaging are discussed and illustrated with examples. For a more detailed description of the principles of elasticity imaging, the reader is referred to other chapters of this book and review papers [16—19].

References

[1] Y.C. Fung, Biomechanics, in: Mechanical Properties of Living Tissues, second ed., Springer, 1993.

[2] L.D. Landau, E.M. Lifshitz, Theory of Elasticity, Pergamon Press, 1959.

[3] G.E. Mase, Theory and Problems of Continium Mechanics, McGraw-Hill Book Company, 1970.

[4] S. Timoshenko, J.N. Goodier, Theory of Elasticity, McGraw-Hill Book Company., 1951.

[4a] A.E.H. Love, A treatise on the mathematical theory of elasticity, 4th ed. New York: Dover Publications; 1944.

[5] J.D. Humphrey, Cardiovascular Solid Mechanics. Cells, Tissues, and Organs, Springer-Verlag, 2002.

[6] P.E. Barbone, A.A. Oberai, in: S. De, F. Guilak, M.R.K. Mofrad (Eds.), Computational Modeling in Biomechanics, Springer Science + Business Media, 2010.

[7] J.N. Goodier, Concentration of stress around spherical and cylindrical inclusions and flaws, Trans. ASME 55 (1933) 39—44.

[8] A.R. Skovoroda, S.Y. Emelianov, M.A. Lubinski, A.P. Sarvazyan, M. O'Donnell, Theoretical analysis and verification of ultrasound displacement and strain imaging, IEEE Trans. Ultrason. Ferroelectr. Freq. Control 41 (1994) 302—313.

[9] A.P. Sarvazyan, O.V. Rudenko, S.D. Swanson, J.B. Fowlkes, S.Y. Emelianov, Shear wave elasticity imaging: a new ultrasonic technology of medical diagnostics, Ultrasound Med. Biol. 24 (1998) 1419—1435.

[10] S.R. Aglyamov, A.B. Karpiouk, Y.A. Ilinskii, E.A. Zabolotskaya, S.Y. Emelianov, Motion of a solid sphere in a viscoelastic medium in response to applied acoustic radiation force: theoretical analysis and experimental verification, J. Acoust. Soc. Am. 122 (2007) 1927—1936.

[11] R.M. Christensen, Theory of Viscoelasticity. An Introduction, Academic Press, 1971.

[12] S. Chen, et al., Assessment of liver viscoelasticity by using shear waves induced by ultrasound radiation force, Radiology 266 (2013) 964—970.

[13] I.Z. Nenadic, et al., Attenuation measuring ultrasound shearwave elastography and in vivo application in post-transplant liver patients, Phys. Med. Biol. 62 (2016) 484—500.

[14] K.R. Nightingale, et al., Derivation and analysis of viscoelastic properties in human liver: impact of frequency on fibrosis and steatosis staging, IEEE Trans. Ultrason. Ferroelectr. Freq. Control 62 (2015) 165–175.

[15] J.F. Doyle, Wave Propagation in Structures. Spectral Analysis Using Fast Discrete Fourier Transforms, Springer, 1997.

[16] K.J. Parker, M.M. Doyley, D.J. Rubens, Imaging the elastic properties of tissue: the 20 year perspective, Phys. Med. Biol. 56 (2010) R1–R29.

[17] J.F. Greenleaf, M. Fatemi, M. Insana, Selected methods for imaging elastic properties of biological tissues, Annu. Rev. Biomed. Eng. 5 (2003) 57–78.

[18] M.M. Doyley, Model-based elastography: a survey of approaches to the inverse elasticity problem, Phys. Med. Biol. 57 (2012) R35–R73.

[19] J. Ophir, S.K. Alam, B.S. Garra, F. Kallel, E. Konofagou, T. Krouskop, T. Varghese, "Elastography: ultrasonic estimation and imaging of the elastic properties of tissues," Proc Inst Mech Eng H: J Eng Med (Special Issue on Acoustic and Ultrasonic Tissue Characterization) 213 (1999) 203–233.

References

Vibration sonoelastography

Kevin J. Parker

William F. May Professor of Engineering, Professor of Electrical and Computer Engineering, of Biomedical Engineering, and of Imaging Sciences (Radiology), University of Rochester, Rochester, NY, United States; Dean Emeritus, School of Engineering & Applied Sciences, University of Rochester, Rochester, NY, United States

1. Early results

As highlighted in Chapter 3(Governing theory of elasticity imaging by Salavat R. Aglyamov), studies of tissue motion were made possible by ultrasound pulse echo, M-mode, and Doppler instruments during the early 1980s. The important transition to *imaging* the biomechanical properties of tissue was launched by sonoelastography techniques. We defined sonoelastography as the application of a continuous low-frequency vibration (between 40 and 1000 Hz) to excite shear waves in tissues [1,2]. The simplest but effective real-time imaging version of sonoelastography pertains to the display of the amplitude of the propagating shear waves in sinusoidal steady state, and this was denoted as vibration-amplitude sonoelastography. If a stiff, inhomogeneous area is present in soft tissues, a disruption in the normal vibration patterns will result. If Doppler detection algorithms are applied, a real-time vibration image can be created to display the shear wave propagation (including modal patterns within regular boundaries). Using the information from these images, the shear wave speed of sound in the organ tissue can be determined [3].

The first vibration-amplitude sonoelastography image is shown in Fig. 3.1 [1,2].

This first-of-a-kind image illustrates the vibration within a sponge and a saline phantom containing a harder region (the dark area). As the phantom was vibrated from below, range-gated Doppler was used to determine the vibration amplitude of the phantom's interior. Researchers at the University of Rochester then used a modified color Doppler instrument to create real-time vibration-amplitude sonoelastography images, wherein vibration above a certain level (~ 2 μm) produced a saturated color (Fig. 3.2).

These initial developments were followed by tissue elastic constant measurements, finite element modeling results, and both phantom and ex vivo tissue sonoelastography [4-6].

By 1990, the field of vibration sonoelastography (also called sonoelastography and sonoelasticity at this time) included real-time imaging techniques and stress-strain analysis of human tissues, such as prostate. Both experimental images and

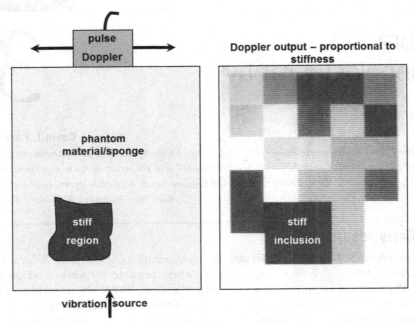

FIGURE 3.1

Original imaging data and schematic of the first known image of relative stiffness, derived from Doppler data in a phantom with applied vibration. The original image was published in 1987 and 1988 [1,2] and marks the emergence of elastographic imaging.

finite element models definitively illustrated that conventional Doppler imaging scanners could detect areas with elevated Young's modulus values.

2. Theory

In addition to the finite element approach [5,6] to model sonoelasticity, additional analytic/numerical techniques were needed to solve theoretic challenges: (1) what effect elastic inhomogeneities might have on the behavior of vibration fields, (2) what image contrast around lesions was produced in a vibration field, (3) how the contrast of these lesions depended on the choice of parameters, and (4) how to use Doppler signal processing (or other similar techniques) to detect and image sinusoidal vibration patterns. A number of journal articles published in the early 1990s addressed these challenges. A seminal result was published in 1992 [3]: it was demonstrated that vibrational eigenmodes could be created in organs, including the liver and kidneys, where surfaces create reflections of sinusoidal steady-state shear waves. A localized disturbance of the eigenmode pattern would be produced by a lesion and, from the patterns at discrete eigenfrequencies, the Young's modulus of the background could be calculated.

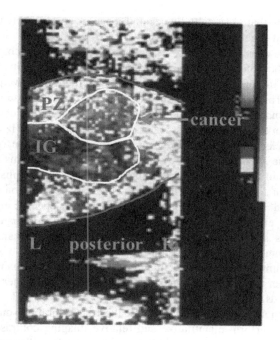

FIGURE 3.2

A representative first-generation image of vibration sonoelastography, circa 1990. Doppler spectral variance is employed as an estimator of vibration in the 2- to −10-μm range and displayed over the B-scan images. No color implies low vibrations below threshold. Shown is the fill-in of vibration within a whole prostate, with a growing cancerous region indicated by the deficit of color within the peripheral zone. *IG*, inner gland; *L*, left; *PZ*, peripheral zone; *R*, right.

Courtesy of Dr. R. M. Lerner.

Simply put, vibration patterns in phantoms and tissues with well-defined boundaries range from a simple, lowest frequency pattern to higher frequency and more complicated natural vibration patterns. The natural vibration patterns of the elastic structure are eigenmodes or simply modal patterns, and the frequencies at which they occur are called eigenfrequencies. When the vibration source is applied at an eigenfrequency, efficient energy coupling to the tissue is realized and the vibrational velocities are high at the antinodes. For a lossless rectangular elastic solid with rigid boundaries and dimensions x_0, y_0, and z_0, the eigenfrequencies occur only at specific frequencies:

$$f_{l,m,n} = \frac{c}{2} \sqrt{ \left(\frac{l}{x_0} \right)^2 + \left(\frac{m}{y_0} \right)^2 + \left(\frac{n}{z_0} \right)^2 }, \qquad 3.1$$

where l, m, and n are integers (1, 2, 3 …) representing the mode number in the x, y, and z directions, respectively, and c is the speed of sound in the rectangular solid. It is important to note that C_l (longitudinal sound speed) is considerably higher than C_s (shear wave sound speed) in biological tissues [7].

As an example consider a cube of tissue 9 cm on an edge, which is subjected to low-frequency vibrations in the 100-Hz regimen, as in sonoelasticity. For a longitudinal wave speed of 1500 m/s, the wavelength for longitudinal wave propagation is 15 m at 100 Hz. This is clearly much larger than the size of the tissue, and even the fundamental resonant mode could not be established in a 9-cm cube. Shear wave speed, however, in tissues is of the order of 300 cm/s, and at 100-Hz excitation frequency, the wavelength for shear wave propagation [7] is of the order of 3 cm. Clearly, this is the appropriate size to set up standing wave patterns within a 9 cm block of tissue, and thus low-order resonance modes may be established.

Eigenmodes of solids can be excited by placing a small vibration source in contact with the solid and tuning the vibration frequency to a specific eigenfrequency. We found that by examining the simpler, lower frequency modal patterns, inhomogeneities such as small stiff tumors of the liver and prostate can be identified [1,2,5,7,8]. We demonstrated that the configuration of eigenmodes, visualized in tissues for the first time in sonoelasticity imaging, can reveal the mechanical properties (for example, shear wave speed) of an organ using Eq. (3.1) and other relations from the theory of solids.

By 1992, when applied in vivo in the clinical setting, vibration sonoelastography could be used to perform quantitative and relative imaging contrast detection. A strategy for simultaneously using multiple frequencies ("chords") and a later review of eigenmodes was described in Ref. [9].

A vibration-amplitude model was created in Refs. [10,11], in which the Born approximation was applied to an isotropic inhomogeneous medium to solve the wave equations. The total wave field inside the medium is

$$\Phi_{\text{total}} = \Phi_i + \Phi_s, \qquad\qquad 3.2$$

where Φ_i is the incident or homogeneous field and Φ_s is the field scattered by the inhomogeneity. Φ_i and Φ_s satisfy the following:

$$\left(\nabla^2 + k^2\right)\Phi_i = 0 \text{ and} \qquad\qquad 3.3$$

$$\left(\nabla^2 + k^2\right)\Phi_s = \delta(x), \qquad\qquad 3.4$$

where k is the wave number, ∇^2 is the Laplacian operator, and $\delta(x)$ is a function of the properties of inhomogeneity. This theory describes how a hard or soft lesion appears as a perturbation in a vibration pattern. A summary of the theory and experimental results is shown in Fig. 3.3.

In studying vibration-amplitude sonoelastography, signal processing estimators were also developed to advance the sensitivity of scanners to shear wave vibrations on a scale of a few micrometers.

Assume that the transmitted or incident signal is

$$s_t(t) = \cos(\omega_0 t), \qquad\qquad 3.5$$

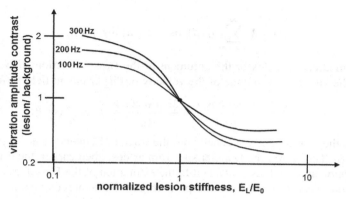

FIGURE 3.3

Theoretical results of the contrast of vibration sonoelastography for soft or hard lesions in a background medium. The image contrast increases with both increasing frequency and with increasing size of the lesion (not shown). On the horizontal axis is the ratio of E_L, the Young's modulus of the lesion, to E_0, the modulus of the background tissue.

and the scatterers are vibrating with the form

$$\xi(t) = \xi_m \sin(\omega_L t + \varphi), \tag{3.6}$$

$$v(t) = \dot{\xi}(t) = v_m \cos(\omega_L t + \varphi), \tag{3.7}$$

where $\xi(t)$ is the displacement of the vibration, $v(t)$ is the velocity of the vibration, ω_L is the vibration frequency, φ is the vibration phase, ξ_m is the vibration amplitude of the displacement field, and $v_m = \omega_L \xi_m$ is the vibration amplitude of the velocity field.

The instantaneous frequency of the received or scattered waves will be shifted to

$$\omega_i = \omega_0 + \Delta\omega_d, \tag{3.8}$$

$$\Delta\omega_d = \Delta\omega_m \cos(\omega_L t + \varphi), \tag{3.9}$$

$$\Delta\omega_m = 2v_m\omega_0 \cos\theta/c_0, \tag{3.10}$$

where ω_i is the instantaneous frequency of the scattered waves, $\Delta\omega_d$ is the Doppler shift, c_0 is the propagation speed of illuminating wave at frequency ω_0, and θ is the angle between the wave propagation and the vibration vectors.

Therefore the received or scattered waves can be written as

$$s_r(t) = A \cos[\omega_0 t + (\Delta\omega_m / \omega_L)\sin(\omega_L t + \varphi)] \tag{3.11}$$

because the instantaneous frequency is, by definition, given by the time derivative of the argument of the carrier cosine wave.

Using trigonometric identities, Eq. (3.11) can be replaced by the series [12].

$$s_r(t) = A \sum_{n=-\infty}^{\infty} J_n(\beta)\cos[\omega_0 t + n(\omega_L t + \varphi)], \qquad 3.12$$

where the modulation index or the argument of the Bessel functions β is directly related to the vibration amplitude of the velocity or displacement field as follows:

$$\beta \equiv \frac{\Delta\omega_m}{\omega_L} = \frac{2v_m\omega_0 \cos\theta}{\omega_L c_0} = \frac{2\xi_m\omega_0 \cos\theta}{c_0} = 4\pi\frac{\xi_m}{\lambda_0}\cos\theta, \qquad 3.13$$

where λ_0 is the wavelength associated with the wave of frequency ω_0 and propagation speed c_0. Thus given the Doppler spectrum as described earlier, the estimation of the vibration amplitude is equivalent to the estimation of the Bessel argument β.

A method to estimate β (a quantity proportional to the target's vibration amplitude) from the Doppler spectral spread (σ_ω) was proposed by Huang et al. [13]:

$$\beta = \sqrt{2}(\sigma_\omega / \omega_L) \qquad 3.14$$

where ω_L is the vibrating target's vibration frequency. This is a very useful and simple property of the Bessel Doppler function. Also studied were sampling, the presence of higher harmonics from source or tissue nonlinearities, and the effect of noise. Later, a family of real-time estimators of vibration amplitude, frequency, and phase was developed. These estimators would be implemented for a number of vibration sonoelastography techniques [14].

A review article [15] places vibration sonoelastography on a biomechanical spectrum with the other imaging techniques that followed (including compression elastography, magnetic resonance elastography, and impulse radiation force excitations). In this view, a master equation provides different solutions as different excitations are applied.

3. Vibration phase gradient sonoelastography

Researchers at the University of Tokyo independently developed a vibration phase gradient approach to sonoelastography by 1990 [16]. Sato and collaborators mapped the phase and amplitude of low-frequency wave propagation in tissues. This enabled them to derive wave propagation velocity and dispersion (linked to tissue elasticity and viscosity).

A pure-tone frequency modulation process is approximated by the phase-modulated Doppler spectrum of a signal returned from sinusoidally oscillating objects (Eq. 3.3). This approximation demonstrates that tissue vibration amplitude and tissue motion phase can be estimated from the ratios of adjacent harmonics. The amplitude ratio between contiguous spectral signal Bessel bands is

$$A_{i+1}/A_i = |J_{i+1}(\beta) / J_i(\beta)|, \qquad 3.15$$

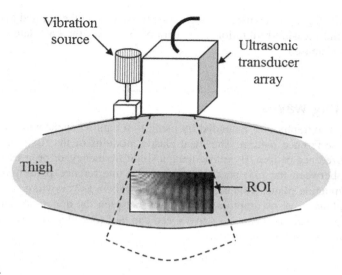

FIGURE 3.4

A depiction of the system developed by Prof. Sato and colleagues at the University of Tokyo, including external vibration and an imaging array with signal processing for estimating the phase of the vibration in tissue. The rate of change of phase can be estimated to yield tissue hardness. *ROI*, region of interest.

where A_i is the amplitude of the ith harmonic, $J_i(\cdot)$ is the ith order Bessel function, and β is the amplitude of vibration (unknown). If A_{i+1}/A_i is calculated as a function of β first, then β can be estimated from experimental data. The vibration phase was calculated from the quadrature signals.

The ability to display wave propagation as a moving image is enabled by the construction of amplitude and phase maps as a function of time (Fig. 3.4).

A minimum squared error algorithm was used to estimate the direction of wave propagation and to calculate amplitude and phase gradients in this direction. The application of this algorithm enabled the amplitude and phase to be computed offline. Sato's group was able to obtain preliminary in vivo results by assuming that shear viscosity has negligible effect at low frequencies [16].

Levinson et al. [17] refined Sato's technique [16] by developing a linear recursive filtering algorithm based on cubic B-spline functions and a more general model of tissue viscoelasticity. They derived the frequency domain displacement equation for a linear, isotropic, homogeneous viscoelastic material by taking the Fourier transform of the wave equation. It was assumed that at low frequencies, viscosity is negligible and shear waves predominate with negligible compression waves. Levinson and colleagues conducted experiments on the human thigh, specifically on the quadriceps muscle group, applying and controlling tension to the muscle and using a pulley device. Calculation of Young's modulus of elasticity was enabled by the phase gradient images of the thighs under conditions of active muscle contraction.

As increasing degrees of contraction were used to counteract the applied load, it was found that the measured vibration propagation speed and the calculated modulus values also increased.

---

4. Crawling waves

Vibration elastography was extended by using two parallel shear wave sources to create an interference pattern. The local elastic modulus of the tissue is revealed by this interference pattern. By introducing a slight frequency, or progressive phase difference, between the two parallel sources, the interference pattern will move across the imaging plane. This moving pattern is known as "crawling waves."

Under plane wave assumption, the relation between the apparent speed of the crawling waves and the true shear wave speed can be deduced as follows. The right traveling waves can be expressed as

$$\exp(ikx - i\omega t). \qquad 3.16$$

The left traveling waves can be expressed as

$$\exp[i(k + \Delta k)(D - x) - i(\omega + \Delta\omega)t], \qquad 3.17$$

where

$$\Delta k = \frac{\Delta\omega}{v_{shear}}. \qquad 3.18$$

The interference patterns are the superposition of the two waves

$$\exp(ikx - i\omega t) + \exp[i(k + \Delta k)(D - x) - i(\omega + \Delta\omega)t]. \qquad 3.19$$

Take only the real part of the terms

$$u(x, t) = \cos(kx - \omega t) + \cos((k + \Delta k)(D - x) - (\omega + \Delta\omega)t)$$

$$= 2\cos\left[\left(k + \frac{\Delta k}{2}\right)x + \frac{\Delta\omega}{2}t\right] * \cos\left[\frac{\Delta k}{2}x + \left(\omega + \frac{\Delta\omega}{2}\right)t\right]. \qquad 3.20$$

Note that D merely introduces a constant phase and is thus omitted for simplicity. Eq. (3.20) factorizes the two waves into rapidly varying carrier signals (the second factor) and slowly varying envelope signals (the first factor). As sonoelastography only images the envelope, it is the absolute value of the first factor that is displayed. Also note that there are both time variables and spatial variables in this factor. It appears to be a traveling wave. The *apparent* speed of these "crawling waves" can be derived by as

$$v_{pattern} = \frac{\Delta\omega}{2\left(k + \frac{\Delta k}{2}\right)}. \qquad 3.21$$

If the frequency difference $\Delta\omega$ is small compared to ω, then Δk is small compared to k. Therefore $v_{pattern} \approx \omega/2k$. This is the typical configuration we used for our experiments. Noting that the speed of the shear wave is $v_{shear} \approx \omega/k$, the relation between the speed of the *crawling waves* and the true shear waves is

$$v_{pattern} \approx \frac{\Delta\omega}{2\omega}v_{shear} \qquad\qquad 3.22$$

given $\Delta\omega \ll \omega$. According to Eq. (3.22), the shear waves can be "slowed down" one or two orders of magnitude by choosing $\Delta\omega$ much smaller than ω, so that a conventional ultrasonic scanner modified for sonoelastography can visualize and track the wave propagation. In practice, $\Delta\omega$ is normally between $\omega/200$ and $\omega/100$ for convenience with respect to the ultrasound scanner's frame rate. Once the apparent speed of the "crawling waves" is measured by analyzing the video sequences, the true shear wave speed can then be obtained from Eq. (3.22). The displayed speed of the crawling waves is controlled by the relative phase between the two shear wave sources [18] (see Fig. 3.5).

There are a number of advantages to the crawling waves method: (1) crawling waves are readily visualized by conventional Doppler imaging scanners at typical Doppler frame rates (no need for ultrafast scanning); (2) the region of interest (ROI) excited between the parallel sources is large; (3) most of the energy in the crawling waves is in the axial/Doppler direction; (4) to derive estimates of local shear wave velocity and Young's modulus, a number of analytic/estimation methods can be applied in an uncomplicated manner; and (5) the crawling waves can be induced by a number of source types, including mechanical line sources, surface applicators, and radiation force excitations of parallel beams [19–21].

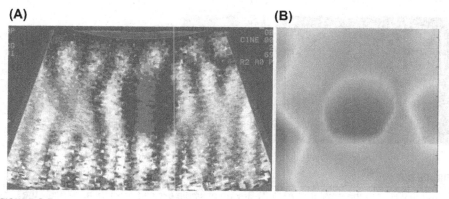

FIGURE 3.5

(A) Image of a crawling wave interference pattern on a Zerdine phantom containing a stiff lesion. External sources (not shown) are located on the right and left sides of the phantom. Stripes represent shear wave interference patterns that are analyzed to determine local shear wave speed and stiffness. (B) Computed shear speed map derived from the crawling wave patterns.

The crawling waves method was introduced at the University of Rochester by Wu et al. [18]. This group was the first to show that crawling waves could be used to identify stiff inclusions and to accurately derive Young's modulus [22–25]. Estimators for shear wave speed and Young's modulus and the shear wave attenuation are derived in Refs. [26–29]. Radiation force excitation along parallel beams was used to implement this method into scanning probes [30–32].

5. Clinical results

Initially, vibration-amplitude sonoelastography was used to characterize tissue properties using eigenmode information [3,33] and to identify stiff lesions by the amplitude contrast effect. A demonstration of its ability to improve the detection of prostate cancer was published in 1995 [34]. This led to in vivo and three-dimensional studies [9,25,35–38], in which it was shown that using sonoelasticity significantly improved the specificity and sensitivity of prostate cancer detection. Furthermore, it was demonstrated that the definition and volumetric measurement of thermal lesions could be markedly improved by implementing sonoelastography [39–42].

Some applications of the crawling waves method are the ex vivo prostate [43–45], in vivo muscle [46], and ex vivo liver [24]. An ex vivo prostate example is given in Fig. 3.6. Other applications of crawling waves include quantification of the elastic properties of muscle [46,47] and thyroid [48]. Crawling waves are

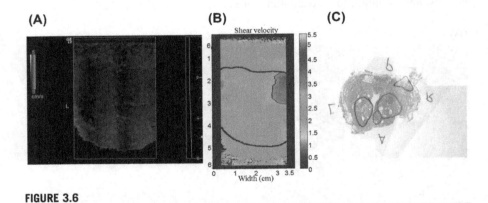

FIGURE 3.6

Crawling waves in the prostate. (A) (left) B-scan of whole excised prostate and (right) the crawling wave frame. (B) Quantitative estimates of shear velocity from crawling-wave analysis, indicating the hard region as a red (dark gray in print version) area (corresponding to the region with cancer). (C) Pathology with labels: black = cancer and blue (gray in print version) = benign prostatic hyperplasia. Estimates from an average of three frequencies are shown. All images are co-registered.

well suited for measuring dispersion (the frequency dependence of shear wave speed in viscoelastic biomaterials, as high signal-to-noise patterns can be generated across relatively large ROIs at discrete frequencies). This enables accurate measurements of phase velocity across an extended range of frequencies and has been utilized to quantify the subtle effects of steatosis in the liver [49–52], capturing an important aspect of viscoelastic behavior.

6. Reverberant shear wave fields

Parker et al. [53] proposed and analyzed a limiting case of a fully reverberant shear wave field in an organ. Mathematically, this limiting case is modeled as the condition where, at an observation point in tissue, shear waves of random amplitude and phase are found to be propagating in all directions as a statistically isotropic distribution across 4π steradians. Practically speaking, all tissue boundaries with reflections, and sources in the vicinity of the observation point, contribute to the overall distribution. Analytic solutions were obtained for the expected value of the autocorrelation function for the vector velocity field as a function of space and time, and then for the projection (or dot product) of this along a single direction, taken as the axis of motion detection of an imaging system. From these analytic solutions, the reverberant or diffuse field approach leads to simple estimators of shear wave speed. The mathematical framework and assumptions of a reverberant field are a departure from previous approaches where directional filters are employed to isolate and characterize one or several principal components of an unknown shear wave field. In contrast, the reverberant or diffuse field explicitly treats a statistically isotropic distribution from all directions, in the imaging plane and out of plane as well, and derives all subsequent processing and estimators from that limiting condition. Thus strategies for identifying principal directions and the use of directional filters are obviated.

Fig. 3.7 shows the reverberant shear wave field properties within a CIRS breast phantom. The displacement patterns using a vibration frequency of 450 Hz is shown in Fig. 3.7A. Fig. 3.7B presents the phase map of the reverberation pattern at 450 Hz; it can be noticed that there are two different wavelengths between 2 and 3 cm depth, which is the region where the inclusion is located (see B-mode image, Fig. 3.7C). The shear wave speed maps (obtained by applying our approach to a correlation window 1.3×1.3 cm^2 in size) overlaying the B-mode image are shown in Fig. 3.7D. An ROI (7×7 mm^2) was taken from the background and the inclusion in order to obtain a mean value for each region: 2.28 ± 0.14 m/s and 3.43 ± 0.18 m/s for 450 Hz, respectively. The shear wave speed result for the background is in agreement with the elasticity properties specified by the CIRS manufacturer: 2.58 ± 0.32 m/s. For the shear wave speed result at the inclusion, the phantom manufacturer only reported that the inclusion stiffness is at least two times higher than the background. Thus the shear wave speed result for the inclusion is in agreement with that information as well.

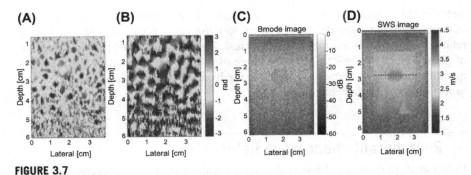

FIGURE 3.7

(A) Reverberant shear wave field displacement pattern at 450 Hz, obtained within a Zerdine breast phantom. (B) Phase map of (A). (C) B-mode image. (D) Shear wave speed (SWS) map calculated from (A) using the second moment algorithm at 450 Hz; the dotted line illustrates the centerline of the lesion, also corresponding to elevated estimates of SWS in the lesion as compared with the surrounding media.

7. Conclusion

There are a number of key advantages to sonoelastographic imaging, crawling waves, and reverberant shear wave imaging. First, in sinusoidal steady-state shear wave excitation, the amplitude contrast effect (of stiff lesions) can be seen with any Doppler imaging platform; it is not necessary to have any synchronization system or specialized platform [54]. Second, by proper choice of shear wave frequency, entire organs can be covered with shear waves, providing large ROIs for analysis [3]. This fact has been used to great advantage in magnetic resonance elastography as well. The crawling waves approach provides data that can be analyzed with a number of simple local estimators, while producing comparable resolution and accuracy to the very best shear wave tracking approaches using radiation force excitations [55]. In addition, the crawling wave patterns can be produced simply by external transducers [19,23] or by an alternating pair of radiation force beams, one on the right and the other on the left of the ROI [20,21]. Multiple excitation sources are also required for producing reverberant shear wave fields; however, the resulting superposition creates favorable conditions for high signal-to-noise estimators across large regions. Thus these techniques are widely applicable and efficacious for studies of tissue elastography in the breast, liver, thyroid, muscle, and other soft tissues.

References

[1] R.M. Lerner, K.J. Parker, Sonoelasticity images derived from ultrasound signals in mechanically vibrated targets, in: Seventh European Communities Workshop, 1987. Nijmegen, The Netherlands.

[2] R.M. Lerner, K.J. Parker, J. Holen, R. Gramiak, R.C. Waag, Sonoelasticity: medical elasticity images derived from ultrasound signals in mechanically vibrated targets, Acoust. Imaging 16 (1988) 317–327.

[3] K.J. Parker, R.M. Lerner, Sonoelasticity of organs: shear waves ring a bell, J. Ultrasound Med. 11 (8) (1992) 387–392.

[4] K.J. Parker, S.R. Huang, R.M. Lerner, F. Lee Jr., D. Rubens, et al., Elastic and ultrasonic properties of the prostate, in: Ultrasonics Symposium, 1993. Proceedings., IEEE 1993, 1993.

[5] R.M. Lerner, S.R. Huang, K.J. Parker, "Sonoelasticity" images derived from ultrasound signals in mechanically vibrated tissues, Ultrasound Med. Biol. 16 (3) (1990) 231–239.

[6] K.J. Parker, S.R. Huang, R.A. Musulin, R.M. Lerner, Tissue response to mechanical vibrations for "sonoelasticity imaging", Ultrasound Med. Biol. 16 (3) (1990) 241–246.

[7] S.R. Huang, Principles of sonoelasticity imaging and its applications in hard tumor detection (Ph.D. thesis), in: Electrical Engineering, University of Rochester, Rochester, NY, 1990.

[8] F. Lee Jr., J.P. Bronson, R.M. Lerner, K.J. Parker, S.R. Huang, et al., Sonoelasticity imaging: results in in vitro tissue specimens, Radiology 181 (1) (1991) 237–239.

[9] L.S. Taylor, B.C. Porter, D.J. Rubens, K.J. Parker, Three-dimensional sonoelastography: principles and practices, Phys. Med. Biol. 45 (6) (2000) 1477–1494.

[10] L. Gao, K.J. Parker, S.K. Alam, R.M. Lerner, Sonoelasticity imaging: theory and experimental verification, J. Acoust. Soc. Am. 97 (6) (1995) 3875–3886.

[11] K.J. Parker, D. Fu, S.M. Graceswki, F. Yeung, S.F. Levinson, Vibration sonoelastography and the detectability of lesions, Ultrasound Med. Biol. 24 (9) (1998) 1437–1447.

[12] A.B. Carlson, Communications Systems, McGraw Hill, New York, 1986, pp. 221–227 (Chapter 8).

[13] S.R. Huang, R.M. Lerner, K.J. Parker, On estimating the amplitude of harmonic vibration from the Doppler spectrum of reflected signals, J. Acoust. Soc. Am. 88 (6) (1990) 2702–2712.

[14] S.R. Huang, R.M. Lerner, K.J. Parker, Time domain Doppler estimators of the amplitude of vibrating targets, J. Acoust. Soc. Am. 91 (2) (1992) 965–974.

[15] K.J. Parker, L.S. Taylor, S. Gracewski, D.J. Rubens, A unified view of imaging the elastic properties of tissue, J. Acoust. Soc. Am. 117 (5) (2005) 2705–2712.

[16] Y. Yamakoshi, J. Sato, T. Sato, Ultrasonic imaging of internal vibration of soft tissue under forced vibration, IEEE Trans. Ultrason. Ferroelectr. Freq. Control 37 (2) (1990) 45–53.

[17] S.F. Levinson, M. Shinagawa, T. Sato, Sonoelastic determination of human skeletal-muscle elasticity, J. Biomech. 28 (10) (1995) 1145–1154.

[18] Z. Wu, L.S. Taylor, D.J. Rubens, K.J. Parker, Sonoelastographic imaging of interference patterns for estimation of the shear velocity of homogeneous biomaterials, Phys. Med. Biol. 49 (6) (2004) 911–922.

[19] A. Partin, Z. Hah, C.T. Barry, D.J. Rubens, K.J. Parker, Elasticity estimates from images of crawling waves generated by miniature surface sources, Ultrasound Med. Biol. 40 (4) (2014) 685–694.

[20] C. Hazard, Z. Hah, D. Rubens, K. Parker, Integration of crawling waves in an ultrasound imaging system. Part 1: system and design considerations, Ultrasound Med. Biol. 38 (2) (2012) 296–311.

[21] Z. Hah, C. Hazard, B. Mills, C. Barry, D. Rubens, et al., Integration of crawling waves in an ultrasound imaging system. Part 2: signal processing and applications, Ultrasound Med. Biol. 38 (2) (2012) 312–323.

[22] Z. Wu, Shear wave interferometry and holography, an application of sonoelastography, in: Electrical & Computer Engineering, University of Rochester, Rochester, NY, 2005, p. 104.

[23] Z. Wu, K. Hoyt, D.J. Rubens, K.J. Parker, Sonoelastographic imaging of interference patterns for estimation of shear velocity distribution in biomaterials, J. Acoust. Soc. Am. 120 (1) (2006) 535–545.

[24] M. Zhang, B. Castaneda, Z. Wu, P. Nigwekar, J.V. Joseph, et al., Congruence of imaging estimators and mechanical measurements of viscoelastic properties of soft tissues, Ultrasound Med. Biol. 33 (10) (2007) 1617–1631.

[25] Z. Wu, L.S. Taylor, D.J. Rubens, K.J. Parker, Shear wave focusing for three-dimensional sonoelastography, J. Acoust. Soc. Am. 111 (1 Pt 1) (2002) 439–446.

[26] K. Hoyt, K.J. Parker, D.J. Rubens, Real-time shear velocity imaging using sonoelastographic techniques, Ultrasound Med. Biol. 33 (7) (2007) 1086–1097.

[27] J. McLaughlin, D. Renzi, K. Parker, Z. Wu, Shear wave speed recovery using moving interference patterns obtained in sonoelastography experiments, J. Acoust. Soc. Am. 121 (4) (2007) 2438–2446.

[28] K. Hoyt, B. Castaneda, K.J. Parker, P4F-4 feasibility of two-dimensional quantitative sonoelastographic imaging, in: Ultrasonics Symposium, 2007. IEEE, 2007.

[29] K. Hoyt, B. Castaneda, K.J. Parker, Two-dimensional sonoelastographic shear velocity imaging, Ultrasound Med. Biol. 34 (2) (2008) 276–288.

[30] Z. Hah, C. Hazard, Y.T. Cho, D. Rubens, K. Parker, Crawling waves from radiation force excitation, Ultrason. Imaging 32 (3) (2010) 177–189.

[31] L. An, D.J. Rubens, J. Strang, Evaluation of crawling wave estimator bias on elastic contrast quantification, in: American Institute of Ultrasound in Medicine Annual Convention, 2010. San Diego, CA.

[32] Y.T. Cho, Z. Hah, L. An, Theoretical investigation of strategies for generating crawling waves using focused beams, in: American Institue for Ultrasound in Medicine Annual Convention, 2010. San Diego, CA.

[33] S.K. Alam, D.W. Richards, K.J. Parker, Detection of intraocular pressure change in the eye using sonoelastic Doppler ultrasound, Ultrasound Med. Biol. 20 (8) (1994) 751–758.

[34] D.J. Rubens, M.A. Hadley, S.K. Alam, L. Gao, R.D. Mayer, et al., Sonoelasticity imaging of prostate cancer: in vitro results, Radiology 195 (2) (1995) 379–383.

[35] L.S. Taylor, T.R. Gaborski, J.G. Strang, D. Rubens, K.J. Parker, Detection and three-dimensional visualization of lesion models using sonoelastography, in: SPIE Medical Imaging 2002: Ultrasonic Imaging and Signal Processing, 2002.

[36] L.S. Taylor, D.J. Rubens, B.C. Porter, Z. Wu, R.B. Baggs, et al., Prostate cancer: three-dimensional sonoelastography for in vitro detection, Radiology 237 (3) (2005) 981–985.

[37] M. Zhang, P. Nigwekar, B. Castaneda, K. Hoyt, J.V. Joseph, et al., Quantitative characterization of viscoelastic properties of human prostate correlated with histology, Ultrasound Med. Biol. 34 (7) (2008) 1033–1042.

[38] B. Castaneda, K. Hoyt, M. Zhang, D. Pasternack, L. Baxter, et al., P1C-9 prostate cancer detection based on three dimensional sonoelastography, in: Ultrasonics Symposium, 2007. IEEE, 2007.

[39] L.S. Taylor, M. Zhang, In-vitro imaging of thermal lesions using three-dimensional vibration sonoelatography, in: Second International Symposium on Therapeutic Ultrasound, 2002.

[40] B. Castaneda, M. Zhang, K. Hoyt, K. Bylund, J. Christensen, et al., P1C-4 real-time semi-automatic segmentation of hepatic radiofrequency ablated lesions in an in vivo porcine model using sonoelastography, in: Ultrasonics Symposium, 2007. IEEE, 2007.

[41] M. Zhang, B. Castaneda, J. Christensen, W.E. Saad, K. Bylund, et al., Real-time sonoelastography of hepatic thermal lesions in a swine model, Med. Phys. 35 (9) (2008) 4132–4141.

[42] B. Castaneda, J.G. Tamez-Pena, M. Zhang, K. Hoyt, K. Bylund, et al., Measurement of thermally-ablated lesions in sonoelastographic images using level set methods, in: Proc SPIE 2008, 2008.

[43] K. Hoyt, B. Castaneda, M. Zhang, P. Nigwekar, P.A. di Sant'agnese, et al., Tissue elasticity properties as biomarkers for prostate cancer, Cancer Biomark. 4 (4–5) (2008) 213–225.

[44] K. Hoyt, K.J. Parker, D.J. Rubens, Sonoelastographic shear velocity imaging: experiments on tissue phantom and prostate, in: Ultrasonics Symposium, 2006. IEEE, 2006.

[45] B. Castaneda, L. An, S. Wu, L.L. Baxter, J.L. Yao, et al., Prostate Cancer Detection Using Crawling Wave Sonoelastography, SPIE, Lake Buena Vista, FL, USA, 2009.

[46] K. Hoyt, T. Kneezel, B. Castaneda, K.J. Parker, Quantitative sonoelastography for the in vivo assessment of skeletal muscle viscoelasticity, Phys. Med. Biol. 53 (15) (2008) 4063–4080.

[47] K. Hoyt, B. Castaneda, K.J. Parker, Muscle tissue characterization using quantitative sonoelastography: preliminary results, in: Ultrasonics Symposium, 2007. IEEE, 2007.

[48] J. Walsh, L. An, B. Mills, Z. Hah, J. Moalem, et al., Quantitative crawling wave sonoelastography of benign and malignant thyroid nodules, Otolaryngol. Head Neck Surg. 147 (2) (2012) 233–238.

[49] C.T. Barry, B. Mills, Z. Hah, R.A. Mooney, C.K. Ryan, et al., Shear wave dispersion measures liver steatosis, Ultrasound Med. Biol. 38 (2) (2012) 175–182.

[50] C.T. Barry, Z. Hah, A. Partin, R.A. Mooney, K.-H. Chuang, et al., Mouse liver dispersion for the diagnosis of early-stage fatty liver disease: a 70-sample study, Ultrasound Med. Biol. 40 (4) (2014) 704–713.

[51] K.J. Parker, A. Partin, D.J. Rubens, What do we know about shear wave dispersion in normal and steatotic livers? Ultrasound Med. Biol. 41 (5) (2015) 1481–1487.

[52] C.T. Barry, C. Hazard, Z. Hah, G. Cheng, A. Partin, et al., Shear wave dispersion in lean versus steatotic rat livers, J. Ultrasound Med. 34 (6) (2015) 1123–1129.

[53] K.J. Parker, J. Ormachea, F. Zvietcovich, B. Castaneda, Reverberant shear wave fields and estimation of tissue properties, Phys. Med. Biol. 62 (3) (2017) 1046.

[54] G. Torres, G.R. Chau, K.J. Parker, B. Castaneda, R.J. Lavarello, Temporal artifact minimization in sonoelastography through optimal selection of imaging parameters, J. Acoust. Soc. Am. 140 (1) (2016) 714–717.

[55] K.J. Parker, J. Ormachea, S.A. McAleavey, R.W. Wood, J.J. Carroll-Nellenback, et al., Shear wave dispersion behaviors of soft, vascularized tissues from the microchannel flow model, Phys. Med. Biol. 61 (13) (2016) 4890.

Introduction to quasi-static elastography

Paul E. Barbone[1], Assad A. Oberai[2], Timothy J. Hall[3]

[1]*Mechanical Engineering, Boston University, Boston, MA, United States;* [2]*Department of Aerospace and Mechanical Engineering, University of Southern California, Los Angeles, CA, United States;* [3]*Department of Medical Physics, University of Wisconsin, Madison, WI, United States*

1. Introduction and background

Palpation has been used for millennia to investigate relative tissue stiffness, especially the isolated "lumps" associated with focal disease. The development of tissue elasticity imaging (*elastography*) provided a method for imaging these variations in tissue stiffness, which can range over orders of magnitude [1,2]. Although elastography can be categorized in a variety of ways [3,4] (e.g., the method of applying deformation, whether the deformation is dynamically applied or not, etc.), quasi-static elastography (defined later) remains the most conceptually simple approach for imaging tissue stiffness. In fact, the successful commercial and clinical application of quasi-static elastography (strain imaging; described later) paved the way for the broad investigations into other clinical applications of elasticity imaging. Implementation and clinical adoption have been sufficiently widespread that the World Federation for Ultrasound in Medicine and Biology, a nonprofit scientific organization of professional societies from around the world, has published consensus statements regarding the appropriate use of elastography in several organ systems [5–8].

In the late 1970s and early 1980s, several groups around the world had reported tracking motion in ultrasound echo signals to infer the underlying motion of tissues [9–12]. By the late 1980s, several additional groups were independently investigating the possibility of using ultrasound to monitor tissue motion and extract information regarding relative tissue (visco)elastic properties [13–16]. In 1991, Céspedes and Ophir [17] reported the formation of strain images obtained by tracking the relative motion between consecutive frames of (radiofrequency [RF]) ultrasonographic images and coined the term "elastography" to describe this process. The term was initially applied to quasi-static elastography but has been extended to encompass all tissue elasticity imaging methods. The *-graph* root in the term elastography refers to the creation of images on the basis of elastic modulus contrast. Increasingly, however, there is emphasis within elastography on quantifying mechanical parameters in vivo, which cannot otherwise be measured directly.

Over the years, strain-based compression elastography has seen an enormous number of potential applications in medicine and biological research. These include

Tissue Elasticity Imaging. https://doi.org/10.1016/B978-0-12-809661-1.00004-2

the detection, diagnosis, and treatment monitoring of diseases such as breast and prostate cancer [18—23], local and diffuse coronary disease [24—28], deep vein thrombosis [29], liver cirrhosis [30,31], and degenerative brain diseases [32,33]. An authoritative review on applications in breast imaging is available in Ref. [34].

Following the initial work in strain imaging, investigations have focused on improving motion tracking performance and robustness of strain images and on extracting more detailed information about the hyperelastic, viscoelastic, and poroelastic properties of tissues. The relationship between the (visco)elastic properties and the underlying fiber network structure of simple polymers is well established [35—37]. However, the inverse problem relating measured response of tissue during a mechanical stimulus to a model for the underlying tissue structure at the micro-, meso-, and macroscales is a topic of current investigation [38,39]. We posit that quasi-static elastography is the most promising approach to investigate and objectively quantify these more detailed descriptions of tissue response to mechanical stimulus.

The remainder of this chapter provides an overview of the steps involved in displacement and strain estimation, interpretation of the relative displacement information, and extensions to interpretation beyond simple linear strain imaging.

2. Deformation application and measurement

At the core of any elastographic system is a means of measuring tissue deformation. In ultrasound elastography, that motion may be intrinsic (e.g., cardiac motion), or extrinsically induced. Acoustic radiation force impulse imaging applies a force to the tissue below the skin surface via a high-intensity ultrasound pulse, applied over a period on the order of 100 μs. This results in a dynamic deformation, which is the subject of Chapter 7. The most commonly used quasi-static deformation in elastography is that resulting from applying a small uniform compression to the tissue. This is accomplished by simply applying gentle pressure to the imaging probe, over a period of roughly 1/4 to 1 s. During the period of tissue deformation, sequential ultrasonographic image data is acquired, from which, the tissue deformation may be measured. The ultrasonographic echo signal (containing phase information) is very sensitive to motion in the direction of sound propagation, and signal processing, described later, can easily detect and accurately measure motions of a just a few micrometers. The high sensitivity of ultrasound to tissue motion makes the pressure required on the probe very small; indeed, the best technique is often to try to hold the probe still.

Measuring the tissue deformation is essentially an advanced problem in image registration, but in a different parameter range than usual. In standard image registration, one seeks a few (say 2—20) transformation parameters in order to align a pair of images to within a millimeter or so. In ultrasound elastography, on the other hand, the image pairs are initially misaligned by at most a millimeter, and usually less than a few hundred micrometers. The goal is then to find tens of 1000s of transformation parameters that align the images to within a few tens of nanometers. Theoretic frameworks to predict optimal strain measurement based on system parameters have been presented, including the strain filter [40] and the Cramer-Rao bound [41].

2.1 Motion tracking algorithms

The phase information available in raw ultrasonographic image data makes it possible to measure deformations with sufficient precision to make elastography possible. When an ultrasound pulse scatters from tissue structures, it carries information from the original pulse and the distribution of subresolution scatterers within the scattering region. The scattered signal serves, therefore, as a kind of "fingerprint" for scattering from that region of tissue, which may be tracked from one ultrasound frame to another as the tissue moves. Standard ultrasonographic images (i.e., B-mode or grayscale images) are log-compressed versions of the envelope of the returned echo signal, whereas elastography usually uses the raw (RF, or in-phase and quadrature components of the analytic signal, I-Q) echo signal in order to maximize the information in the tissue "fingerprint."

Most motion tracking algorithms work from a pair of images, one representing a "predeformation image" and the other a "postdeformation image." (Large deformations may be tracked by estimating deformation increments through a sequence of images, as discussed later.) There are two main classes of methods to measure tissue deformation from a pair of ultrasonographic images: block-matching methods and phase-based methods.

In block matching, a small window (the fingerprint) is selected in the predeformation image. This window is alternately called the correlation window by some authors, or correlation kernel by others. Choosing the size of the kernel involves a trade-off in resolution versus precision. Generally speaking, a smaller kernel permits higher spatial resolution estimates of displacement, at the expense of estimate precision. The block matching "kernel" can be one dimensional (1D), two dimensional (2D), or three dimensional (depending on the imaging system). Multidimensional tracking kernels allow compact "fingerprints" that minimize the echo signal decorrelation that occurs with (postdeformation) axial strain inside the "fingerprint" [42]. To estimate the displacement from the predeformation image to the postdeformation image, the "fingerprint" from the first image is compared to similarly sized windows in the postdeformation image, until the best match is found [43].[1] In this way the new location of the "fingerprint" is found by the best match in the postdeformation echo field. Knowing both the initial position (X) and the final position (x) of the tissue imaged in the kernel, we compute the displacement ($\boldsymbol{u}$) of that tissue as

$$\boldsymbol{u} = x - X \qquad\qquad (4.1)$$

By computing the displacement at a large number of densely spaced kernels in the predeformation image, a "displacement image" can be created that represents

---

[1] Judging the "best match" may be done by a number of different metrics, including sum of absolute differences, sum of squared differences, or normalized cross-correlation [44]. These typically apply at the precision of individual samples and result in displacement estimates that are too imprecise for the purposes of elastography. Subsample displacement estimates may be obtained by fitting the matching function to a model (e.g., quadratic, cosine, spline) and finding the peak of the model [45–48].

the displacement of each region of tissue between the predeformation image and the postdeformation image (see Fig. 4.1).

The phase-based methods compute the phase difference between the predeformation and postdeformation kernels. For phase-based tracking, these kernels are 1D and are aligned with the acoustic beam. This approach may be thought of as a generalization of tissue Doppler [49], which applies traditional Doppler signal processing to tissue rather than blood. Other popular variants of the phase-based motion tracking method include zero-phase crossing estimation [50,51] and the Loupas algorithm [52,53]. The phase-based methods are more computationally efficient than the correlation-based methods, but they have aliasing when displacements exceed half an ultrasound wavelength. Although phase-unwrapping for large displacements is conceptually reasonable and would allow the use of those methods for tracking relatively large deformations, phase unwrapping is troublesome and adds computational complexity. Thus phase-tracking methods are ideally suited to very small displacements or two-step methods that track large-scale motion, deform one RF echo field into the coordinate system of the other, and then perform fine-scale motion tracking, such as the "companding" methods [54–56] and the "combined autocorrelation method" [57].

Numerous other methods for motion tracking in ultrasound echo signals have also been investigated. Several approaches to regularized motion tracking have been implemented, including those that are highly efficient [59–62]. Alternatively, there are methods that directly estimate material strain by, for example, temporal stretching [58,63,64]; by warping a spline-interpolation of one RF echo field (pre- or postdeformation) to fit the sampled RF echo field of the other [46]; or by comparing pre- and postdeformation echo signal power spectra [65,66]. All displacement estimation methods depend on the implicit assumption that the acoustic properties of the material are constant with loading, but they might not be [67].

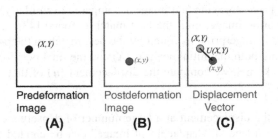

Predeformation Image **(A)** Postdeformation Image **(B)** Displacement Vector **(C)**

FIGURE 4.1

(A) A small region of tissue appears at the location (X, Y) in the predeformation image. (B) That same region of tissue appears at location (x, y) in a subsequent image, collected after a small overall deformation is applied. (C) By matching the echos in the region around (X, Y) in (A) with those in the region around (x, y) in (B), the displacement of that small region of tissue can be measured.

2.2 Motion tracking performance and error

Initial implementations of strain imaging were compared, effectively, as a "beauty contest"; the "best" algorithms were thought to be those whose strain images had high correlation with the associated B-mode images while also providing high contrast and spatial resolution. This approach works reasonably well in simple phantoms, but tissues are often too heterogeneous to expect smoothly varying strain images. So the observer must consider, for example, was the local variation in strain values due to resolvable spatial variation in elastic properties or was it simply displacement estimate noise?

Metrics that objectively evaluate the quality of a strain image help select high-quality strain images, provide feedback when learning strain imaging techniques, and help determine if the information in a strain image can be trusted. One such measure, the displacement quality metric (DQM) [68,69], is the product of two normalized cross-correlation coefficients. The first is the cross-correlation between the predeformation RF echo field and the motion-compensated postdeformation echo field, which provides a measure of motion tracking accuracy. (If the estimated deformation correctly compensates for motion, the two RF fields should align well, within the limits imposed by the decorrelation with the pulse volume that cannot be recovered.) The second cross-correlation is between consecutive motion-compensated strain images. (If the incremental strain is relatively small [< 2% frame-average strain], the differences between consecutive strain images should be small; this provides a measure of strain image consistency.) DQM values exceeding about 0.8 suggest high-quality consecutive strain images (e.g., cross-correlation values for each component above about 0.9). For reference, the DQM values are 0.8 and 0.7 for the strain images in Fig. 4.2A and B, respectively. The DQM illustrates a trade-

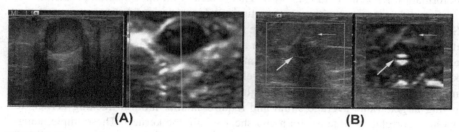

(A)   (B)

FIGURE 4.2

Example of breast B-mode and strain image pairs from a commercial ultrasonographic system. (A) The image pair shows a relatively high-quality strain image of a fibroadenoma. Although the background surrounding the tumor is not homogeneous, variations in strain do not interfere with the interpretation of the lesion size, shape, margins, etc. The image depth is 4.0 cm. (B) The strain image pair, also a fibroadenoma, initially appears quite noisy. On closer inspection, the variations in strain correlate well with structures seen in (A). For example, the dark band in (B) separating two bright areas (*large arrows*) corresponds to a thin band in (A). The triangular area near the top region of interest in (B) corresponds to a similar-shaped structure in (A). The image depth is 4.0 cm. The commercial system used here does not report the strain range in these images (no quantitative color bar).

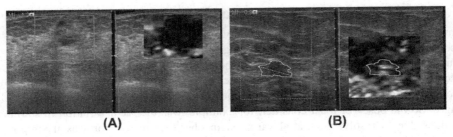

(A) **(B)**

FIGURE 4.3

Example of breast strain images from a commercial ultrasonographic system. (A) A relatively low-quality strain image of an invasive ductal carcinoma (displacement quality metric [DQM] = 0.55). The area in the strain image corresponding to the abnormality seen in the B-mode image is considerably larger in strain than the B-mode image, and the background is heterogeneous. Similarly, in (B) (also an invasive ductal carcinoma; DQM = 0.7), the region above the apparent abnormality in the B-mode image appears quite stiff, and the background appears heterogeneous. In both cases, a larger region of interest would have allowed better investigation into the extent of the increased stiffness region and a likely increase in the DQM values.

off in strain imaging: motion-compensated cross-correlation in the pre- and postdeformation echo signals is maximized when there is no deformation. However, in that case, there is also no information provided in relative stiffness variations in the tissue (little useful information in a strain image). Strain image contrast increases with increasing deformation (up to a point), but with increasing decorrelation in the postdeformation RF echo field (Fig. 4.3). This is clearly an optimization problem.

Strain imaging algorithms are optimized for a desired frame-average strain. Some commercial implementations are designed for use with minimal incremental deformation with the transducer. Such an implementation is relatively easy to use and allows for relatively large motion tracking kernels (little strain within the kernel), but as a result provides relatively low detail in strain images. Other implementations require continuously changing surface pressure with the transducer targeting larger frame-average strain and require smaller motion tracking kernels to avoid decorrelation (due to strain and shear within the kernel). These implementations tend to have a steeper learning curve to master but also provide greater detail (spatial resolution) in the deformation field.

2.3 Tracking large deformations

The "image fingerprint" of a region of tissue results from a combination of subresolution tissue microstructures and the point spread function (PSF) of the imaging system. Because the PSF is different in different parts of the image, the tissue fingerprint will change if the tissue moves very far within the image. It will also change if the microstructure is significantly distorted from one image to another, as occurs with tissue strain. (Strain is essentially a measure of this tissue distortion.) Therefore in order to measure large displacements (more than a few percentage strain or a few millimeters of displacement), the displacement must be measured in several small increments. This is known as "accumulation" or "multicompression" [42,70–75].

An important practical consideration in the application of a large displacement multicompression is that the same tissue region be imaged throughout the deformation sequence. Motion out of the imaging plane may cause a region of tissue imaged in the first few frames of a sequence to leave the imaging plane in subsequent frames. Operators working with 1D arrays and 2D imaging systems should be cognizant of such issues when performing elasticity imaging.

3. **Interpretation of the measured deformation**

Once tissue deformation is measured, the measurements need to be presented to the operator in a manner convenient for interpretation, usually in the form of an image. In quasi-static elastography, this is most commonly a *strain image* (see Section 3.5). Elastic modulus (relative or calibrated) images are also possible, as are elastic nonlinearity images. This section will focus on the physics of strain imaging in quasi-static elastography, with the aim of putting it into the context of other approaches described in other chapters.

3.1 **The quasi-static approximation**

The deformation of soft tissues and other continuous media is governed by two fundamental conservation laws: conservation of mass and conservation of momentum. For solid materials, the more informative of these two is conservation of momentum.

The differential form of the equation for conservation of momentum in a continuous medium takes the form:

$$\sum_{j=1}^{3} \frac{\partial \sigma_{ij}}{\partial x_j} - \rho a_i = 0 \qquad (4.2)$$

Here, σ_{ij} is the component of the (Cauchy) stress tensor.[2] The stress distribution represents the internal force distribution within the tissue. The number σ_{ij} represents the force in the x_i direction (say x, y, or z) on a face perpendicular to the x_j direction. The numbers a_1, a_2, and a_3 (equivalently a_x, a_y, and a_z) are the three components of the acceleration vector of the tissue during its motion from its position before deformation to its position after deformation. ρ is the mass density (mass per unit

---

[2] We are neglecting the body force term here, representing, e.g., the effects of gravity on the tissue. In doing so, it is supposed that the body is in the same gravitational field before, during, and after any elastography experiment, and so the deformation caused by gravity is implicitly accounted for by starting the experiment from the gravitationally deformed state.

volume), typically about 1 g per cubic centimeter for soft tissue. Eq. (4.2) is essentially an expression of Newton's second law of motion, $f = ma$, applied to an infinitesimal volume of tissue. The first term represents f and summarizes the net force on the infinitesimal volume from neighboring tissues. The second term represents ma, with ρ providing the mass part (per unit volume).

We can estimate the order of magnitude of these terms in a typical quasi-static elastography experiment. To that end, we suppose we compress a 10-cm-high region of tissue 1 mm over a period of about 1 s. A compression of 1 mm over a tissue depth of 10 cm gives a strain $\varepsilon \approx 1$ mm/10 cm = 1 mm/100 cm = 0.01. As $\sigma = E\varepsilon$, for tissue with Young's modulus $E \approx 10$ kPa (see later discussion), we find $\sigma \approx O(100 \text{ Pa})$.   From   this,   we   estimate $\frac{\partial \sigma_{ij}}{\partial x_j} \approx O(\sigma/L) = O(100 \text{ Pa}/10 \text{ cm}) = 10^3 \text{ N/m}^3$.   On   the   other   hand,   the   inertia   term $\rho a_i = O(1000 \text{ kg/m}^3 \times 1 \text{ mm}/1 \text{ s}^2) = O(1) \text{ N/m}^3$. Therefore in quasi-static elastography, the inertia term is about a 1000 times smaller than the stress term, and hence may be neglected. This results in the approximation:

$$\sum_{j=1}^{3} \frac{\partial \sigma_{ij}}{\partial x_j} = 0 \qquad (4.3)$$

Eq. (4.3) is called the "quasi-static approximation" because inertia effects are neglected.

3.2 Strain

There are a variety of definitions of strain that are useful in solid mechanics. For small strains, however, all various strain measures reduce to the so-called *linearized strain tensor*:

$$\varepsilon_{ij} = \frac{1}{2} \left[\frac{\partial u_j}{\partial x_i} + \frac{\partial u_i}{\partial x_j} \right] \qquad (4.4)$$

The distinctions between different strain measures are roughly the same size as the strain themselves. So as a rule of thumb, the different strain measures are within roughly 10% of each other at a strain magnitude of 10%.

The interpretation of these numbers is as follows. If $i = j$, then ε_{ii} (no sum) represents $\Delta L/L$, i.e., the change in length divided by the initial length, of a line segment originally oriented along the x_i direction. This is the most commonly used strain measure in quasi-static elastography to measure tissue compression in the direction of the applied load. So, for example, if y is the direction in the ultrasound axial direction, then a strain $\varepsilon_{yy} = 1\%$ at a particular point in the image indicates that layers of tissue immediately above and below that point got 1% closer together as a result of the deformation. If $i \neq j$, then ε_{ij} represents $\frac{1}{2}\Delta\phi_{ij}$, where ϕ_{ij} is the angle (measured in radians) between two lines initially parallel to the x_i and x_j axes, respectively. As coordinate axes are mutually perpendicular, the initial angle is always $\phi_{12} = \phi_{13} = \phi_{23} = \pi/2$ radians, or 90 degrees.

3.3 The elastic approximation

Stress and strain are usually recognized to be correlated within a material, and this is also true in soft tissues. Generally speaking, the more a tissue is strained, i.e., the greater it is deformed, the greater is the internal stress within the tissue. In modeling the tissue's response to an imposed strain, it is common to make an "elastic assumption," which is to say that the stress is a function of the strain:

$$\sigma_{ij} = f_{ij}(\boldsymbol{\varepsilon}) \equiv f_{ij}(\varepsilon_{11}, \varepsilon_{12}, \varepsilon_{13}, \varepsilon_{21}, \ldots) \tag{4.5}$$

Eq. (4.5) seems rather general, as the functional form of f_{ij} is left unspecified at this stage. However, it neglects many potential confounding factors, such as fluid volume and pressure, deformation rate, and prestress. Eq. (4.5) represents an "elastic material." Accounting for fluid volume and pressure effects requires modeling the tissue as a "poroelastic material." Accounting for strain rate requires modeling the tissue as a "viscoelastic material."

Over small deformations, the tissue behavior may be often accurately modeled as linear elastic. In that case, Hooke's law relating stress to linearized strain in an isotropic material is written as

$$\sigma_{ij} = -p\delta_{ij} + 2\mu\varepsilon_{ij} \tag{4.6}$$

where $p = -\lambda(\varepsilon_{11} + \varepsilon_{22} + \varepsilon_{33})$ and λ and μ are the first and second Lamé constants, respectively. This is a special case of Eq. (4.5) and is a generalized form of Hooke's law. Here, δ_{ij} is the Kronecker delta defined by $\delta_{11} = \delta_{22} = \delta_{33} = 1$ and $\delta_{12} = \delta_{21} = \delta_{13} = \delta_{31} = \delta_{23} = \delta_{32} = 0$. If the material is elastic, but the relation between stress and strain does not follow a straight line over the range of deformations considered, then the tissue should be modeled as nonlinear elastic. This possibility is included in Eq. (4.5) but is excluded in Eq. (4.6).

One way to understand the physical difference between the first and second Lamé constants, λ and μ, respectively, is to consider, briefly, elastic wave propagation. An isotropic elastic material can support two main classes of acoustic waves. P-waves, or pressure waves, tend to be longitudinally polarized and travel at the speed c_p. S-waves, or shear waves, tend to be transversely polarized and travel at the speed c_s. These speeds are related to the tissue material properties through the following equations:

$$c_s^2 = \frac{\mu}{\rho} \tag{4.7}$$

$$c_p^2 = \frac{\lambda + 2\mu}{\rho} \approx \frac{\lambda}{\rho} \tag{4.8}$$

In soft tissues, high-frequency P-waves tend to travel at $c_p \approx O(10^3 \text{m/s})$, whereas shear waves tend to travel at $c_s \approx O(1 - 10 \text{ m/s})$. This implies that $\lambda \gg \mu$, and so for deformations related to elastography, λ may be taken as practically infinite. Hence there is only one elastic material parameter, μ, that controls the elastic response of the tissue. Therefore we consider μ in more detail.

The elastic parameter μ is commonly known as the "shear modulus," although its traditional and more descriptive name is the "modulus of rigidity." In engineering literature, it is often given the symbol G. It is sometimes confused with Young's modulus (symbols E or Y; definition below) to which it is related through the following formula:

$$E = \frac{\mu(3\lambda + 2\mu)}{\lambda + \mu} \approx 3\mu \qquad (4.9)$$

The second approximation results from considering the limiting case of $\lambda \to \infty$.

Although G, μ, and E are all called moduli of *elasticity*, their numerical interpretations are not necessarily consistent with one's intuitive notion of the meaning of "elastic." Generally speaking, in common parlance, a rubber elastic band is considered "more elastic" than a loop of steel wire. Nevertheless, the steel in the wire has a modulus of elasticity roughly a million times larger than that of rubber. Thus a higher value of E or G or μ does not translate to a material feeling more "elastic" than another, but rather more rigid than the other. Therefore the traditional name "modulus of rigidity" for μ (and G) is more accurately descriptive of the physical interpretation of μ and less likely to be misinterpreted than the name "modulus of elasticity." In this chapter, we will adopt the term modulus of rigidity for μ (and G) and Young's modulus for E.

3.4 The one-dimensional assumption

In the typical application of quasi-static elastography, the tissue is compressed by pressing the transducer gently against the patient's skin. This results in a slight compression of the tissue immediately below the transducer. As the tissue is compressed axially, it is natural for it to expand in the perpendicular directions to preserve the overall volume of the tissue. Assuming that nothing restricts its horizontal expansion is tantamount to assuming that there is no confining stress in the perpendicular directions. This is the "uniaxial stress" approximation that is central to the interpretation of strain images.

The mathematical derivation of the uniaxial stress assumption proceeds as follows. Let us suppose that y is the direction of compression, which is the same as the ultrasound axial direction, i.e., the direction of sound propagation. We call the ultrasound lateral direction the x direction and z as the ultrasound elevation (i.e., out of plane) direction. Then the uniaxial stress assumption reads in mathematical terms as

$$\sigma_{xx} = \sigma_{zz} = 0; \quad \sigma_{yy} = \sigma_0 \neq 0 \qquad (4.10)$$

Similarly, $\sigma_{xy} = \sigma_{yx} = \sigma_{xz} = \sigma_{zx} = \sigma_{yz} = \sigma_{zy} = 0$. As in the uniaxial stress assumption there is only one nonzero stress component (the σ_{yy} component), we denote that one component by σ_0. Substituting Eq. (4.10) into Eq. (4.6) yields $p = -\frac{1}{3}\sigma_0$, and

$$\varepsilon_{yy} = \frac{1}{E}\sigma_0 \equiv \varepsilon_0 \qquad (4.11)$$

$$\varepsilon_{xx} = \varepsilon_{zz} = -\frac{1}{2}\varepsilon_0 \qquad (4.12)$$

Here, parallel with the stress nomenclature, we introduce ε_0 to denote the dominant strain component. We note, however, that there are three nonzero components of strain, whereas there is only one nonzero component of stress. From Eq. (4.12), we may solve for E to find

$$E = \frac{\sigma_0}{\varepsilon_0} \qquad (4.13)$$

In the uniaxial stress case the momentum Eq. (4.2) reduces to

$$\frac{\partial \sigma_0}{\partial y} = 0 \qquad (4.14)$$

From this, we can conclude that σ_0 is independent of y. In other words, the stress σ_0 is constant throughout the depth of the tissue. If we make the further assumption that σ_0 is constant in the lateral and elevation directions as well, then Eq. (4.13) shows that strain ε_0 is inversely proportional to Young's modulus, E. This interpretation has been computationally and experimentally evaluated [76,77]. See also the following example.

3.5 Strain image interpretation

We have just seen that in the uniaxial approximation, strain is inversely proportional to the tissue Young's modulus. This appeals to one's intuition: the more rigid a region of tissue is, the less easily it can be deformed. Therefore in quasi-static elastography, it is common to create images that show tissue strain, which are interpreted as being inversely related to Young's modulus.

The strain image in Fig. 4.4C may be interpreted with the help of Fig. 4.5, which shows a simple conceptual model for the interpretation of strain images.

Consistent with Hooke's law, the tissue is assumed to behave as a collection of linear springs, possibly with different stiffness values. When compressed, the heavy bold spring compresses less (i.e., shows less strain) than the surrounding springs. This is consistent with the dark circle representing the rigid inclusion in Fig. 4.4. The springs above and below the heavy spring compress more than their equally stiff neighbors, to the right and left, to compensate for the lack of compression in the stiffer spring. This phenomenon is also evident in the strain image in Fig. 4.4C and corresponds to the brighter regions immediately above and below the more rigid dark inclusion. Finally, we note that horizontal lines linking the ends of springs are angled around the inclusion. Real tissues resist this type of deformation, called shear deformation. It is the cause of the brighter regions immediately to the right and left of the inclusion. This "plus" sign pattern of high strain surrounding a more rigid

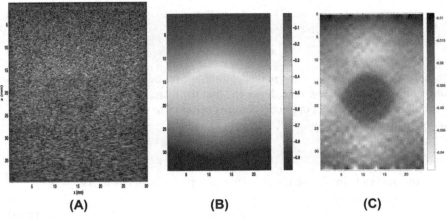

FIGURE 4.4

Example of (A) B-mode image, (B) axial displacement image, and (C) the corresponding strain image. The measured deformation is a quasi-uniform compression applied from the top. The B-mode image shows a faint, slightly hypoechoic circular inclusion of about 12 mm diameter. An overall compression of about 1 mm is applied to the top of the phantom, which causes the distant structures to appear to move toward the transducer by the same amount. The displacement image (B) thus shows the lower region moving about 1 mm (blue [gray in print version]), whereas the upper layer (red [black in print version]) remains approximately motionless relative to the probe (at the top; not shown). The presence of the stiff inclusion disturbs the otherwise uniform compression of this object. The inclusion is easily visualized in the strain image (C), which represents the vertical spatial derivative of the displacement image. The grayscale bar indicates percentage strain. Note the low value of strain within the inclusion and the high value of strain in the background. (Negative implies compression.) Comparing the ratio of strain within the inclusion to that of the background yields a strain contrast of approximately 1.9, which implies that the inclusion is about 1.9 times more rigid than the background. Note the artifacts in the strain image (C) dominated by short horizontal lines of white-black-white. Similar-appearing artifacts are described as "worm" artifacts, as described in Ref. [2].

Data collected and processed by Congxian Jia; Phantom courtesy of Marvin Doyley.

inclusion with low strain is typical of strain images. It is often called an artifact, but it represents an accurate measurement of the physical strain distribution around a stiff inclusion. The "artifact" is not in the measurement, but rather in the interpretation of the strain image as a reciprocal modulus image [2]. The strain is not the reciprocal of the modulus image and this demonstrates that even at small strains, the uniaxial stress assumption is not perfectly satisfied. The review by Ophir et al. [2] contains a rather complete discussion of artifacts, including signal-processing artifacts as well as similar "interpretation artifacts." For example, if the compression was applied nonuniformly so that the left side was more compressed than the right,

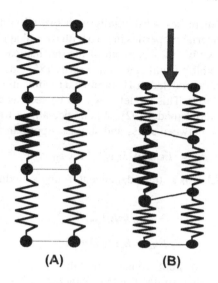

FIGURE 4.5

Interpretation of strain image. The tissue behaves as a collection of linear springs, possibly with different stiffness values. When compressed, the heavy bold spring compresses less (shows less strain) than its neighbors. Here, for clarity, only springs oriented in the direction of applied compression are shown. In principle, isotropic tissue has equal "springiness" in all directions. Some of the horizontal links in (A) are seen to have rotated in (B), indicating that nonuniform compression is usually accompanied by shear strain.

then the strain would be (on average) higher on the left side than the right, independent of the tissue stiffness distribution.

4. Beyond linear elastic imaging: biomechanical imaging

The technologic capability of ultrasonography to make precise strain measurements in vivo enables the interrogation of properties far beyond linear elastic modulus. Indeed, it has been recognized that soft tissues exhibit nonlinear elastic, viscoelastic, and even poroelastic behaviors. Therefore some of the current limitations of quasi-static elastography stem not from the ability to measure tissue deformation but from a lack of appropriate modeling and inverse solvers to take full advantage of the available data. Here we will mention two directions beyond linear elastic modulus imaging that are discussed in other chapters in this volume.

4.1 Nonlinear elastic imaging

It has long been recognized that strain contrast between a lesion and its background depends on the total applied deformation of the tissue [78−83]. In shear wave elastography, an equivalent phenomenon has been reported, where it has been observed that shear wave speed increases substantially with overall compression [84]. The observed variation in material properties in vivo with overall applied strain is consistent with the fact that ex vivo measurement of the mechanical properties of breast tissues shows them to exhibit significant nonlinearity [85–88].

This may be accurately modeled in 1D through a modification of the Blatz model [89] (see also Refs. [90,91]). This model may also be seen as a special case of the Veronda-Westmann material model [92]. When specialized to uniaxial stress and small strain, the relation between stress and strain is approximately

$$\sigma_0 \approx E\varepsilon_0 \exp\left[3\gamma\varepsilon_0^2\right] \quad \text{for } \varepsilon_0 \ll 1 \tag{4.15}$$

This may be interpreted as a strain-dependent Young's modulus by rewriting Eq. (4.15) as

$$\sigma_0 = E_{NL}\varepsilon_0 \tag{4.16}$$

$$E_{NL} = E \exp\left[3\gamma\varepsilon_0^2\right] \tag{4.17}$$

This shows that the apparent Young's modulus, E_{NL}, increases exponentially with the square of the applied strain. The rate of increase, γ, is a material parameter that differs from tissue to tissue. The increase in apparent modulus is sometimes measured as a tangent modulus, with overall tissue compression being a confounding factor in the interpretation of all elastograms, whether compression elastogram (quasi-static) or shear wave elastogram [79,84]. For $\gamma \approx 10$, which is a reasonable estimate for normal breast tissue [93], Eq. (4.17) suggests that E_{NL} will be within 10% of E, provided the overall applied strain is less than 5%. Therefore it is advisable that elasticity images be collected with minimal possible preload.

The fact that the nonlinear phenomenon just described is observable and quantifiable, however, also presents an opportunity. The ability to image γ in phantoms and in vivo has been studied and demonstrated in Refs. [90,91,93–97]. In particular, it is shown in Ref. [93] to be higher in invasive ductal carcinomas (a malignant breast tumor) than in fibroadenomas (a benign breast tumor), and higher than in normal breast tissue. One contribution of the nonlinear mechanical behavior of breast and other tissues may be traced back to the fibrillar structure of the extracellular matrix, and indeed, initial studies demonstrate the possibility of inferring the extracellular matrix structure from macroscopic observations of tissue mechanical behavior [39].

4.2 Poroelastic imaging

Soft tissues are often modeled as incompressible, with the acknowledgment that they have a large water content. Although the incompressible behavior of soft tissue is evident at short or early times, longer observation periods can reveal local volumetric changes as fluid moves from one region of the tissue into another. The effects of this fluid motion are observable in vivo [98]; as fluid redistributes within the tissue, local patterns of strain in the tissue are altered. The observability of these effects leads to an opportunity to measure the material parameters that govern the fluid exchange. In particular, microvessel density (MVD) has exciting possibilities in the differential diagnosis of tumors. MVD is more strongly predictive of prognosis than tumor grade, size, estrogen-receptor positivity, or other prognostic markers [99]. Because of this and other motivating applications, the development of effective methods for imaging, interpreting, and quantifying the multiphase effects in soft tissues is an area of active interest [98,100–115].

Several multiphase models of tissue mechanics have been proposed to capture the link between fluid flow and tissue deformation. The primary application of such models has traditionally been to cartilage, starting with the "biphasic model" [116], which may be regarded as a specialization of Biot's theory to incompressible phases. When modeling the mechanics of vascularized soft tissues, however, one must account for different compartments within which the fluids may reside. For example, fluids within the microvasculature move relatively freely compared with motile fluids in the extravascular compartment. Furthermore, local fluid redistribution or fluid pressure can give rise to nonlinear phenomena, akin to those mentioned earlier [117]. This gives rise to different phases that may be, in fact, the same constituent (e.g., water or lymphatic fluid) but which currently reside in different locations within the tissue. Thus water within the microvasculature is accounted for separately from water within the extravascular compartment, and so gives rise to two distinct phases in the model. The simplest model that takes such an approach is a generalized poroelastic continuum model for fluid transport in solid tumors [91,104,118]. The inverse problem to determine vascular permeability from strain relaxation measurements was considered in Ref. [110].

5. Summary

In this chapter, we reviewed the physical basis for quasi-static elastography, which in its common implementation is called compression elastography. To generate a compression elastogram, the ultrasound probe is simply pressed gently against the patient's skin in order to very slightly compress the tissue immediately below the probe. The ideal level of compression is typically a few hundred micrometers (a compression of approximately 0.5 percent–1 percent), and therefore best results can often be achieved by attempting to hold the probe almost entirely still.

We discussed how the compression is measured from the ultrasound signal via either block matching or phase-based methods. We showed why tissue inertia is negligible in the interpretation of quasi-static elastographic measurements and how that interpretation depends on the uniaxial stress assumption. These lead us to the conclusion that an image of tissue strain approximately represents the inverse of the tissue rigidity measured in terms of Young's modulus. The Young's modulus is the initial slope of the graph of stress versus strain.

Finally, we briefly discussed applications of compression elastography beyond linear elastic imaging. We briefly discussed the confounding effects of tissue nonlinearity and how these might be quantified to provide new tissue information linked to the structure of extracellular matrix proteins. We also discussed the confounding effects of strain relaxation caused by the redistribution of fluid within a tissue that is subject to compression. Inasmuch as the fluid redistribution is dominated by microvasculature, there is the potential to infer information about the microvasculature by quantifying strain relaxation. The relationship is quite complicated, however, and may best be accomplished through model-based inversion [110,112–115].

In summary, quasi-static elastography offers opportunities to use ultrasonography to interrogate a number of different mechanical properties of soft tissues. These properties are dominated by features of the tissue microstructure, including properties of the extracellular matrix and of the microvasculature. Compression strain imaging barely scratches the surface of the full potential of quasi-static elastography, which offers many exciting prospects ahead.

Acknowledgments

Research reported in this publication was partially supported by the National Institutes of Health through Grants R01 CA140271 and R01 CA195527 from the National Cancer Institute. The content is solely the responsibility of the authors and does not necessarily represent the official views of the National Institutes of Health.

References

[1] A.P. Sarvazyan, O.V. Rudenko, S.D. Swanson, J.B. Fowlkes, S.Y. Emelianov, Shear wave elasticity imaging: a new ultrasonic technology of medical diagnostics, Ultrasound Med. Biol. 24 (9) (1998) 1419–1435.

[2] J. Ophir, S.K. Alam, B. Garra, F. Kallel, E. Konofagou, T. Krouskop, T. Varghese, Elastography: ultrasonic estimation and imaging of the elastic properties of tissues, Proc. Inst. Mech. Eng. H J. Eng. Med. 213 (H3) (1999) 203–233.

[3] J. Bamber, D. Cosgrove, C.F. Dietrich, J. Fromageau, J. Bojunga, F. Calliada, V. Cantisani, J.-M. Correas, M. Donofrio, E.E. Drakonaki, et al., EFSUMB guidelines and recommendations on the clinical use of ultrasound elastography. Part 1: basic principles and technology, Ultraschall Med. 34 (02) (2013) 169–184.

[4] T. Shiina, K.R. Nightingale, M.L. Palmeri, T.J. Hall, J.C. Bamber, R.G. Barr, L. Castera, B.I. Choi, Y.-H. Chou, D. Cosgrove, C.F. Dietrich, H. Ding, D. Amy, A. Farrokh, G. Ferraioli, C. Filice, M. Friedrich-Rust, K. Nakashima, F. Schafer, I. Sporea, S. Suzuki, S. Wilson, M. Kudo, WFUMB guidelines and recommendations for clinical use of ultrasound elastography: part 1: basic principles and terminology, Ultrasound Med. Biol. 41 (5) (2015) 1126–1147.

[5] R.G. Barr, K. Nakashima, D. Amy, D. Cosgrove, A. Farrokh, F. Schafer, J.C. Bamber, L. Castera, B.I. Choi, Y.-H. Chou, C.F. Dietrich, H. Ding, G. Ferraioli, C. Filice, M. Friedrich-Rust, T.J. Hall, K.R. Nightingale, M.L. Palmeri, T. Shiina, S. Suzuki, I. Sporea, S. Wilson, M. Kudo, WFUMB guidelines and recommendations for clinical use of ultrasound elastography: Part 2: Breast, Ultrasound Med. Biol. 41 (5) (2015) 1148–1160.

[6] G. Ferraioli, C. Filice, L. Castera, B.I. Choi, I. Sporea, S.R. Wilson, D. Cosgrove, C.F. Dietrich, D. Amy, J.C. Bamber, R.G. Barr, Y.-H. Chou, H. Ding, A. Farrokh, M. Friedrich-Rust, T.J. Hall, K. Nakashima, K.R. Nightingale, M.L. Palmeri, F. Schafer, T. Shiina, S. Suzuki, M. Kudo, WFUMB guidelines and recommendations for clinical use of ultrasound elastography: part 3: Liver, Ultrasound Med. Biol. 41 (5) (2015) 1161−1179.

[7] D. Cosgrove, R. Barr, J. Bojunga, V. Cantisani, M.C. Chammas, M. Dighe, S. Vinayak, J.-M. Xu, C.F. Dietrich, WFUMB guidelines and recommendations on the clinical use of ultrasound elastography: part 4. thyroid, Ultrasound Med Biol 43 (1) (2017 Jan) 4−26.

[8] R.G. Barr, D. Cosgrove, M. Brock, V. Cantisani, J. Michel Correas, A.W. Postema, G. Salomon, M. Tsutsumi, H.-X. Xu, C.F. Dietrich, WFUMB guidelines and recommendations on the clinical use of ultrasound elastography: part 5. prostate, Ultrasound Med Biol 43 (1) (2017 Jan) 27−48.

[9] J.C. Gore, S. Leeman, N.J. Plessner, K. Willson, C. Metreweli, Dynamic autocorrelation analysis of a-scans in vivo, in: M. Linzer (Ed.), Ultrasonic Tissue Characterisation II, National Bureau of Standards, Washington, DC, USA, 1979, pp. 275−280.

[10] R.J. Dickinson, C.R. Hill, Measurement of soft tissue motion using correlation between a-scans, Ultrasound Med. Biol. 8 (3) (1982) 263−271.

[11] L.S. Wilson, D.E. Robinson, Ultrasonic measurement of small displacements and deformations of tissue, Ultrason. Imaging 4 (1) (1982) 71−82.

[12] A. Eisensher, E. Schweg-Toffler, G. Pelletier, G. Jacquemard, La palpation échographique rythmée-echosismographie, J. Radiol. 64 (1983) 255−261.

[13] M. Tristam, D.C. Barbosa, D.O. Cosgrove, D.K. Nassiri, J.C. Bamber, C.R. Hill, Ultrasonic study of in vivo kinetic characteristics of human tissues, Ultrasound Med. Biol. 12 (12) (1986) 927−937.

[14] T.A. Krouskop, D.R. Dougherty, F.S. Vinson, et al., A pulsed Doppler ultrasonic system for making noninvasive measurements of the mechanical properties of soft tissue, J. Rehabil. Res. Dev. 24 (2) (1987) 1−8.

[15] R.M. Lerner, K.J. Parker, J. Holen, R. Gramiak, R.C. Waag, Sono-elasticity: medical elasticity images derived from ultrasound signals in mechanically vibrated targets, in: Acoustical Imaging, Springer, 1988, pp. 317−327.

[16] M. Bertrand, J. Meunier, M. Doucet, G. Ferland, Ultrasonic biomechanical strain gauge based on speckle tracking, in: Ultrasonics Symposium, 1989. Proceedings, IEEE, IEEE, 1989, pp. 859−863.

[17] J. Ophir, I. Cespedes, H. Ponnekanti, Y. Yazdi, X. Li, Elastography: a quantitative method for imaging the elasticity of biological tissues, Ultrason. Imaging 13 (2) (1991) 111−134.

[18] T.L. Chenevert, A.R. Skovoroda, M. O'Donnell, S.Y. Emelianov, Elasticity reconstructive imaging by means of stimulated echo MRI, Magn. Reson. Med. 39 (3) (1998) 482−490.

[19] C.R. Hill, M. Tristam, J.C. Bamber, et al., Ultrasonic remote palpation (URP): use of shear elastic modulus to differentiate pathology, J. Ultrasound Med. 7 (1988) S129.

[20] B.S. Garra, I. Cespedes, J. Ophir, S. Spratt, R.A. Zuurbier, C.M. Magnant, M.F. Pennanen, Elastography of breast lesions: initial clinical results, Radiology 202 (1997) 79−86.

[21] J.B. Weaver, E.E.W. Van Houten, M.I. Miga, F.E. Kennedy, K.D. Paulsen, Magnetic resonance elastography using 3D gradient echo measurements of steady-state motion, Med. Phys. 28 (8) (2001) 1620−1628.

[22] A.R. Skovoroda, S.Y. Emelianov, M. O'Donnell, Tissue elasticity reconstruction based on ultrasonic displacement and strain images, IEEE Trans. Ultrason. Ferroelectr. Freq. Control 42 (1995) 747−765.

[23] C. Sumi, A. Suzuki, K. Nakayama, Estimation of shear modulus distribution in soft tissue from strain distribution, IEEE Trans. Biomed. Eng. 42 (1995) 193−202.

[24] L.K. Ryan, F.S. Foster, Ultrasonic measurement of differential displacement strain in a vascular model, Ultrason. Imaging 19 (1) (1997) 19−38.

[25] C.L. deKorte, E.I. Cespedes, A.F.W. vanderSteen, C.T. Lancee, Intravascular elasticity imaging using ultrasound: feasibility studies in phantoms, Ultrasound Med. Biol. 23 (5) (1997) 735−746.

[26] C.L. de Korte, A.F.W. van der Steen, E.I. Cespedes, G. Pasterkamp, Intravascular ultrasound elastography in human arteries: initial experience in vitro, Ultrasound Med. Biol. 24 (3) (1998) 401−408.

[27] C.L. de Korte, E.I. Cespedes, A.F.W. van der Steen, Influence of catheter position on estimated strain in intravascular elastography, IEEE Trans. Ultrason. Ferroelectr. Freq. Control 46 (3) (1999) 616−625.

[28] C.L. de Korte, A.F. van der Steen, E.I. Cepedes, G. Pasterkamp, S.G. Carlier, F. Mastik, A.H. Schoneveld, P.W. Serruys, N. Bom, Characterization of plaque components and vulnerability with intravascular ultrasound elastography, Phys. Med. Biol. 45 (6) (2000) 1465−1475.

[29] S.Y. Emelianov, X. Chen, M. O'Donnell, B. Knipp, D. Myers, T.W. Wakefield, J.M. Rubin, Triplex ultrasound: elasticity imaging to age deep venous thrombosis, Ultrasound Med. Biol. 28 (6) (2002) 757−767.

[30] K.R. Nightingale, R.W. Nightingale, D. Stutz, G.E. Trahey, Acoustic radiation force impulse imaging of in vivo vastus medialis muscle under varying isometric load, Ultrason. Imaging 24 (2) (2002) 100−108.

[31] K.R. Nightingale, M.S. Soo, R.W. Nightingale, G.E. Trahey, Acoustic radiation force impulse imaging: in vivo demonstration of clinical feasibility, Ultrasound Med. Biol. 28 (2) (2002) 227−235.

[32] S.A. Kruse, G.H. Rose, K.J. Glaser, A. Manduca, J.P. Felmlee, C.R. Jack Jr., R.L. Ehman, Magnetic resonance elastography of the brain, Neuroimage 39 (1) (2008) 231−237.

[33] M.C. Murphy, K.J. Glaser, A. Manduca, J.P. Felmlee, J. Huston III, R.L. Ehman, Analysis of time reduction methods for magnetic resonance elastography of the brain, Magn. Reson. Imaging 28 (10) (2010) 1514−1524.

[34] G.B. Richard, Breast Elastography, Thieme Medical Publishers, 2014.

[35] P.E. Rouse Jr., A theory of the linear viscoelastic properties of dilute solutions of coiling polymers, J. Chem. Phys. 21 (7) (1953) 1272−1280.

[36] A.V. Tobolsky, Stress relaxation studies of the viscoelastic properties of polymers, J. Appl. Phys. 27 (7) (1956) 673−685.

[37] J.D. Ferry, Viscoelastic Properties of Polymers, John Wiley & Sons, 1980.

[38] B. Lee, X. Zhou, K. Riching, K.W. Eliceiri, P. J Keely, S.A. Guelcher, A.M. Weaver, Y. Jiang, A three-dimensional computational model of collagen network mechanics, PLoS One 9 (11) (2014) e111896.

[39] T. Liu, T.J. Hall, P.E. Barbone, A.A. Oberai, Inferring spatial variations of microstructural properties from macroscopic mechanical response, Biomech. Model. Mechanobiol. (2016) 1−18.

[40] T. Varghese, J. Ophir, A theoretical framework for performance characterization of elastography: the strain filter, IEEE Trans. UFFC 44 (1) (1997) 164–172.

[41] W.F. Walker, G.E. Trahey, A fundamental limit on delay estimation using partially correlated speckle signals, IEEE Trans. Biomed. Eng. 42 (2) (1995).

[42] M. Bayer, T.J. Hall, L.P. Neves, A.A.O. Carneiro, Two-dimensional simulations of displacement accumulation incorporating shear strain, Ultrason. Imaging 36 (1) (2014) 55–73.

[43] Y. Zhu, T.J. Hall, A modified block matching method for real-time freehand strain imaging, Ultrason. Imaging 24 (3) (2002) 161–176.

[44] F. Viola, W.F. Walker, A comparison of the performance of time-delay estimators in medical ultrasound, IEEE Trans. Ultrason. Ferroelectr. Freq. Control 50 (4) (2003) 392–401.

[45] I. Céspedes, Y. Huang, J. Ophir, S. Spratt, Methods for estimation of subsample time delays of digitized echo signals, Ultrason. Imaging 17 (2) (1995) 142–171.

[46] F. Viola, W.F. Walker, A spline-based algorithm for continuous time-delay estimation using sampled data, IEEE Trans. Ultrason. Ferroelectr. Freq. Control 52 (1) (2005) 80–93.

[47] F. Viola, R.L. Coe, K. Owen, D.a Guenther, W.F. Walker, Multi-Dimensional Spline-Based Estimator (MUSE) for motion estimation: algorithm development and initial results, Ann. Biomed. Eng. 36 (12) (2008) 1942–1960.

[48] J. Jiang, T.J. Hall, A coupled subsample displacement estimation method for ultrasound-based strain elastography, Phys. Med. Biol. 60 (21) (2015) 8347.

[49] K. Miyatake, M. Yamagishi, N. Tanaka, M. Uematsu, N. Yamazaki, Y. Mine, a Sano, M. Hirama, New method for evaluating left ventricular wall motion by color-coded tissue Doppler imaging: in vitro and in vivo studies, J. Am. Coll. Cardiol. 25 (3) (1995) 717–724.

[50] A. Pesavento, C. Perrey, M. Krueger, H. Ermert, A time-efficient and accurate strain estimation concept for ultrasonic elastography using iterative phase zero estimation, IEEE Trans. Ultrason. Ferroelectr. Freq. Control 46 (5) (1999) 1057–1067.

[51] A. Pesavento, A. Lorenz, H. Ermert, Phase root seeking and the Cramer-Rao-Lower bound for strain estimation, in: 1999 IEEE Ultrasonics Symposium. Proceedings. International Symposium (Cat. No.99CH37027), vol. 2, 1999, pp. 1669–1672.

[52] T. Loupas, J.T. Powers, R.W. Gill, An axial velocity estimator for ultrasound blood flow imaging, based on a full evaluation of the Doppler equation by means of a two-dimensional autocorrelation approach, IEEE Trans. Ultrason. Ferroelectr. Freq. Control 42 (4) (1995) 672–688.

[53] G.F. Pinton, J. J Dahl, G.E. Trahey, Rapid tracking of small displacements with ultrasound, IEEE Trans. Ultrason. Ferroelectr. Freq. Control 53 (6) (2006) 1103–1117.

[54] P. Chaturvedi, M.F. Insana, T.J. Hall, 2-D companding for noise reduction in strain imaging, IEEE Trans. Ultrason. Ferroelectr. Freq. Control 45 (1) (1998) 179–191.

[55] P. Chaturvedi, M.F. Insana, T.J. Hall, Testing the limitations of 2-D companding for strain imaging using phantoms, IEEE Trans. Ultrason. Ferroelectr. Freq. Control 45 (4) (1998) 1022–1031.

[56] M.F. Insana, P. Chaturvedi, T.J. Hall, M. Bilgen, 3-D companding using linear arrays for improved strain imaging, in: 1997 IEEE Ultrasonics Symposium, 1997, pp. 1435–1438.

[57] T. Shiina, N. Nitta, E.I. Ueno, J.C. Bamber, Real time tissue elasticity imaging using the combined autocorrelation method, J. Med. Ultrason. 29 (3) (2002) 119–128.

[58] S.K. Alam, J. Ophir, Reduction of signal decorrelation from mechanical compression of tissues by temporal stretching: applications to elastography, Ultrasound Med. Biol. 23 (1) (1997) 95−105.

[59] H. Rivaz, E. Boctor, P. Foroughi, R. Zellars, G. Fichtinger, G. Hager, Ultrasound elastography: a dynamic programming approach, IEEE Trans.Med. Imaging 27 (10) (2008) 1373−1377.

[60] J. Jiang, T.J. Hall, A generalized speckle tracking algorithm for ultrasonic strain imaging using dynamic programming, Ultrasound Med. Biol. 35 (11) (2009) 1863.

[61] J. Jiang, T.J. Hall, A fast hybrid algorithm combining regularized motion tracking and predictive search for reducing the occurrence of large displacement errors, IEEE Trans. Ultrason. Ferroelectr. Freq. Control 58 (4) (2011) 730−736.

[62] H. Rivaz, E.M. Boctor, M.A. Choti, G.D. Hager, Real-time regularized ultrasound elastography, IEEE Trans. Med. Imaging 30 (4) (2011) 928−945.

[63] S.K. Alam, J. Ophir, E.E. Konofagou, An adaptive strain estimator for elastography, IEEE Trans. Ultrason. Ferroelectr. Freq. Control 45 (2) (1998) 461−472.

[64] S.K. Alam, F.L. Lizzi, T. Varghese, E.J. Feleppa, S. Ramachandran, Adaptive spectral strain estimators for elastography, Ultrason. Imaging 26 (3) (2004) 131−149.

[65] T. Varghese, E.E. Konofagou, J. Ophir, S.K. Alam, M. Bilgen, Direct strain estimation in elastography using spectral cross-correlation, Ultrasound Med. Biol. 26 (9) (2000) 1525−1537.

[66] M.K. Hasan, E.M.A. Anas, S.K. Alam, S.Y. Lee, Direct mean strain estimation for elastography using nearest-neighbor weighted least-squares approach in the frequency domain, Ultrasound Med. Biol. 38 (10) (2012) 1759−1777.

[67] M.F. Insana, T.J. Hall, P. Chaturvedi, C. Kargel, Ultrasonic properties of random media under uniaxial loading, J. Acoust. Soc. Am. 110 (6) (2001) 3242−3251.

[68] J. Jiang, T.J. Hall, A.M. Sommer, A novel performance descriptor for ultrasonic strain imaging: a preliminary study, IEEE Trans. Ultrason. Ferroelectr. Freq. Control 53 (6) (2006) 1088−1102.

[69] J. Jiang, T.J. Hall, A.M. Sommer, A novel image formation method for ultrasonic strain imaging, Ultrasound Med. Biol. 33 (4) (2007) 643−652.

[70] S.Y. Yemelyanov, A.R. Skovoroda, M.A. Lubinski, B.M. Shapo, M. O'Donnell, Ultrasound elasticity imaging using Fourier based speckle tracking algorithm, in: Proceedings of the IEEE Ultrasonics Symposium, 1992, pp. 1065−1068.

[71] M. O'Donnell, A.R. Skovoroda, B.M. Shapo, S.Y. Emelianov, Internal displacement and strain imaging using ultrasonic speckle tracking, IEEE Trans. Ultrason. Ferroelectr. Freq. Control 41 (3) (1994) 314−325.

[72] T. Varghese, J. Ophir, Performance optimization in elastography: multicompression with temporal stretching, Ultrason. Imaging 18 (3) (1996) 193−214.

[73] M.A. Lubinski, S.Y. Emelianov, M. O'Donnell, Speckle tracking methods for ultrasonic elasticity imaging using short-time correlation, IEEE Trans. Ultrason. Ferroelectr. Freq. Control 46 (1) (1999) 82−96.

[74] H. Du, J. Liu, C. Pellot-Barakat, M.F. Insana, Optimizing multicompression approaches to elasticity imaging, IEEE Trans. Ultrason. Ferroelectr. Freq. Control 53 (1) (2006) 90−99.

[75] M. Bayer, T.J. Hall, Variance and covariance of accumulated displacement estimates, Ultrason. Imaging 35 (2) (2013) 90−108.

[76] M.M. Doyley, S. Srinivasan, S.A. Pendergrass, Z. Wu, J. Ophir, Comparative evaluation of strain-based and model-based modulus elastography, Ultrasound Med. Biol. 31 (6) (2005) 787–802.

[77] S. Srinivasan, T. Krouskop, J. Ophir, A quantitative comparison of modulus images obtained using nanoindentation with strain elastograms, Ultrasound Med. Biol. 30 (7) (2004) 899–918.

[78] A.R. Skovoroda, M.A. Lubinski, S.Y. Emelianov, M. O'Donnell, Reconstructive elasticity imaging for large deformations, IEEE Trans. Ultrason. Ferroelectr. Freq. Control 46 (3) (1999) 523–535.

[79] T. Varghese, J. Ophir, T.A. Krouskop, Nonlinear stress-strain relationships in tissue and their effect on the contrast-to-noise ratio in elastograms, Ultrasound Med. Biol. 26 (5) (2000) 839–851.

[80] R.Q. Erkamp, S.Y. Emelianov, A.R. Skovoroda, X. Chen, M. O'Donnell, Exploiting strain-hardening of tissue to increase contrast in elasticity imaging, in: IEEE Ultrasonics Symposium vol. 2, 2000, pp. 1833–1836.

[81] R.Q. Erkamp, S.Y. Emelianov, A.R. Skovoroda, M. O'Donnell, Nonlinear elasticity imaging, in: IEEE Ultrasonics Symposium vol. 2, 2002, pp. 1891–1894.

[82] T.J. Hall, Y. Zhu, C.S. Spalding, *In vivo* real-time freehand palpation imaging, Ultrasound Med. Biol. 29 (3) (2003) 427–435.

[83] R.Q. Erkamp, S.Y. Emelianov, A.R. Skovoroda, M. O'Donnell, Nonlinear elasticity imaging: theory and phantom study, Ultrason. Ferroelectr. Freq. Control, IEEE Trans. on 51 (5) (2004) 532–539.

[84] R.G. Barr, Sonographic breast elastography: a primer, J. Ultrasound Med. 31 (5) (2012) 773–783.

[85] T.A. Krouskop, T.M. Wheeler, F. Kallel, B.S. Garra, T. Hall, Elastic moduli of breast and prostate tissues under compression, Ultrason. Imaging 20 (4) (1998) 260–274.

[86] P.S. Wellman, E.P. Dalton, D. Krag, K.A. Kern, R.D. Howe, Tactile imaging of breast masses: first clinical report, Arch. Surg. 136 (2) (2001) 204–208.

[87] A. Samani, D. Plewes, A method to measure the hyperelastic parameters of ex vivo breast tissue samples, Phys. Med. Biol. 49 (18) (2004) 4395–4405.

[88] J.J. O'Hagan, A. Samani, Measurement of the hyperelastic properties of 44 pathological ex vivo breast tissue samples, Phys. Med. Biol. 54 (2009) 2557–2569.

[89] P.J. Blatz, B.M. Chu, H. Wayland, On the mechanical behavior of elastic animal tissue, J. Rheol. 13 (1969) 83–102.

[90] S. Goenezen, P. Barbone, A.A. Oberai, Solution of the nonlinear elasticity imaging inverse problem: the incompressible case, Comput. Methods Appl. Mech. Eng. 200 (13) (2011) 1406–1420.

[91] P.E. Barbone, A.A. Oberai, J.C. Bamber, G.P. Berry, J.F. Dord, E.R. Ferreira, S. Goenezen, T.J. Hall, Nonlinear and poroelastic biomechanical imaging: elastography beyond Young's modulus, in: C.P. Neu, G.M. Genin (Eds.), Handbook of Imaging in Biological Mechanics, CRC Press, 2014, pp. 199–215.

[92] D.R. Veronda, R.A. Westmann, Mechanical characterization of skin—Finite deformations, J. Biomech. 3 (1) (1970) 111–122.

[93] S. Goenezen, J.-F. Dord, Z. Sink, P.E. Barbone, J. Jiang, T.J. Hall, A.A. Oberai, Linear and nonlinear elastic modulus imaging: an application to breast cancer diagnosis, IEEE Trans. Med. Imaging 31 (8) (2012) 1628–1637.

[94] A.A. Oberai, N.H. Gokhale, S. Goenezen, P.E. Barbone, T.J. Hall, A.M. Sommer, J. Jiang, Linear and nonlinear elasticity imaging of soft tissue in vivo: demonstration of feasibility, Phys. Med. Biol. 54 (5) (2009) 1191.

[95] T.J. Hall, A.A. Oberait, P.E. Barbone, A.M. Sommer, N.H. Gokhale, S. Goenezen, J. Jiang, Elastic nonlinearity imaging, in: Engineering in Medicine and Biology Society, 2009. EMBC 2009. Annual International Conference of the IEEE, IEEE, 2009, pp. 1967−1970.

[96] T.J. Hall, P. Barbone, A.A. Oberai, J. Jiang, J. Francois Dord, S. Goenezen, T.G. Fisher, Recent results in nonlinear strain and modulus imaging, Curr. Med. Imag. Rev. 7 (4) (2011) 313.

[97] E.R. Ferreira, A.A. Oberai, P.E. Barbone, Uniqueness of the elastography inverse problem for incompressible nonlinear planar hyperelasticity, Inverse Probl. 28 (6) (2012) 065008.

[98] G.P. Berry, J.C. Bamber, P.S. Mortimer, N.L. Bush, N.R. Miller, P.E. Barbone, The spatio-temporal strain response of oedematous and non-oedematous tissue to sustained compression *in vivo*, Ultrasound Med. Biol. 34 (4) (2008) 617−629.

[99] J. Folkman, Clinical applications of research on angiogenesis, N. Engl. J. Med. 333 (26) (1995) 1757−1763.

[100] E.E. Konofagou, T.P. Harrigan, J. Ophir, T.A. Krouskop, Poroelastography: imaging the poroelastic properties of tissues, Ultrasound Med. Biol. 27 (10) (2001) 1387−1397.

[101] R. Righetti, J. Ophir, S. Srinivasan, T.A. Krouskop, The feasibility of using elastography for imaging the Poisson's ratio in porous media, Ultrasound Med. Biol. 30 (2) (2004) 215−228.

[102] R. Righetti, J. Ophir, B.S. Garra, R.M. Chandrasekhar, T.A. Krouskop, A new method for generating poroelastograms in noisy environments, Ultrason. Imaging 27 (4) (2005) 201−220.

[103] R. Righetti, J. Ophir, T.A. Krouskop, A method for generating permeability elastograms and Poissons ratio time-constant elastograms, Ultrasound Med. Biol. 31 (6) (2005) 803−816.

[104] R. Leiderman, P.E. Barbone, A.A. Oberai, J.C. Bamber, Coupling between elastic strain and interstitial fluid flow: ramifications for poroelastic imaging, Phys. Med. Biol. 51 (2006) 6291−6313.

[105] G.P. Berry, J.C. Bamber, C.G. Armstrong, N.R. Miller, P.E. Barbone, Towards an acoustic model-based poroelastic imaging method: I. Theoretical foundation, Ultrasound Med. Biol. 32 (4) (2006) 547−567.

[106] G.P. Berry, J.C. Bamber, C.G. Armstrong, N.R. Miller, P.E. Barbone, Towards an acoustic model-based poroelastic imaging method: ii. Experimental investigation, Ultrasound Med. Biol. 32 (12) (2006) 1869−1885.

[107] R. Righetti, B.S. Garra, L.M. Mobbs, C.M. Kraemer-Chant, J. Ophir, T.A. Krouskop, The feasibility of using poroelastographic techniques for distinguishing between normal and lymphedematous tissues in vivo, Phys. Med. Biol. 52 (21) (2007) 6525.

[108] R. Righetti, S. Srinivasan, A.T. Kumar, J. Ophir, T.A. Krouskop, Assessing image quality in effective poisson's ratio elastography and poroelastography: I, Phys. Med. Biol. 52 (5) (2007) 1303.

[109] R. Righetti, J. Ophir, A.T. Kumar, T.A. Krouskop, Assessing image quality in effective Poisson's ratio elastography and poroelastography: II, Phys. Med. Biol. 52 (5) (2007) 1321.

[110] R. Leiderman, A.A. Oberai, P.E. Barbone, Theory of reconstructing the spatial distribution of the filtration coefficient in vascularized soft tissues: exact and approximate inverse solutions, Compt. Rendus Mec. 338 (7-8) (2010).

[111] P.R. Perriñez, F.E. Kennedy, E.E.W. Van Houten, J.B. Weaver, K.D. Paulsen, Magnetic resonance poroelastography: an algorithm for estimating the mechanical properties of fluid-saturated soft tissues, IEEE Trans.Med. Imaging 29 (3) (2010) 746−755.

[112] B.A. Galaz, R.H. Acevedo, Optimization of a pixel-to-pixel curve-fitting method for poroelastography imaging, Ultrasound Med. Biol. 43 (1) (2016).

[113] J.J. Pitre, W.F. Weitzel, J.L. Bull, Evaluation of a model-based poroelastography algorithm for edema quantification, J. Acoust. Soc. Am. 140 (4) (2016) 3310.

[114] A. Chaudhry, I.K. Yazdi, R. Kongari, E. Tasciotti, R. Righetti, A new class of phantom materials for poroelastography imaging techniques, Ultrasound Med. Biol. 42 (5) (2016) 1230−1238.

[115] A. Chaudhry, N. Kim, G. Unnikrishnan, S. Nair, J.N. Reddy, R. Righetti, Effect of interstitial fluid pressure on ultrasound axial strain and axial shear strain elastography, 0161734616671713, Ultrason. Imaging (2016).

[116] V.C. Mow, S.C. Kuei, W.M. Lai, C.G. Armstrong, Biphasic creep and stress relaxation of articular cartilage in compression: theory and experiments, J. Biomech. Eng. 102 (1980) 73−84.

[117] N. Nazari, P.E. Barbone, Shear wave speed in pressurized soft tissue, J. Mech. Phys. Solids 119 (2018) 60−72.

[118] P.A. Netti, L.T. Baxter, Y. Boucher, R. Skalak, R.K. Jain, Macro- and microscopic fluid transport in living tissues: application to solid tumors, AIChE J. Bioeng. Food Nat. Prod. 43 (3) (1997) 818−834.

Acoustic radiation force and shear wave elastography techniques

5

Arsenii V. Telichko, Carl D. Herickhoff, Jeremy J. Dahl

Department of Radiology, Stanford University, Stanford, CA, United States

1. Introduction

In 1990, Sugimoto et al. [1] proposed the use of acoustic radiation force to probe the hardness of tissue as an alternative to manual palpation. Manual palpation has been widely utilized by practicing clinicians for hundreds of years for the purpose of disease identification and diagnosis, based on the concept that diseased tissue is different in stiffness than the surrounding normal tissue. Although manual palpation is a useful diagnostic tool today, it is limited to tissues near the surface and to disease processes that are large enough to be detected by human hands and senses.

The technique proposed by Sugimoto et al. is an example of remote palpation, allowing clinicians to probe the mechanics of deep tissues previously unavailable by manual palpation. By dynamically exciting tissue with acoustic radiation force, for example, a response can be elicited from tissues deep under the surface of the skin, from which the elastic or viscoelastic properties can be derived. This initial remote palpation technique touched of a variety of acoustic radiation force methods in the decades following, which seek to remotely and noninvasively extract quantitative and qualitative properties of tissues not previously available to clinicians. These methods have attempted to differentiate tissues based on their mechanical properties, including stiffness, Young's modulus, shear modulus, viscoelasticity, and vibroacoustic resonance. The ability to noninvasively characterize the mechanical properties of tissues has many potential applications in both disease identification and procedure monitoring.

This chapter will explore the physical basis for the generation of radiation force from acoustic waves and the response of tissues to the radiation force. Several of the many imaging and measurement techniques that utilize acoustic radiation force will be introduced, such as acoustic radiation force impulse (ARFI) imaging, shear wave elasticity imaging (SWEI), vibroacoustography, harmonic motion imaging, viscoelastic response (VisR) imaging, and others. Both preliminary applications and potential future applications of these acoustic radiation force methods will be reviewed.

Tissue Elasticity Imaging. https://doi.org/10.1016/B978-0-12-809661-1.00005-4

2. Physical basis for acoustic radiation force from ultrasonography

Acoustic radiation force is a unidirectional force that arises when the absorbing or scattering targets are in the propagation path of an acoustic wave. Radiation force is the result of a transfer of momentum from an acoustic wave to the propagation medium or target. In the following derivation, we begin with the assumption that the acoustic radiation force is largely a product of the absorption of the acoustic wave due to the weak scattering of ultrasound from tissues. We combined the classic derivations of radiation force from Eckart [2] and Nyborg [3] and attempted to reduce the complexity of the derivation so that readers of all backgrounds may follow the discussion. In addition, we introduce later the full definition of acoustic radiation force, including those terms arising from the scattering of ultrasound. More rigorous and thorough derivations of radiation force can be found in the works by Eckart [2], Westervelt [4], Nyborg [3], Beyer [5], and Sarvazyan et al. [6].

2.1 The acoustic radiation force

Consider a small element of homogeneous isotropic fluid (i.e., a fluid with the same acoustic and mechanical properties throughout and in all directions). This small element of fluid is much smaller than the wavelength of the acoustic wave yet much larger than the interatomic spacing of the fluid. The density, pressure, and particle velocity of the small element of the fluid are described by $\rho(x, y, z, t)$, $\mathbf{p}(x, y, z, t)$, and $\mathbf{u}(x, y, z, t)$, respectively. $\mathbf{p}$ and $\mathbf{u}$ are both vector quantities with unit vectors $\hat{\mathbf{x}}$, $\hat{\mathbf{y}}$, and $\hat{\mathbf{z}}$. For brevity, the x, y, z, and t notation will be dropped from the vectors $\mathbf{p}(x, y, z, t)$ and $\mathbf{u}(x, y, z, t)$ and the pressure and particle velocity will be referred to as p and u, respectively. The particle velocity describes the velocity of this small element under the influence of the pressure wave and should not be confused with the longitudinal speed of sound, c_0, which is a scalar quantity that describes the speed at which the pressure wave travels through a medium.

The derivation of the acoustic radiation force begins with Newton's second law, $\mathbf{F} = m\mathbf{a}$, the well-known relation that describes a force resulting from the acceleration of a mass. Expressing acceleration as the time derivative of velocity and density as a mass per unit volume, Newton's second law is written in the form

$$\mathbf{f} = \rho d\mathbf{u}/dt, \tag{5.1}$$

where $\mathbf{f}$ is the force per unit volume. Recall that the components of the vector $\mathbf{u}$ are functions of x, y, and z, so the derivative of the particle velocity requires the *material derivative*, which includes differentiation with respect to the spatial dimensions as well as the time derivative. Applying the material derivative to Newton's second law provides the momentum balance equation:

$$\mathbf{f} = \rho \frac{\partial \mathbf{u}}{\partial t} + \rho (\mathbf{u} \cdot \nabla)\mathbf{u}, \tag{5.2}$$

where ∇ is the vector operator defined as

$$\nabla = \frac{\partial}{\partial x}\widehat{\mathbf{x}} + \frac{\partial}{\partial y}\widehat{\mathbf{y}} + \frac{\partial}{\partial z}\widehat{\mathbf{z}}.$$

The first term in Eq. (5.2) is the well-known mass times acceleration of Newton's second law, and the second term describes the force as a result of the spatial dependencies of $\mathbf{u}$. In the case of a plane acoustic wave propagating in the z direction, Eq. (5.2) becomes

$$\mathbf{f} = \left[\rho \frac{\partial u_z}{\partial t} + \rho \frac{\partial u_z}{\partial z}\frac{\partial z}{\partial t} \right]\widehat{\mathbf{z}}. \tag{5.3}$$

where u_z is the component of $\mathbf{u}$ in the direction $\widehat{\mathbf{z}}$.

The continuity equation says that the rate at which mass enters the small volume of fluid under a pressure wave must be equal to the rate at which mass exits the volume. This is described mathematically as

$$\frac{\partial \rho}{\partial t} + \nabla \cdot (\rho \mathbf{u}) = 0. \tag{5.4}$$

By multiplying Eq. (5.4) by the vector $\mathbf{u}$ and applying the chain rule to $\frac{\partial(\rho\mathbf{u})}{\partial t}$ and rearranging, the following equation is obtained:

$$\rho \frac{\partial \mathbf{u}}{\partial t} = \frac{\partial(\rho \mathbf{u})}{\partial t} + \mathbf{u}\nabla \cdot (\rho \mathbf{u}). \tag{5.5}$$

Inserting Eq. (5.5) into Eq. (5.2) obtains

$$\mathbf{f} = \frac{\partial(\rho \mathbf{u})}{\partial t} + \mathbf{u}\nabla \cdot (\rho \mathbf{u}) + \rho(\mathbf{u} \cdot \nabla)\mathbf{u}. \tag{5.6}$$

In fluid dynamics, the $\mathbf{f}$ in Eq. (5.6) is often replaced by a sum of forces to achieve the Navier-Stokes equation for fluid dynamics. This equation has been applied to determine the streaming velocity of a fluid resulting from acoustic radiation force [3]. However, the force induced by a propagating acoustic wave is the desired solution here, so it is necessary to only carry out the derivation of the radiation force vector described in Eq. (5.6). Unfortunately, the true velocities, pressures, and densities of tissues are complicated functions, making it difficult to derive the exact solution to Eq. (5.6). To reduce complexity, the density, pressure, and velocity can be approximated by a sum of terms of increasing order:

$$\rho = \rho_0 + \rho_1 + \rho_2 + \cdots \tag{5.7a}$$

$$\mathbf{p} = \mathbf{p}_0 + \mathbf{p}_1 + \mathbf{p}_2 + \cdots \tag{5.7b}$$

$$\mathbf{u} = \mathbf{u}_0 + \mathbf{u}_1 + \mathbf{u}_2 + \cdots \tag{5.7c}$$

Each successive higher order term in these equations is smaller than the preceding term and has a smaller impact on the overall solution. This approximation allows the use of perturbation methods, such as Stokes' expansion [7], to obtain an approximate solution for $\mathbf{f}$.

To determine the most influential terms of Eq. 5.7 to the solution of Eq. (5.6), a nondimensionalized version of Eq. (5.7) is needed. Let U be an arbitrary value in units of meters per second, T be an arbitrary value in units of seconds, and Z be an arbitrary value in units of meters. The variables in Eqs. (5.4) and (5.6) can then replaced by the nondimensionalized variables: $\upsilon = \mathbf{u}/U$, $\tau = t/T$, and $\zeta = z/Z$. The respective nondimensionalized versions of Eqs. (5.4) and (5.6) are thus

$$\frac{\partial \rho}{\partial \tau} + N\nabla \cdot (\rho \upsilon) = 0 \tag{5.8}$$

$$\mathbf{f} = \frac{\partial (\rho \upsilon)}{\partial \tau} + N\upsilon\nabla \cdot (\rho \upsilon) + N_\rho (\upsilon \cdot \nabla)\upsilon, \tag{5.9}$$

where the density term, ρ, has been left as the only dimensionalized parameter and the x and y components of ∇ have also been nondimensionalized by Z. N is the so-called dimensionless perturbation parameter, defined as $N = UT/Z$. The pressure, density, and velocity approximations are then written, respectively, to include the perturbation parameter:

$$\rho = \rho_0 + N\rho_1 + N^2\rho_2 + \cdots \tag{5.10a}$$

$$\mathbf{p} = \mathbf{p}_0 + N\mathbf{p}_1 + N^2\mathbf{p}_2 + \cdots \tag{5.10b}$$

$$\upsilon = \upsilon_o + N\upsilon_1 + N^2\upsilon_2 + \cdots. \tag{5.10c}$$

In the Stokes' perturbation analysis, the number of higher order terms one wishes to include in the analysis is inserted into the nondimensionalized equations. Here, it will suffice to include only those terms up to the first order in Eq. (5.10). These first-order terms are then inserted into Eqs. (5.8) and (5.9). The nondimensionalized continuity equation (Eq. 5.8) in this case becomes

$$0 = \frac{\partial \rho_0}{\partial \tau} + N\frac{\partial \rho_1}{\partial \tau} + N\nabla \cdot (\rho_0 + N\rho_1)(\upsilon_o + N\upsilon_1)$$

$$= \frac{\partial \rho_0}{\partial \tau} + N\frac{\partial \rho_1}{\partial \tau} + N\nabla \cdot (\rho_0\upsilon_o) + N^2\nabla \cdot (\rho_0\upsilon_1) + N^2\nabla \cdot (\rho_1\upsilon_o) + N^3\nabla \cdot (\rho_1\upsilon_1).$$
$$\tag{5.11}$$

The terms according to the power of the perturbation parameter N are collected to form zeroth- and first-order equations, and the nondimensionalized parameters can then be replaced with their original dimensionalized parameters. The respective zeroth- and first-order continuity equations are therefore

$$N^0 : \frac{\partial \rho_0}{\partial t} = 0 \tag{5.12}$$

$$N^1 : \frac{\partial \rho_1}{\partial t} + \rho_0\nabla \cdot \mathbf{u}_0 = 0. \tag{5.13}$$

The zeroth-order equation states that the steady-state density of the propagation medium (ρ_0) is a constant. By combining the first-order terms from a similar perturbation of Eq. (5.9) with the first-order continuity Eq. (5.13), the first-order solution for the instantaneous force is obtained:

$$\mathbf{f} = \rho_0 \frac{\partial \mathbf{u}_1}{\partial t} + \frac{\partial (\rho_1 \mathbf{u}_0)}{\partial t} + \rho_0 (\mathbf{u}_0 \nabla \cdot \mathbf{u}_0 + \mathbf{u}_0 \cdot \nabla \mathbf{u}_0). \tag{5.14}$$

The acoustic radiation force arises from the time average of the instantaneous force and is described as

$$\mathbf{f_{rad}} = \langle \mathbf{f} \rangle = \frac{1}{T} \int\limits_{-T/2}^{T/2} \mathbf{f} \, dt. \tag{5.15}$$

In the case of radiation force, T is selected to be larger than the period of the acoustic wave, but not so long as to allow T to approach infinity. Because most acoustic waves are assumed to be sinusoidal in nature, the time average eliminates the time-dependent derivatives in Eq. (5.14) and reduces Eq. (5.15) to

$$\mathbf{f_{rad}} = \rho_0 \langle \mathbf{u}_0 \nabla \cdot \mathbf{u}_0 + \mathbf{u}_0 \cdot \nabla \mathbf{u}_0 \rangle. \tag{5.16}$$

Solutions to Eq. (5.16) can be complicated for three-dimensional waves. However, in the case of an exponentially decaying plane acoustic wave, an easy analytic solution can be found. A plane wave traveling in the $\hat{\mathbf{z}}$ direction is described as $\mathbf{u_0} = U_0 \exp(-\alpha z) \cos(2\pi f t - kz)\hat{\mathbf{z}}$, where f is the frequency of the wave, $k = 2\pi/\lambda$ is the wave number, λ is the wavelength, and α is the absorption coefficient, which is a function of the bulk and shear viscosities of the fluid. Applying this plane wave representation to Eq. (5.16), the acoustic radiation force is reduced to

$$\mathbf{f_{rad}} = 2\rho_0 \left\langle u_0 \frac{\partial u_0}{\partial z} \right\rangle \hat{\mathbf{z}}. \tag{5.17}$$

Carrying out the partial derivatives, the following solution is obtained:

$$\mathbf{f_{rad}} = -2\rho_0 \alpha U_0^2 \exp(-2\alpha z) \langle \cos^2(2\pi f t - kz) + k \sin(4\pi f t - 2kz) \rangle \hat{\mathbf{z}}. \tag{5.18}$$

In the time average of Eq. (5.18), the sinusoidal term on the right goes to zero and the $\cos^2(\cdot)$ term is equal to 1/2. The acoustic radiation force is then given by

$$\mathbf{f_{rad}} = -\alpha \rho_0 U_0^2 \exp(-2\alpha z) = -\frac{2\alpha I}{c_0} \hat{\mathbf{z}}, \tag{5.19}$$

where I is the time-averaged intensity of the acoustic wave given by $I = \frac{1}{2} \mathbf{p} \mathbf{u} = \frac{1}{2} \rho_0 c_0 U_0^2 \exp(-2\alpha z)$ and c_0 is the longitudinal steady-state speed of sound of the propagating medium. In practice, the attenuation coefficient (i.e., the decrease in the wave's amplitude due to both scattering and absorption) is typically used for α rather than the absorption coefficient because the effects of scattering and absorption are difficult to separate empirically. In addition, this derivation neglects the radiation force associated with scattering. In most applications to human tissues, however, scattering is weak and absorption dominates the attenuation coefficient, so the attenuation coefficient is a good approximation for the absorption coefficient and Eq. (5.19) is a good approximation for the radiation force.

In the presence of objects with high reflection coefficients, however, the radiation force due to scattering at angles off-axis cannot be neglected. In this case, the radiation force is described by

$$\mathbf{f}_{\mathbf{rad,scatt}} = \mathbf{d}_r S \langle E \rangle, \tag{5.20}$$

where S is the projected area of the object, $\langle E \rangle$ is the time-averaged energy density of the incident acoustic wave, and $\mathbf{d}_r$ is a drag coefficient. The drag coefficient is a vector value with components in the direction of and transverse to the wave propagation and represents both the scattering and absorbing properties of the object. The drag coefficient is given by [4]

$$\mathbf{d}_r = \frac{1}{S} \left(\Pi_a + \Pi_s - \int \gamma r \cos\theta \, dr \, d\theta \right) \widehat{\mathbf{z}} - \frac{j}{S} \int \gamma r \sin\theta \, dr \, d\theta \, \widehat{\boldsymbol{\theta}}, \tag{5.21}$$

where Π_aa is the total power absorbed by the target, Π_s is the total power scattered by the target, γ is the magnitude of the scattered intensity, θ is the scattering angle, and $rdrd\theta$ is an area element of the projection of the target onto the axial/lateral plane. Compared to the contribution of absorption to the radiation force, which is in the direction of wave propagation, the contribution of scattering to the radiation force depends on the angular scattering properties of the object. If the scattering object has an axis of symmetry perpendicular to the direction of wave propagation (e.g., a spherical object), the radiation force from scattering is in the direction of wave propagation only. In this situation, the radiation force is given by

$$\mathbf{f}_{\mathbf{rad,scatt}} = \left(\Pi_a + \Pi_s - \int \gamma \cos\theta r \, dr \, d\theta \right) \langle E \rangle \widehat{\mathbf{z}}. \tag{5.22}$$

Eq. (5.22) provides a much more generalized solution to the radiation force than the absorptive solution (Eq. 5.16) and can take into account the shape of the object. This equation is useful for describing the radiation force impinging on objects that have significant scattering properties, such as those that have large acoustic impedance differences from the surrounding propagation medium.

3. Acoustic radiation force imaging techniques
3.1 Acoustic radiation force applied to tissues
Although plane waves provide a simple solution for the acoustic radiation force in Eq. (5.19), they are not particularly useful in medical ultrasonographic applications.

First, a plane wave cannot probe finely enough with radiation force to provide the detailed resolution of a medium. Second, greater radiation force can be achieved with less transmit power by focusing the ultrasonic wave. Eq. (5.19) implies that conventional diagnostic ultrasonography used in soft tissue imaging can generate radiation force. This is often apparent when imaging objects in a water tank, where air bubbles in the tank that oat into the imaging region accelerate away from the transducer. In soft tissues, this effect is not apparent because the radiation force is relatively small.

The objective of radiation-force-based imaging methods is to elicit a response from tissues from which the elastic or viscoelastic properties can be measured. Generally, the desired response from tissue is a movement or displacement of the tissue. To obtain displacements that are detectable by ultrasonic instrumentation (i.e., displacements greater than a micrometer), the pulse-averaged intensity of the acoustic wave must be increased to well above the limits for diagnostic B-mode ultrasonographic imaging [8]. To maintain the spatial-peak temporal-average intensity (I_{spta}) under the FDA limit of $720 \, \mathrm{mW/cm^2}$, the frame rate for a conventional ARFI imaging sequence must be low, with a mechanical index not exceeding 1.9. Such high pulse-averaged intensities can be achieved by increasing the length or pressure of the acoustic pulse. Generally, increasing the pulse length is preferred because increasing the pressure may lead to large rarefactional pressures, which can induce bioeffects such as cavitation or heating.

In medical imaging applications of radiation force, the radiation force field is a function of the spatial distribution of the transmitted wave's intensity as well as the tissue properties, including speed of sound and attenuation. The spatial distribution of acoustic wave intensity is determined by the transmitted pulse characteristics, such as aperture size, pulse length, and frequency. Fig. 5.1 demonstrates the radiation force field created in a medium with an attenuation coefficient of 0.7 and 2.0 dB/cm/MHz by a linear ultrasound transducer array. Higher attenuation will result in increased acoustic energy losses in the propagation path before the region of interest, thus making the total displacements lower than those in media with less attenuation. In such cases, effective use of ARFI imaging will be limited to shallow depths only.

The radiation force pattern is primarily determined by the intensity of the transmitted pressure field. For focused ultrasound waves transmitted from a linear transducer array, the field has a shape similar to a flattened pyramid. As shown in Fig. 5.1, the amount of attenuation (or absorption in this case) also impacts the distribution of intensity and the radiation force—a higher attenuation will reduce the relative intensity at the focal region and shift the maximum forces toward the transducer.

In the following sections, we introduce a variety of acoustic radiation force imaging techniques. These techniques are qualitative imaging techniques that use radiation force to probe the stiffness of tissues.

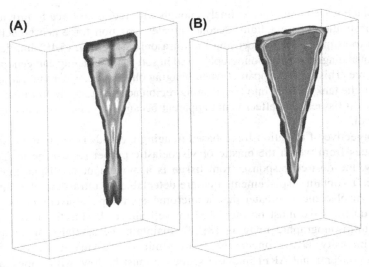

FIGURE 5.1

Image of the displacements from radiation force created by a linear transducer array for media having an attenuation coefficient of (A) 0.7 dB/cm/MHz and (B) 2.0 dB/cm/MHz (red [black in print version] is high displacement, blue [dark gray in print version] is low displacement). The shape of the radiation force field (and thus induced displacements) is determined by both the intensity of the pressure distribution and the attenuation coefficient. At 0.7 dB/cm/MHz, the stronger radiation force occurs at the focal depth of the ultrasound beam. At 2.0 dB/cm/MHz, the force is distributed more evenly in the region before the focal depth.

3.2 Acoustic radiation force impulse imaging

In 2001, Nightingale et al. [9,10] introduced ARFI imaging as a derivative of their work in acoustic streaming [11]. ARFI imaging is a radiation-force-based imaging method that uses relatively short-duration acoustic pulses to generate localized regions of radiation force in tissues. The acoustic radiation force pulses are longer (in the order of a few hundred cycles, i.e., hundreds of microseconds) than the diagnostic imaging pulses, which are less than a few microseconds, but are very short relative to the response of the tissue elicited by the radiation force. This relatively short timescale is what characterizes this method as an "impulsive" application of radiation force.

For ARFI imaging, the response of the tissue is typically a rapid acceleration away from its original position, followed by a recovery back to its original position over several milliseconds. Because the response of the tissue to the radiation force impulse is much slower than the longitudinal speed of sound of acoustic waves in the tissue, techniques from conventional pulse-echo ultrasonic imaging can be used to monitor the response in the localized region where radiation force is applied (also referred to as the excitation region). Before the application of the radiation force, an echo signal is acquired along the intended line of force as a reference of the initial tissue position within the excitation region. The ARFI is then transmitted to displace the region of interest. After application of the ARFI "push," a series of echo "tracking" signals are acquired in order to observe the response of the tissue in the excitation region over time. These tracking signals are then compared to the reference signal using cross-correlation or phase-shift estimation techniques to determine the displacement of the tissue from its original location, over a range of depths that includes the excitation region [12]. By applying this "reference-push-track" method many times over a wide field of view, images of the tissue response can be made.

Typically, an ARFI image is created from the displacements of the tissue at a fixed time relative to the radiation force push. In general, softer tissues respond with greater displacements than stiffer tissues under the same applied force; thus the contrast mechanism in ARFI imaging is the difference in displacement between tissue types, and more indirectly, the tissue stiffness or Young's modulus. However, because the absorption coefficient and speed of sound in the interrogated tissue may vary spatially and the amplitude and distribution of the ARFI push beam intensity cannot be precisely determined, the applied radiation force is unknown and only the relative differences in tissue stiffness can be observed, i.e., no quantitative values of stiffness can be obtained.

In addition, the relative difference in tissue stiffness can only be reasonably compared to nearby or neighboring tissues in ARFI images. As observed in Fig. 5.1, the radiation force from a focused ultrasound transmission, such as that used in ARFI imaging, has a spatial distribution that is defined by the intensity of the ultrasound beam and is impacted by the local variation in tissue types and the associated acoustic parameters. Therefore the applied radiation force can vary across the imaging field of view, which affects the assessment of relative stiffness. For example, a subcutaneous fat layer of varying thickness lying over the liver will produce a different amount of attenuation, thereby impacting the radiation force achieved at the focal depth at two spatial locations situated far apart. This difference in radiation force produces different displacements, thus creating an apparent difference in stiffness in the liver at the focal depth.

3.2.1 Displacement tracking

The displacements generated by an ARFI push are usually on the order of a 10th to a 100th of the ultrasound wavelength, and techniques such as cross-correlation are used to compare signals and detect and track such displacements over time. Conventional normalized cross-correlation techniques can be used on radio frequency (RF) data to estimate the ARFI displacements; however, as shown by Pinton et al. [12], a sampling frequency as high as 40 MHz corresponds to a sample spacing of approximately 20 μm, whereas the typical axial displacements induced by ARFI are roughly an order of magnitude smaller. Consequently, interpolation is necessary to resolve these small displacements using cross-correlation-based methods, which significantly increases the computational burden over that of phase-shift estimation algorithms.

In order to reduce the data processing requirements for ARFI imaging, instead of the conventional RF echo signals, down-sampled in-phase and quadrature (I/Q) components are commonly acquired. While the down-sampled I/Q data can be remodulated back to RF data, most autocorrelation tracking techniques typically compute displacements directly from phase-shift estimations at each depth in I/Q data. Phase-shift estimation algorithms can detect a maximum displacement of one-half of the acoustic wavelength; larger displacements introduce phase-wrapping artifacts. Kasai et al. [13] proposed a normalized autocorrelator that determines displacements between the reference and displaced I/Q data by measuring the average phase shift with respect to the central frequency of the transmitted pulse.

This normalized autocorrelator is considered a one-dimensional approach, as the estimator operates in the phase domain only.' The phase-shift estimator of Loupas et al. [14] relies on a two-dimensional autocorrelation approach to obtain estimates of both the mean Doppler frequency and RF of the signal inside the observation window. These estimates permit a full evaluation of the Doppler equation and, consequently, address the unrealistic assumption that the RF of the backscattered signal remains constant and is equal to the central frequency of the transmitted pulse.

Because the tracking algorithms described earlier use a phase-shift estimation approach, they have limitations similar to those found in Doppler imaging techniques. These techniques tend to be accurate for axial displacement but cannot directly measure lateral displacement. The most direct way to obtain ARFI-induced lateral displacement estimates would be to utilize speckle tracking or multiple-beam techniques, in which two or more separate angled tracking beams each estimate the displacement at the respective angle, and the geometric relationship between the beams is taken into account to estimate the displacement in two dimensions. Another approach to lateral tracking does not require multiple beams, but rather applies receive apodization functions to estimate the lateral displacement components, such that a two-dimensional displacement vector can be estimated [15,16].

3.2.2 High speed tracking techniques for acoustic radiation force impulse

Although ARFI imaging introduces new capabilities for detecting diseases, there are several challenges to its practical implementation. One of the downsides of ARFI imaging (and other radiation force elastography techniques) is that their frame rates are relatively slow than those of conventional B-mode and Doppler imaging because the response of tissue to the radiation force requires several milliseconds of observation. In addition, the increased acoustic intensity transmitted by the ultrasonographic system creates the potential for undesirable thermal and nonthermal bioeffects.

There have been many solutions proposed to increase the frame rate of radiation force imaging methods [17−19]. Many of these methods rely on parallel receive beamforming and interleaving push beams with track beams. Dahl et al. [17]

introduced a method to interrogate a wide field of view with ARFI imaging by using fewer radiation force "pushes" than previous ARFI imaging methods. This method relied on parallel receive beamforming of the reference and tracking signals to expand the local field of view and required less overlap of the radiation force fields. This method was used to reduce the heat generated by the transducer at the surface as well as the heat generated by absorption of the ultrasonic waves.

The methods proposed by Hsu et al. [18] and Bouchard et al. [19] take advantage of the long observation time of the tissue response necessary to correct for the tissue motion described previously. Although the total observation time is long, it is not necessary to track the tissue response at high temporal sampling during the recovery phase. Hsu et al. [18] and Bouchard et al. [19] interleaved push and track beams at multiple locations. Combined with parallel tracking, this allowed ARFI imaging to achieve substantial reductions in heat generation [19], enabling not only real-time ARFI imaging of up to 20 frames per second [20] but also real-time sequences of ARFI imaging combined with other imaging modalities, such as B-mode and Doppler imaging [21].

In contrast to the interleaving and parallel tracking methods, another approach to increase the ARFI frame rates is to utilize simultaneous push beams, known as the parallel transmit method [22]. In this case, multiple push beams are transmitted simultaneously from an aperture in a fan or comb-type pattern. Interleaved or parallel tracking beams are then used to observe the response to the simultaneous excitations. This significantly improves the frame rate in a manner similar to that of the interleaving methods proposed by Bouchard et al. [19] and Hsu et al. [18].

3.2.3 Harmonic tracking
The presence of inhomogeneities in the acoustic properties of tissues in the layers between the transducer and the desired region of displacement tracking can create significant errors in the displacement estimates. In particular, diffuse and coherent reverberation from these tissues can possibly appear in the region of displacement estimation, corrupting the displacement estimates. Doherty et al. [23] proposed a "fully sampled" technique utilizing a pulse inversion (PI) harmonic sequence to remove clutter (signal artifact due to multipath scattering and reverberation [24]) and maintain the temporal sampling rate of conventional ARFI tracking techniques.

Conventional PI techniques, such as that used in contrast imaging and tissue harmonic imaging, use a transmit pulse sequence whereby two sequential transmits are identical in pulse shape and focal location, but have alternating polarity. The echoes from the two transmitted pulses (an "echo pair") are summed to eliminate the fundamental frequency components of the signal while maintaining the second harmonic components. The second harmonic components have higher resolution and lack the reverberation clutter found at the fundamental frequency [25]. When this is applied to tracking in ARFI imaging, the sampling rate is decreased by a factor of two because the two transmit pulses are used to create one harmonic tracking signal.

In the fully sampled technique, the sum of the negative polarity echo of the first pair with the positive polarity echo of the next pair is computed in addition to the

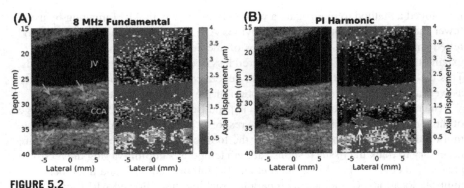

FIGURE 5.2

Images of a human carotid artery and its associated plaque. Shown are the
(A) fundamental B-mode (left) and acoustic radiation force impulse (ARFI) (right) images
and the (B) harmonic B-mode (left) and ARFI (right) images. CCA, common carotid
artery; JV, jugular vein; PI, pulse inversion.

Courtesy of Josh Doherty.

sum of positive and negative polarity echoes as in the conventional case. This essentially creates a two-sample "sliding window" across the positive- and negative-polarity echo signals to obtain a harmonic tracking sample rate equal to the pulse repetition frequency of the system. This technique demonstrated significant improvement in displacement estimation of in vivo signals due to the reduction in reverberation clutter from the subcutaneous tissue layers.

Fig. 5.2 shows an example of harmonic tracking applied to ARFI imaging of an atherosclerotic plaque in the carotid artery. In the 8-MHz fundamental B-mode image (left image in Fig. 5.2A), a plaque (arrows) is visible on the wall of the common carotid artery (CCA). Low-amplitude image clutter is present inside the lumen of the CCA and jugular vein (JV). The image clutter creates a false tissue signal that appears stationary and produces an apparent seeming stiff region of tissue with little or no displacement in the ARFI image (right image of Fig. 5.2A). In the ARFI image, the lumen appears similar to the surrounding stiff plaque and normal vascular tissue, as indicated by dark blue. With PI harmonic tracking (4 MHz transmit; Fig. 5.2B), the stationary clutter signal inside the lumen of the CCA and JV is reduced, thereby eliminating the "stiff plaque" artifact in the ARFI image. In addition, finer details are now visible that were hidden in the fundamental image, such as the soft plaque region indicated by the arrow in the right image of Fig. 5.2B.

3.2.4 Motion filters

Because the displacements induced by radiation force are small, the displacements can be obscured by nearby physiologic motion and motion introduced by the sonographer or clinician scanning the subject [26,27]. Nightingale et al. [10] utilized a simple linear filter to eliminate the errors caused by these motions. This filter assumes that the physiologic or transducer motion is linear during the observation window and that the tissue completely recovers to its original position within that observation window. The filter computes a linear displacement profile between the first and last sample and subtracts this displacement profile from the estimated displacements. Similarly, Hsu et al. [18] introduced a quadratic version of this filter to allow for greater flexibility in the assumed physiologic motion.

Other motion filters, including model-based [28] and blind source separation [29] filters, have been proposed to eliminate physiologic motion in ARFI imaging, however, the most widely used filter in current ARFI imaging applications is an adaptation of the quadratic motion filter proposed by Giannantonio et al. [30]. In this filter, called the quadratic extrapolation filter, multiple reference signals are obtained before the radiation force beam. These "pretrack" signals are used to estimate the trajectory of tissue motion before the radiation force application, and this extrapolated motion trajectory is included in the quadratic filter to obtain a more accurate fit to the actual physiologic motion.

3.3 Preliminary applications of acoustic radiation force impulse imaging

The first proposed applications of ARFI imaging included distinguishing benign from malignant breast lesions [10,31], based on the likelihood that breast cancer would have greater stiffness than benign lesions. Similarly, ARFI imaging was also proposed to differentiate hepatocellular carcinoma from metastatic disease in the liver [32]. Because ARFI imaging has recently been made available on commercial ultrasound scanners, clinical studies have recently demonstrated the use of ARFI imaging in breast and liver tumors with some success [33–35].

Other applications of ARFI imaging include characterizing the mechanical composition of carotid plaques [36–38], guiding prostate tumor biopsy [39,40], monitoring ablation nodules in cardiac and other tissues [41–43], and observing natural and pathologic changes in the stiffness in myocardial tissues [18].

Carotid plaques have the potential to rupture, resulting in stroke and potentially death, and therefore identifying such vulnerable plaques is important to stroke management. ARFI imaging of atherosclerosis has demonstrated the ability to detect soft regions within vascular plaques, which are linked to potential lipid regions of the plaque [36–38]. A study comparing ARFI imaging to histology of carotid endarterectomy samples demonstrated good agreement between the low ARFI displacements and calcified tissue and high ARFI displacements in lipid/necrotic cores [44].

ARFI imaging has also demonstrated, in both in vivo [40,45,46] and ex vivo [39] studies, the ability to identify various prostate pathologic conditions that correlate with histologic findings and magnetic resonance imaging. Although ultrasound-guided needle biopsy is the gold standard for determining prostate cancer diagnosis, the method is problematic because it relies on a grid of biopsy cores and includes additional biopsy locations only if a known suspicious region has been identified. In particular, ARFI imaging of the prostate shows high contrast of prostate tumors, making it easier to apply biopsies to suspected tumor regions.

ARFI imaging has also been proposed to monitor therapies for liver and cardiac diseases. Radio frequency ablation (RFA) procedures are often guided by imaging, and ultrasonography is a popular modality for guidance due to its real-time capabilities. However, cancerous lesions are sometimes inadequately visible to guide electrode placement. In addition, the amount of time needed to ablate tissues to an appropriate margin varies between patients and protocols. Active monitoring of the ablation lesion size can result in more accurate and potentially safer procedures. Because the resulting ablation lesion is much stiffer than either the healthy or cancerous tissue, ablation lesions are easily observed in ARFI imaging. ARFI imaging has shown the ability to observe both RFA and cryoablation lesions in liver tissues [32,47].

Similarly, in cardiac RFA, a line of RFA lesions are used to isolate or destroy the circuit pathways of supraventricular tachyarrhythmias. Any gaps in the line of lesions will facilitate conduction of residual arrhythmias, so accurate placement and size of the ablation lesions are necessary for successful treatment. Identification of these lesion gaps with ARFI imaging has been shown in vitro with an intracardiac echocardiography transducer and in an in vivo animal model [41,42,48,49]. ARFI imaging was employed in a human clinical study in patients undergoing catheter ablation for atrial utter and atrial fibrillation [50]. This study showed that the ablation lesions could be visualized to confirm the size and placement of the lesion and that this could enable optimization of RF energy deployment. The main challenge found by this study was the need for the probe to be within 2.5 cm of the ablation site, when a distance of 4–6 cm would be preferred.

ARFI imaging has also been employed to measure and visualize normal and pathologic changes in cardiac tissue stiffness during contraction of the heart [22,51,52]. Hsu et al. [52] showed that during artificial pacing of a canine heart, the propagation of heart muscle contraction could be observed by measuring the tissue's stiffness with ARFI imaging. With this capability, it is possible to detect infarcted regions of the heart based on the tissue's ability or inability to contract and induce changes in the cardiac muscle stiffness.

3.4 Vibroacoustography

A modulating ARFI can also be used to elicit a mechanical response from tissues. In this case, a modulating particle velocity is applied to the plane wave model in Eq. (5.17):

$$\mathbf{u}_0 = U_0 \exp(-\alpha z)\cos(2\pi f t - kz)\cos\left(\frac{2\pi \Delta f_0 t}{2}\right)\widehat{\mathbf{z}}, \tag{5.23}$$

where $\Delta f_0/2$ is the modulating frequency and $\Delta f_0 \ll f$. It is also assumed that the period T in the time average of the radiation force is much longer than the period of the acoustic frequency f, but much shorter than the period of the modulating frequency Δf_0, or $1/f \ll T \ll 1/\Delta f_0$. This assumption allows the modulating term to escape the time average of Eq. (5.23). By inserting Eq. (5.23) into Eq. (5.18) and taking the time average, the radiation force function due to amplitude modulation is shown to be

$$\mathbf{f_{rad}} = -2\rho_0 \alpha U_0^2 \exp(-2\alpha z)\cos^2\left(\frac{2\pi\Delta f_0 t}{2}\right)\left\langle \cos^2(2\pi f t - kz) + k\sin(4\pi f t - 2kz)\right\rangle\hat{\mathbf{z}}$$

$$= -\rho_0 \alpha U_0^2 \exp(-2\alpha z)\cos^2\left(\frac{2\pi\Delta f_0 t}{2}\right)\hat{\mathbf{z}}.$$

(5.24)

Using the trigonometric identity $\cos^2(\theta) = (1 + \cos 2\theta)/2$, and substituting in the definition of time-averaged intensity from Section 2.1, Eq. (5.24) can be rewritten as

$$\mathbf{f_{rad}} = -\left(\frac{\alpha I}{c_0} + \frac{\alpha I}{c_0}\cos\Delta f_0 t\right)\hat{\mathbf{z}}.$$

(5.25)

The first term in Eq. (5.25) is a steady-state term resulting from the time-averaged intensity of the acoustic wave and is identical to Eq. (5.19) except for a factor of two. The second term in Eq. (5.25) is a modulating term that varies the total radiation force between 0 and $-2\alpha I/c_0$. The resulting modulation force is a vibration at a frequency of Δf_0 within the region of excitation.

Ultrasound-stimulated vibroacoustography (USVA) is a technique that utilizes this modulating radiation force function to image tissues [53,54]. The modulation of tissue acts like a piston, from which the tissue creates an acoustic emission. This acoustic emission is dependent on the tissue's viscoelastic properties, modulated beam's cross section, and the shape and size of the objects present in the region of excitation. In USVA, a focused modulating radiation force is scanned over the xy plane in tissue and the amplitude and phase of the acoustic emissions from the tissue are recorded using a hydrophone placed on the surface of the medium (see Fig. 5.3).

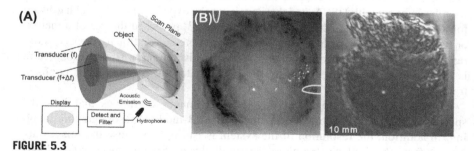

FIGURE 5.3

(A) A vibroacoustography system. A confocal transducer is used to create a modulating radiation force that is scanned in a raster format. Acoustic emissions are received by a nearby hydrophone and displayed. (B) A radiograph (left) and vibroacoustographic (right) image of an excised human prostate. The images show a large calcification in the center of the image and a cluster of smaller calcifications on the right side of the images. The vibroacoustography image agrees well with the radiograph.

An image is then constructed from the magnitude of the recorded acoustic emissions at every point in the scan plane.

The modulating radiation force can be created by an amplitude-modulated ultrasound beam; however, this strategy creates a modulating radiation force field that extends over a very large range in the medium. A large-range modulating force field is problematic from an imaging prospective because the origin of tissue response cannot be precisely determined in the range dimension, thus limiting the achievable image quality.

A better strategy involves the use of two unmodulated ultrasound beams that differ in frequency by Δf_0, where Δf_0 is in the order of tens of kilohertz. In this approach, the two beams are crossed so that they interfere for only a short range. The resulting interference of the two beams is an amplitude-modulated acoustic beam. The smaller the region over which these two beams interfere, the better the resolution of the imaging system.

There are several strategies for creating an amplitude-modulated beam by crossing two unmodulated beams [55]. The simplest method is to use two piston transducers to form crossing beams at a specified focal depth into the medium. Although this can be effective, aligning two piston transducers to a desired location requires precise control of the transducers. A similar strategy involves the use of a confocal arrangement, in which a ring-shaped transducer surrounds a piston transducer. This method allows two beams to be crossed at a specified focal depth while minimizing positioning errors and expensive setups for two separate transducers. Conventional diagnostic transducers can also be used to generate the two unmodulated beams [56]. This configuration may be the most ideal, particularly for clinical applications, because the linear array of elements in a diagnostic transducer would allow the two beams to be steered electronically over a field of view without requiring the mechanical steering system necessary for the dual or confocal transducer approaches. Furthermore, utilizing a diagnostic transducer to generate the radiation force modulation enables multimodality imaging, including B-mode or Doppler imaging with USVA, and does not incur the expense of additional hardware for mechanical steering. Mehrmohammadi et al. [57] proposed the use of a matrix, or two-dimensional, transducer array to achieve more conventional vibroacoustography planes with a clinical imaging system. In this system, a square subaperture is created with a pattern that closely approximates the dual-element confocal transducer.

Vibroacoustography makes highly detailed images of tissues that lack the speckle signature that accompanies conventional B-mode ultrasonographic imaging. In addition, the resolution of the system is isotropic because the images are created from the C-plane (the plane perpendicular to the transducer axis). USVA excels particularly in identifying hard tissues, such as microcalcifications, because these tissues create high-amplitude acoustic emissions under the modulating radiation force. These acoustic emissions are an indirect measure of the viscoelastic properties of a tissue and therefore, like ARFI imaging, USVA images are qualitative in nature. However, USVA images are obtained in a *xy*-plane of constant range from the transducer applying the radiation force and thus will not exhibit the depth-dependent variation (of radiation force and the measured response) present in conventional xz-plane ARFI imaging. Therefore relative differences in viscoelastic properties can be compared across the image. The main disadvantage of vibroacoustography systems is that the scans currently require several minutes to complete. This makes vibroacoustography susceptible to patient and physiologic motion and prevents it from being used on targets such as the heart.

3.5 Preliminary applications of vibroacoustography

Vibroacoustography has demonstrated potential applications in breast imaging [58,59], prostate imaging [60—62], observing calcifications in carotid and peripheral arterial stenosis [63,64], and hip replacement [65]. In breast imaging, the major advantage of vibroacoustography is its ability to detect microcalcifications [58,59]. Microcalcifications are small hard crystals with different acoustic properties than typical soft tissues. They are high-scattering targets and are thus subject to significant modulating radiation force, making them bright and easily detectable targets in USVA images. In addition, the images produced by vibroacoustography are very similar to conventional X-ray mammography, making the images easily recognizable by radiologists.

Like microcalcifications, brachytherapy seeds are also hard and highly scattering compared with normal human tissue, and thus make excellent objects for visualization with USVA. The placement of brachytherapy seeds in the prostate is often performed under the guidance of transrectal ultrasonography; however, there are no ideal diagnostic imaging modalities for confirming seed placement. In conventional ultrasonography, the seed is often invisible due to its small reflecting surface and the surrounding speckle. Mitri et al. [61] used USVA as a method to detect brachytherapy seeds, independent of seed orientation. This method could potentially be complementary to B-mode ultrasonography and allow for immediate and quick determination of seed placement. Other applications of vibroacoustography in the prostate include detection of cryotherapy lesions [60] and prostate calcifications [62].

Another potential target for vibroacoustography is the characterization of carotid and peripheral arterial plaque morphology [63,64]. Calcification is often present in these plaques, which, like microcalcifications, are hard structures that vibrate significantly under modulating radiation force. Pislaru et al. [63] demonstrated the visualization of calcified plaques from in vivo and in situ porcine femoral arteries. Bone and joint replacement materials are also significant scatterers of ultrasound, making the surfaces of these tissues and materials suitable for vibroacoustography. Kamimura et al. [65] demonstrated the use of three-dimensional vibroacoustography to image a model of a total hip arthroplasty and was able to show that the exposed metal of the implant could be accurately detected and visualized.

3.6 Harmonic motion imaging

Harmonic motion imaging (HMI) is a method similar to USVA in that it utilizes a modulating radiation force field as a forcing function [66]. However, instead of measuring acoustic emissions resulting from the vibrating tissue, the oscillation of the tissue is observed directly using conventional ultrasonographic methods. The measurement of the vibrating tissue is accomplished much in the same way that the tissue response is tracked in ARFI imaging, except that in HMI, the tracking occurs during the application of the modulating radiation force rather than after excitation. In HMI the image is created from the amplitude of the modulating tissue displacements at each location in the image. A key difference between HMI and USVA is that the Δf used in vibrating the tissue for HMI must be in the order of hundreds of hertz so that tissue motion can be observed with conventional ultrasound, whereas Δf in USVA is in the order of tens of kilohertz.

A key element of the HMI system is the placement of the imaging transducer for tracking the modulating tissue. In order to effectively measure vibrations, the imaging transducer must be placed on the same axis where the modulating radiation force is applied. This can be accomplished by crossing the beams of two piston transducers to generate radiation force and placing the imaging transducer between the two piston transducers [66]. Although this configuration creates viable HMI images, the setup is cumbersome for practical uses in clinics.

HMI, however, has demonstrated potential use in the area of high-intensity focused ultrasound (HIFU) [67,68]. An HIFU transducer is a large concave transducer that is used to generate high-intensity ultrasound to a targeted region in tissues to induce ablation; because it is designed for high intensity, it can easily be used to generate radiation force. The most common approach for HMI uses a dual-transducer setup, incorporating a smaller imaging array with a larger HIFU transducer that has a hole in the center. The imaging array is placed in this hole and precisely aligned with the axis of the radiation force beam in order to monitor the tissue displacements. The imaging transducer is activated while the amplitude-modulated radiation force is applied [67].

Like ARFI and USVA, HMI is a qualitative imaging mode that shows images of relative tissue stiffness. In HMI, there is a spatial distribution of the applied modulating force (as well as a spatial distribution of the acoustic properties in tissues), and thus, like ARFI, objects and tissues in HMI images can only reasonably be compared when they are close to each other.

HMI was primarily developed to monitor ablation lesions during focused ultrasound surgery. The HIFU transducer serves a dual purpose in HMI by applying different pulses to ablate and apply radiation force to the tissue. HMI has successfully demonstrated the visualization of ablation lesions in preclinical studies of mammary tumors in mice [68] and thigh muscle in rabbits [69].

3.7 Viscoelastic response imaging and model-based aproaches

The limitation of many of the radiation force imaging methods described previously is that they are all qualitative; that is, only relative comparisons of elastic or viscoelastic properties could be made from the measurements and images. Using methods similar to ARFI imaging, however, Selzo and Gallippi [70] demonstrated the ability to produce quantitative measurements and images of the viscoelastic properties of tissues. This method, called VisR imaging, is a model-based radiation force approach to estimate the viscoelastic stress relaxation time constant, τ_σ. It is a derivative of other model-based radiation force approaches, including kinetic acoustic vitreoretinal examination (KAVE) [71] and monitored steady-state excitation and recovery (MSSER) [72], that estimate viscoelastic properties using radiation force.

In the model-based approaches of KAVE and MSSER, continuous acoustic radiation forces are used to compress the tissue in localized regions. The tissue is held in place by the steady-state radiation force. During application of the radiation force, the tissue displacements are measured using conventional ultrasonography in the same way that tissue displacements are measured in ARFI imaging. The displacement of the tissue is then fit to the displacements expected from the solution to a viscoelastic model with a unit step or Heaviside function as an input.

Although model-based methods such as KAVE and MSSER are capable of estimating viscoelastic properties, the required steady-state radiation forces induce significant heating of the tissue, require long observation times, and are problematic with the FDA limitations on acoustic exposure [73] if used for imaging purposes. VisR was proposed as a model-based approach to avoid these problems and enable real-time imaging of viscoelastic properties.

VisR uses the Voigt model as the basis for the viscoelastic properties of tissues. Given a radiation force $F(t)$, the Voigt model relates the radiation force to the tissue displacement, $x(t)$, by

$$F(t) = Ex(t) + E\tau_\sigma \frac{dx(t)}{dt}. \tag{5.26}$$

In (Eq. 5.26), E relates to the spring constant in newtons per meter and $\tau_\sigma = \eta/E$ in seconds, where η is the viscosity [70]. The time application of the radiation force used in this method is slightly longer than that used in ARFI imaging. Therefore the radiation force input, $F(t)$, is modeled as a rectangular pulse rather than an impulse function.

In the VisR method, the objective is to estimate τ_σ from the response of the tissue to the radiation force input to system. A short-duration rectangular pulse of width t_{ARF} (or "on-time" for the radiation force) is applied to the tissue and the displacements D_1 and D_2 of the tissue are recorded at time points t_1 and t_2, respectively, during the recovery of the tissue. A second rectangular pulse is then applied at t_2, and the displacement D_3 is measured at the time point t_3. From the solution of the modeled rectangular pulse inputs to the Voigt model and the measured times and displacements, the stress relaxation time constant can be estimated as

$$\tau_\sigma = \frac{-t_1}{\ln\left(\dfrac{D_3 - D_1}{D_2}\right)} \tag{5.27}$$

Although τ_σ could be estimated from a single rectangular pulse using this method, the single-pulse approach has been shown to have significant errors away from the focal depth of the radiation force pulse [70]. A "two-push" approach has been shown to produce much more stable measurements of τ_σ with depth [70,74].

Selzo et al. [74] later improved the VisR method by incorporating two key missing elements from the Voigt model. First, the Voigt model lacks material mass, so it does not account for the inertial displacement of tissues that is often seen with short-duration radiation force pulses and impulses. Second, the ultrasonic tracking of tissue displacements with conventional ultrasonography often underestimates the actual displacement induced by the radiation force [75]. By accounting for these two missing elements, a more accurate estimation of τ_σ is obtained. To account for inertia and tracking underestimation, a mass-spring-damper model was used with finite element modeling [74]. This approach achieved greater accuracy in the measurement of τ_σ.

4. Shear wave elastography techniques
4.1 Shear wave generation by radiation force

The radiation force field that is used in the previous section to generate displacement of tissue creates shear stresses at the boundaries of the region of excitation. These stresses pull the tissue back toward its original location, thereby inducing low-frequency transverse oscillations in the form of a shear wave. The shear wave propagates away from the region of excitation in a direction normal to the direction of applied radiation force. The shear wave oscillations are described by [76]

$$\mathbf{F_z} = \frac{\partial^2 s_z}{\partial t^2} - c_t^2 \Delta_\perp s_z + \frac{\eta}{\rho} \frac{\partial}{\partial t} \Delta_\perp s_z, \tag{5.28}$$

where $\mathbf{F_z}$ is the applied radiation force in the z direction, η is the shear viscosity, s_z is the $\hat{\mathbf{z}}$ component of the displacement vector $\mathbf{s}$, and $\Delta_\perp$ is the Laplacian operator for the x and y components. c_t is the shear wave velocity and is defined as

$$c_t = \sqrt{\mu/\rho}, \tag{5.29}$$

Where μ is the shear modulus, and the tissue is assumed to be an isotropic, elastic, and quasi-incompressible medium. In Eq. (5.28) the first two terms describe the linear wave equation with forcing function F_z and the last term on the right is a dissipative term. Eq. (5.28) essentially describes a low-frequency (typically hundreds of hertz) shear wave that propagates with speed c_t away from the axis of the acoustic beam. The shape of the shear wave in the xy-plane is much like an enlarging ring, similar to what one would observe by dropping a pebble into a pool of water. By observation of these shear waves, properties of the medium in which they propagate can be determined. In the following sections, we explore a few of the techniques used for measuring and imaging the viscoelastic properties of tissues, based on shear waves.

4.2 Shear wave imaging techniques

SWEI is a radiation-force-based method that derives the shear modulus of tissues from the propagating shear waves [76]. In SWEI, the tissue is assumed to be an isotropic, elastic, and quasi-incompressible medium. This allows the shear modulus of the wave to be estimated from Eq. (5.29) by measuring the speed of the shear wave and assuming a density of 1000 kg/m^3. The shear wave can be produced from any type of radiation force application, including impulses (as in Fig. 5.4) and forcing functions.

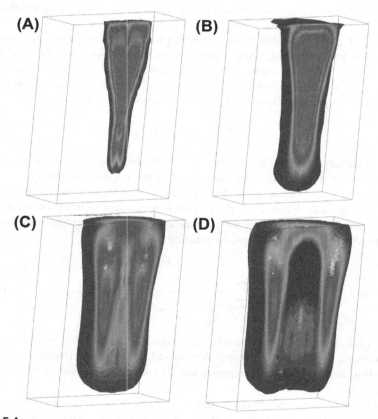

FIGURE 5.4

Shear wave propagation from a region of excitation due to an impulsive radiation force. At 0.15 ms, the displacement field is similar to the radiation-force field (dark blue [dark gray in print version] is zero displacement; red [black in print version] is maximum displacement of 4 μm). As the shear forces pull the tissue back to its neutral position (1.0 ms), a shear wave is formed that propagates away from the region of excitation (2.25 and 3.25 ms). This image shows a center plane cutting through the middle of the region of excitation showing a well-defined shear wave propagating outward (appearing as "two" waves propagating from the region of excitation).

4.2.1 Shear modulus by inversion of the shear wave equation

Bercoff et al. [77] proposed an SWEI method based on a supersonic shear source to create a shear wave with a large extent in range. This was accomplished by applying many radiation force impulses at increasingly greater depths. The "supersonic" name is given to this shear wave source because the compressional waves of ultrasound used to create the radiation force impulses travel at velocities orders of magnitude greater than those of the shear waves, allowing multiple radiation force impulses to be applied over a large range before the shear waves propagate a substantial distance (Fig. 5.5). Comparatively, compressional waves travel at speeds around 1500 m/s in tissues, whereas shear waves travel at speeds of approximately 1–5 m/s.

Bercoff et al. [77] used a plane wave transmit with massive parallel receive beamforming to continuously image and measure tissue displacements induced by the propagating shear wave over the entire field of view at high frame rates. The shear modulus is estimated from the measured displacements (s) by inversion of the shear wave equation. The wave equation for the shear wave is given by

$$\rho \frac{\partial^2 s_z(x,z)}{\partial t^2} = \mu \nabla^2 s_z(x,z), \tag{5.30}$$

where x is the lateral direction perpendicular to the beam axis z. Bercoff's method assumes that the displacements of interest are in the xz-plane (the imaging plane) and assumes that the Laplacian in the y direction is negligible compared with the x and z components. Because the shear wave is a function of time, the solution to the shear modulus at all points in the xz-plane can be written in terms of the Fourier transform, $\mathcal{FT}\{\,\cdot\,\}$, of the shear wave equation:

$$\mu(x,z) = \frac{\rho}{N} \sum_{\omega} \frac{\mathcal{FT}\left\{\dfrac{\partial^2 s_z(x,z)}{\partial t^2}\right\}}{\mathcal{FT}\left\{\dfrac{\partial^2 s_z(x,z)}{\partial x^2} + \dfrac{\partial^2 s_z(x,z)}{\partial z^2}\right\}}. \tag{5.31}$$

In this solution for μ, the problem is broken up into several solutions at N discrete frequencies, ω. The inverse problem is then solved N times in the frequency domain. The resulting solutions at each frequency are then summed together to complete the estimate of μ.

Like most inverse problems, the inversion process is very sensitive to noise, which comes in the form of jitter errors from displacement estimation. However, this method obtains high resolution because the inversion can be solved for every point in the measured displacement field. An example SWEI image using Bercoff's method is shown in Fig. 5.6.

Many of the applications of the shear wave inversion technique are similar in scope to that of ARFI imaging, but they provide quantitative information about the elastic properties of the tissue. The technique may be applied for several purposes, including differentiating benign and malignant thyroid nodules [80], broad

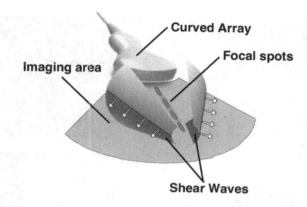

FIGURE 5.5

Diagram of the generation of a supersonic shear wave source.

Reproduced from E. Bavu, J.L. Gennisson, M. Couade adn, J. Vercoff, V. Mallet, M. Fink, A. Badel, A. Vallet-Pichard, B. Nalpas, M. Tanter, S. Pol. Noninvasive in vivo liver fibrosis evaluation using supersonic shear imaging: a clinical study on 113 Hepatitis C virus patients, Ultrasound Med. Biol. 37 (9) (2011):1361–1373 with permission from Elsevier.

mapping of liver stiffness for fibrosis staging [81,82], cervical stiffness [83], and the monitoring of thermally induced lesions [79] among a host of other applications. This technique may have challenges due to attenuation causing difficulty generating sufficient shear wave displacements at greater depths. The amplitude of the shear wave decreases owing to cylindrical spreading as it propagates, and in deep applications, the small initial displacements will only be detectable for a limited extent of lateral propagation before measurements become too noisy. This can effectively limit the field of view for the shear wave inversion from a single supersonic excitation, such that multiple excitations and inversion image frames may be needed. This may be solved by inducing shear waves at multiple locations and solving each displacement field individually with Eq. (5.31).

4.2.2 Shear wave speed by time-to-peak displacement

Inversion problems are inherently susceptible to noise and require significant filtering. As an alternative to this solution, time-of-flight-based algorithms that can estimate the shear wave speed, and thus shear modulus, are becoming increasingly popular. Palmeri et al. [84] proposed the lateral time-to-peak (TTP) displacement as an alternative to the inversion method of imaging shear wave properties. The TTP method is based on the shear waves generated by an ARFI [85] and uses parallel tracking lines [17] located adjacent to the region of excitation to monitor the propagating shear wave in the xz-plane. The time for the wave to reach its peak displacement at each of these lateral locations is then estimated and the shear wave speed is computed from the inverse of the slope of the line between the time to reach peak displacement and lateral location. Using the shear wave speed,

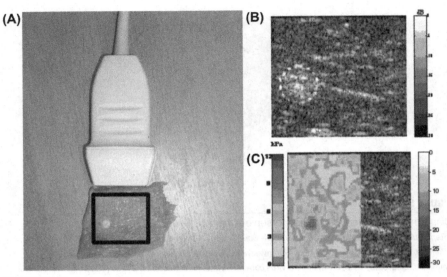

FIGURE 5.6

(A) A high-intensity focused ultrasound (HIFU)-induced lesion in a tissue sample and corresponding 40 × 40 mm² imaged area. (B) B-mode image just after HIFU treatment. (C) Map of the shear modulus achieved using the supersonic shear wave inversion technique, superimposed on the B-mode image after necrosis.

Reprinted from J. Bercoff, M. Pernot, M. Tanter, M. Fink, Monitoring thermally-induced lesions with supersonic shear imaging, Ultrason. Imaging 26 (2) (2004):71–84 with permission from SAGE Publications.

the shear modulus can be estimated from Eq. (5.29) (see Fig. 5.7). Although the shear modulus can be computed from the shear wave speed, the current systems developed by manufacturers report the shear wave speed in their shear wave elastographic imaging mode. Because the applied radiation force is an impulse, this method is sometimes referred to as ARFI or ARFI imaging by the medical community, but it should not be confused with the ARFI imaging method in Section 3.2.

The TTP method is a direct measure of the speed of the shear wave, so it is less susceptible to noise and errors in the displacement estimations. However, because the TTP algorithm requires observation of the shear wave over a finite lateral extent, the resolution of TTP is determined by the size of this observation region. Therefore the TTP method has worse resolution than the inversion method.

To mitigate this loss in resolution, Palmeri et al. [86] proposed an alternative to the lateral TTP method that depended on tracking shear wave displacements only from the "region of excitation," or along the beam axis where the radiation force is applied. This method takes advantage of the spatial extent of the radiation force field shown in Fig. 5.1. The basis for this method is somewhat similar in concept introduced by Sandrin et al. [87] in monitoring the mechanically induced shear waves from a piston impulse.

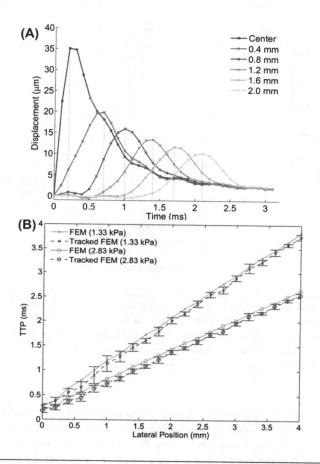

FIGURE 5.7

(A) The displacements observed by several tracking lines in a lateral time-to-peak (TTP) shear wave configuration. At locations further away from the excitation region, the shear wave is observed at a later time. (B) The shear wave velocity, and hence shear modulus, is estimated from the reciprocal of the slope of the line plotting the tracking location versus the time to reach peak displacement. *FEM*, finite element model.

Reproduced from M.L. Palmeri, M.H. Wang, J.J. Dahl, K.D. Frinkley, K.R. Nightingale, Quantifying hepatic shear modulus in vivo using acoustic radiation force, Ultrasound Med. Biol. 34 (4) (2008):546–558 with permission from Elsevier.

The shear wave at the focal region is narrow in its spatial extent, whereas the shear wave in the region between the focus and the transducer is broad and triangular. As the shear wave propagates away from the region of excitation, there is overlap of the shear wave with itself in the region before the focal depth. The overlap in the shear wave produces an interference pattern that "propagates" along the beam axis toward the transducer. The method proposed by Palmeri et al. [86] measures the

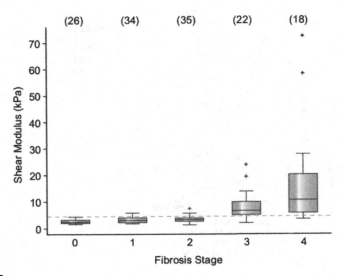

FIGURE 5.8

The lateral time-to-peak method applied to staging liver fibrosis. Stages F0—F2 are differentiable from stages F3 and F4 on the basis of their shear modulus.

Reproduced from M.L. Palmeri, M.H. Wang, N.C. Rouze, M.F. Abdelmalek, C.D. Guy, B. Moser, A.M. Diehl, K.R. Nightingale, Noninvasive evaluation of hepatic fibrosis using acoustic radiation force-based shear stiffness in patients with nonalcoholic fatty liver disease, J. Hepatol. 55 (3) (2011):666—672 with permission from Elsevier.

time for this interference wave to reach peak displacement in this "region of excitation" and is therefore called the region of excitation TTP method. The propagation speed of this interference wave can be directly related to the shear wave speed. The disadvantage of this approach is that the shear wave estimates are far noisier than the lateral TTP method. Walsh et al. [88] proposed a Bayesian displacement estimation approach to the shear wave displacement measurements in order to reduce this error with good success.

The lateral TTP method was initially proposed as a method to quantify liver fibrosis in fibrosis staging [84,89] (see Fig. 5.8). Palmeri et al. [89] demonstrated significant differences in shear wave modulus between stages F0—F2 and stages F3—F4. With the implementation of this method on a commercial scanner, the technique has since been applied to a wide number of clinical studies staging liver fibrosis as well as a host of other clinical applications, including kidney [90], spleen [91], and thyroid [92]. New applications of the lateral TTP method include a host of applications to measure the mechanical contrast in cardiac tissues, including measurement of stiffness changes in cardiac ablation lesions [93], the impact of coronary perfusion on cardiac compliance [94], cardiac function [95], breast tumor staging [96], monitoring response of cancers to therapy [97], observation of the contractility of myocardial tissue under normal and diseased states [98,99], monitoring HIFU lesions in cardiac tissues [100], and monitoring temperature changes for thermal therapy guidance [101,102].

Temperature monitoring of thermally induced lesions [101] is an interesting application of shear waves. Because the speed of a shear wave in tissues is temperature dependent, the relative temperature change of the tissue can be calculated from the change in estimated shear modulus. Arnal et al. [101] demonstrated a temperature dependence of the shear modulus in tissues that was linear and repeatable in the range of 30–45°C. In this temperature range, proteins are not denatured and the shear modulus is not permanently affected by ablation of the tissue. However, because the shear modulus has a different temperature dependence for each tissue type, the tissue type needs to be known in order to compute temperature. The advantage of this method over ultrasound- and magnetic resonance–based thermometry techniques is that it is not as susceptible to motion artifacts, making it ideal for in vivo temperature monitoring.

4.2.3 Comb-push ultrasound elastography

Although the shear wave inversion technique provides an elegant solution for solving the shear modulus at every location, the cylindrical spreading of shear waves and the viscosity of the tissue can limit the effective lateral range over which shear waves can be accurately tracked.

In addition, a limitation of the TTP method is that the shear wave velocity is an average of the shear wave measured over a finite extent and that the shear wave velocity cannot be measured at the region of excitation.

Song et al. [103] proposed the comb-push ultrasound elastography (CUSE) method to address the issues of cylindrical spreading and resolution encountered in the previous methods. The application of radiation force in the CUSE method is similar to that used in the parallel transmit method with ARFI imaging. In the CUSE approach, the transmitting aperture is divided into several subapertures to provide simultaneous transmission of multiple focused radiation force beams at the same depth but different lateral locations, giving the appearance of a "comb-push" (this is in contrast to the use of a single aperture to push along one axis at a single lateral location at a time, as is commonly done in ARFI and shear wave inversion methods). This provides a propagating shear wave near every point in the image where the shear modulus is reconstructed. Song et al. [104] later proposed modifications to the unfocused beams to produce focused (F-CUSE) and marching focused (M-CUSE) beams. In the case of M-CUSE, the apertures used to generate the radiation force beams overlap, and so the radiation force is applied in sequence rather than simultaneously, much like the interleaving method proposed by Bouchard et al. [19].

Displacements from the shear waves are tracked using parallel beamforming in the manner described by Bercoff et al. [77] or Montaldo et al. [105]. In order to separate the complex wave interaction between the propagating shear waves from each simultaneous push, a direction filter is applied in the left-to-right and right-to-left directions to isolate the waves propagating in each respective direction [106].

Images of shear wave velocity are then produced from shear wave velocity estimates by lateral cross-correlation of the shear wave displacement images with lateral kernel sizes of eight ultrasound wavelengths.

Nabavizadeh et al. [107] proposed a modification to the CUSE method, called axicon CUSE (AxCUSE), in which the push beams were angled with respect to the transducer normal. This allows both axial and lateral spatial frequency information to be included in the shear wave velocity estimate and produces a sharper cross-correlation peak, thereby improving shear wave velocity image quality compared with conventional CUSE. CUSE imaging has recently been used to image benign and malignant masses in the breast [108] and thyroid [109].

4.2.4 Viscoelastic properties by shear wave dispersion ultrasound vibrometry

The shear wave inversion, TTP, and CUSE methods make the assumption that the tissue is a purely elastic medium. However, this is an oversimplification because a tissue has viscous behavior as well. In a medium that contains both elastic and viscous properties, the shear modulus is a complex value [110]. In the frequency domain, the complex shear modulus derived from the Voigt model is

$$G(\omega) = \mu(\omega) + j\eta(\omega) = \frac{\rho\omega^2}{k^2(\omega)} \qquad (5.32)$$

where k is the complex wave number, $k(\omega) = k_1(\omega) + jk_2(\omega)$, and μ and η are the shear modulus and shear viscosity, respectively, defined previously. The shear wave velocity, $c_s(\omega)$, and shear wave attenuation, $\alpha_s(\omega)$, are shown to be functions of μ and η by the relationships [111]

$$c_s(\omega) = \sqrt{\frac{2(\mu^2 + \omega^2\eta^2)}{\rho\left(\mu + \sqrt{\mu^2 + \omega^2\eta^2}\right)}} \qquad (5.33)$$

$$\alpha_s(\omega) = \sqrt{\frac{\rho\omega^2\left(\sqrt{\mu^2 + \omega^2\eta^2} - \mu\right)}{2(\mu^2 + \omega^2\eta^2)}}. \qquad (5.34)$$

Eqs. 5.33 and 5.34 can be used to solve for the shear modulus and shear viscosity if the shear wave speed and attenuation can be measured at discrete frequencies. However, it is often difficult (but not impossible [112]) to obtain accurate measurements of the shear wave attenuation. One approach to obtain these parameters is the shear wave dispersion ultrasound vibrometry (SDUV) [113].

The SDUV method utilizes a modulating radiation-force forcing function, much like that used in USVA and HMI, but measures the resulting shear wave at two detection points adjacent to the region of excitation and along a vector orthogonal to the direction of the applied radiation force (see Fig. 5.9). A practical implementation to this process can be accomplished by using a linear array transducer to implement both the modulating radiation force and the detection or tracking beams

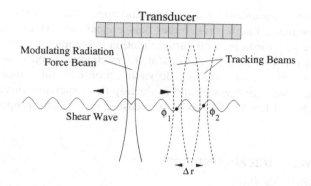

FIGURE 5.9

The SDUV method. A modulating radiation force beam is applied to an object or medium to create a continuous shear wave at a discrete frequency that travels outward from the region of excitation. Two tracking beams are used to observe the tissue displacement at some distance from the region of excitation. The phases of the shear wave are estimated and used to compute the shear wave velocity.

(see Fig. 5.9) [114,115]. Because the shear wave is sinusoidal in this method, the shear wave speed can be computed from the phase difference of the shear wave between the two measurement locations [114]:

$$c_s(\omega) = \frac{\omega \Delta r}{\Delta \phi} \tag{5.35}$$

where Δr is the distance between the two measurement locations and $\Delta \phi = \phi_2 - \phi_1$ is the phase difference of the shear wave between the two locations. Generally, the measurements are performed over a frequency range of 200–800 Hz. The shear wave speed values over this range of frequencies are then fit to the model described in Eq. (5.33) using nonlinear least squares [113]. The shear modulus and viscosity are then obtained from the model fit.

Urban and Greenleaf [116] modified the SDUV method to obtain shear wave velocity measurements at multiple frequencies by utilizing a "square wave radiation" force function. In this technique, the radiation force is turned on and off, rather than using an amplitude-modulated burst. The "square wave" application of radiation force creates a set of harmonic frequencies in the shear wave for which measurements of shear wave speed can be obtained. For example, a 100-Hz square wave will yield harmonic frequencies at 200, 300, 400 Hz, etc. In this manner, the shear wave speed can be measured using a single application of radiation force.

SDUV has been applied to measure the viscoelastic parameters of a variety of tissues, including prostate [117], kidney [118], Achilles tendon [119], and liver [120]. For tissues consisting of a thin layer, such as arteries and the bladder wall, the focal length of the radiation force beam may significantly exceed the layer thickness, such that the tissue layer's response is best modeled as a viscoelastic plate surrounded by fluid. Thus the application of radiation force in such tissues induces a Lamb wave, rather than a shear wave (Lamb waves travel along a surface of a medium, with particle motion perpendicular to that surface). Lamb waves are considered oscillations of the medium as a bulk substrate, and Lamb waves will also have a velocity dispersion profile characteristic of the medium. By generating Lamb waves and evaluating the dispersion curves, Nenadic et al. [121] adapted the SDUV model to measure viscoelastic properties of the bladder wall.

4.3 Shear wave tracking methods
4.3.1 Multitracking lines
Early methods of shear wave tracking with ultrasound utilized a synthesis approach to measure shear wave velocity, as described in Ref. [85]. In the synthesis approach, one tracking location is selected adjacent to the shear wave excitation region and the displacements are measured over the observation period. The tracking location is then moved to the next position and the radiation force push and displacement measurement is obtained. This process is repeated until a sufficient number of tracking locations are obtained. A single shear wave is synthesized from the recorded data by plotting the displacement data as a function of depth, tracking location, and observation time relative to the radiation force push. The shear wave velocity is then estimated from the synthesized displacement data. The methods proposed in Refs. [17,77] utilize parallel receive beamforming to reduce the number of radiation force applications per image line and thereby decrease the time necessary to obtain shear wave images and measurements. These methods are called multitracking line (MTL) methods because they use observations from (multiple) separate positions to estimate the shear wave velocity.

Many of the tracking methods employed by ARFI imaging, such as interleaved and multitrack approaches [19], can be employed with shear wave displacement tracking. Similar methods have been employed by other techniques to improve shear wave velocity estimation and imaging frame rate, such as the time-aligned sequential tracking technique employed with CUSE [122].

4.3.2 Single tracking line
McAleavey et al. [123] recognized that the MTL tracking methods are subject to a bias of the arrival time of the shear wave due to bright reflectors in the path of the propagating shear wave. These bright reflectors typically correspond to the brighter parts of the speckle pattern and create an apparent increase or decrease in the shear wave speed, which increases the variance and accuracy of the shear wave velocity measurement. McAleavey referred to this phenomenon as the "speckle bias." Because the reflectors are bright, they dominate the cross-correlation and phase-shift estimation techniques used in displacement estimation. Therefore a bright reflector close to the track location will correspond to the movement of the bright reflector rather than the movement at the desired location, resulting in an artificially

soon or late arrival time of the shear wave peak, depending on whether the bright reflector appears before or after the track location, respectively.

McAleavey et al. [123] proposed a spatially modulated ultrasonic radiation force pattern that creates multiple simultaneous pushes by the use of an apodization function, and uses a single track location (STL) or line to observe the shear wave displacement. The shear wave velocity can be estimated by the known distance between the radiation force pushes (by the Fraunhofer approximation to the acoustic pressure field) and the measured time difference between the arrival times of the shear waves at the STL. This approach eliminates the speckle bias caused by bright reflectors because the shear wave speed depends only on the difference in arrival times at a single location and not the absolute arrival times. This method demonstrates a dramatic improvement in shear wave imaging and measurements.

Further adaptation of this technique showed that a spatially modulated pushing field was not necessary [124,125] and that the shear wave velocity images and measurements could be synthesized by moving the location of the radiation force beam rather than the tracking beam. This is akin to the acoustic reciprocity principle, but is applied to shear waves. Hollender et al. [124] showed that these STL techniques could achieve superior resolution and contrast compared with the traditional MTL techniques (see Fig. 5.10).

The kernel size of the tracking technique determines the lateral resolution of the shear wave image. In both MTL and STL approaches, the kernel size is determined by the width of the region encompassing all push and track beams. The STL

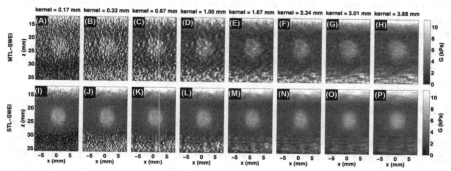

FIGURE 5.10

(A—H) Multitracking line shear wave elasticity imaging (MTL-SWEI) compared to the (I—P) single tracking line (STL)-SWEI of a 6-mm lesion in a phantom. The lesion has a nominal shear modulus of 10.7 kPa and the background has a nominal shear modulus of 2.7 kPa. The STL images show much higher resolution and lower shear wave speed variance than the MTL images for a given kernel size.

Image reproduced from P.J. Hollender, S.J. Rosenzweig, K.R. Nightingale, G.E. Trahey, Single and multiple-track-location shear wave and acoustic radiation force impulse imaging: Matched comparison of contrast, contrast-to-noise ratio and resolution, Ultrasound Med. Biol. 41 (4)(2015):1043—1057 with permission from Elsevier.

approach achieves better resolution because of the reduction in speckle bias [124]. While the smallest kernel shows the optimal resolution, the variance in the shear wave speed estimate is greater. A balance is needed between kernel size and resolution in order to reduce variance, avoid artifacts, and maintain visibility of small structures.

4.3.3 Ultrafast tracking of shear waves

The method employed by Bercoff et al. [77] was designed as a fast tracking technique that allowed the authors to obtain high temporal sampling of the shear wave displacements. In this approach, a plane wave was transmitted and, using parallel beamforming techniques, hundreds of focused receive beams were formed by delay-and-sum beamforming. This allowed for temporal sampling of the shear wave displacements at the pulse repetition frequency of the imaging system. For example, at 4 cm imaging depth, upward of 12 kHz could be achieved. However, this method traded off image resolution for high frame rates.

Montaldo et al. [105] proposed a modification of this method to achieve high spatial resolution with high temporal sampling. The method relies on sequential transmission of multiple plane waves, each at a different multiple angle, to form the tracking beams. The tracking beam is formed by the summation of the echoes from the plane waves corresponding to the same reflection point. This method, called ultrafast coherent compounding, is a form of virtual source synthetic transmit aperture [126–128], in which the transducer replicates the acoustic wave from a virtual source located behind or in front of the transducer. In the case of ultrafast coherent compounding, the virtual sources are placed in the far field of the transducer. To achieve high temporal sampling, the number of plane waves is kept lower than the number of parallel receive beams, but the temporal sampling of the displacements is decreased compared with Bercoff's method [77]. However, the entire field of view is computed with high resolution, thereby enabling real-time shear wave imaging.

4.3.4 Harmonic tracking of shear waves

Shear wave imaging techniques, such as ARFI, are subject to the reverberation clutter by subcutaneous and other tissue layers, which decrease shear wave tracking performance. Song et al. [104] utilized PI tracking of the shear waves generated with a radiation force impulse. This method lacked the fully sampled capabilities employed by Doherty et al. [23] but showed that the method was capable of providing shear wave velocity estimates when none were available by the fundamental component.

Similarly, Correia et al. [129] proposed a combination of the ultrafast tracking technique [105] with the fully sampled PI technique [23] to improve shear wave velocity estimates. Here, diverging-wave virtual sources placed closely behind the transducer are employed rather than plane waves because a phased-array transducer was used; however, plane waves can also be easily used with this technique on linear array transducers.

4.3.5 Three-dimensional shear wave imaging

Although ultrasonography is well known as a two-dimensional imaging modality, the shear waves generated from acoustic radiation force are three-dimensional in nature. Typically, the radiation force beam generates a shear wave that is cylindrical in extent and travels outward from the push location. The observation plane intersects the cylindrical shear wave, giving it the appearance of two separate waves traveling away from the excitation region. In some tissues, the propagation speed of shear waves is dependent on the direction of propagation, i.e., the mechanical properties of the tissue are anisotropic. Conventional two-dimensional imaging of shear wave speed in such tissues will be dependent on initial transducer placement and its spatial orientation, with inconsistent data from one measurement to another. For example, tissues such as cardiac and skeletal muscles are composed of thin fibers, which are often aligned approximately parallel to each other, and shear waves will tend to travel faster along the direction the fibers are oriented and slower in directions more normal to the fiber orientation. In such tissues, muscle fiber orientation is generally not precisely known and shear modulus reconstruction techniques cannot assume an isotropic medium. The issue of unknown muscle fiber orientation can be solved by accurately estimating the shear wave speed in multiple directions using a transducer mounted on a rotation device [130] or by tracking shear wave propagation in three dimensions. Three-dimensional shear wave tracking is useful for anisotropic tissues because it can observe differences in shear wave speed values in different directions and can be used to construct a more accurate model of shear modulus in the medium.

In anisotropic, fibrous muscle tissues, a more specific approach is required to estimate shear wave velocity as a function of propagation direction. In a linear approximation of Hooke's law for anisotropic media, the elastic tensor C_{mnkl} for transverse isotropic media (muscle tissues) will have only five independent values (the Voigt notation is used here to simplify indexing):

$$C_{ij} = \begin{pmatrix} C_{11} & C_{12} & C_{13} & 0 & 0 & 0 \\ & C_{11} & C_{13} & 0 & 0 & 0 \\ & & C_{33} & 0 & 0 & 0 \\ & & & C_{44} & 0 & 0 \\ & & & & C_{44} & 0 \\ & & & & & C_{66} \end{pmatrix}, \qquad (5.36)$$

where $C_{66} = 1/2(C_{11}-C_{12})$. Generally, the propagation of three ultrasound waves is considered for anisotropic media. To obtain the ultrasound wave velocities and directions of propagation, the eigenvalues ($\rho V^2 \delta_{ml}$) and eigenvectors (U_l) of the Christoffel equation should be calculated [131]:

$$\left(\Gamma_{ml} - \rho V^2 \delta_{ml} \right) U_l = 0. \qquad (5.37)$$

In Eq. (5.37), $\Gamma_{ml} = C_{mnkl}N_nN_k$ is the Christoffel tensor, N_k are components of the unit propagation direction vector, U_l are components of the unit particle displacement vector (polarization of the wave), and δ_{ml} is the Kronecker delta. It should be noted that in the current C_{ij} orientation, the z axis (or [001]) is the axis of symmetry that can be treated as the fibrous muscle direction.

For an acoustic wave that propagates along the z direction in a transverse isotropic media, i.e., $N = [001]$, the Christoffel Eq. (5.37) becomes

$$\begin{pmatrix} C_{13} - \rho V^2 & 0 & 0 \\ 0 & C_{13} - \rho V^2 & 0 \\ 0 & 0 & C_{33} - \rho V^2 \end{pmatrix} \cdot \begin{pmatrix} U_1 \\ U_2 \\ U_3 \end{pmatrix} = 0. \qquad (5.38)$$

One of the eigenvalues of Eq. (5.38) would be $U = [001]$, corresponding to a longitudinal wave propagating in the z direction. Obtaining the eigenvalue for this case $(C_{33} - \rho V_L^2 = 0)$ results in a longitudinal speed of sound of $V_L = \sqrt{C_{33}/\rho}$. However, for the other eigenvalue, $V_S = \sqrt{C_{13}/\rho}$, the eigenspace for the associated eigenvector is two dimensional. This means that any vector lying in the xy-plane would be an eigenvector for Eq. (5.38), and the shear wave propagating in the z direction would have a speed of $V_S = \sqrt{C_{33}/\rho}$ and can have any polarization in the xy-plane.

In a more general case, the directionality of material properties about x, y, and z directions is arbitrary and equivalent for any choice of axes. Therefore without any loss in generality, we can set $N = [n_1, 0, n_3]$ and only consider wave propagation in the xz-plane. Thus the Christoffel tensor would become

$$\begin{pmatrix} n_1^2 C_{11} + n_3^2 C_{44} & 0 & n_1 n_3 (C_{13} + C_{44}) \\ 0 & n_1^2 C_{66} + n_3^2 C_{44} & 0 \\ n_1 n_3 (C_{13} + C_{44}) & 0 & n_1^2 C_{44} + n_3^2 C_{33} \end{pmatrix}. \qquad (5.39)$$

One of the eigenvectors of Eq. (5.39) is $U = [010]$ with the corresponding wave velocity $\rho V_S^2 = n_1^2 C_{66} + n_3^2 C_{44}$. As the wave propagation direction is $N = [n_1, 0, n_3]$, this would be a pure shear wave. Two other solutions of the Christoffel equation are obtained by setting $U = [u_1 0 u_2]$. The expressions for the velocities of these two modes can be found in Ref. [132].

Owing Due to the complexity of the propagating shear waves, three-dimensional tracking of shear waves with a matrix transducer is usually required. Wang et al. [133] followed an approach described previously using a HIFU transducer to generate a radiation force push and then tracking the shear wave displacements using a matrix array placed in the center of an HIFU transducer (similar to the experimental setup used in HMI). In these experiments the transducers were placed perpendicular and oblique to fiber orientation. With this arrangement, Wang et al. were able to observe different shear wave velocities (computed using Eq. 5.39) at propagation directions parallel and perpendicular to skeletal muscle fibers. The anisotropic behavior of this tissue is apparent in both the phase and group velocity of the shear wave.

Using a transducer configuration similar to that used by Wang, Hollender et al. [134] adapted the STL technique for three-dimensional shear wave tracking and was able to observe the cylindrical nature of the shear wave in a homogeneous, isotropic, elastic medium. The TTP shear wave velocity estimation technique was used to obtain three-dimensional volumes of shear wave speed.

Similarly, Gennisson et al. [135] utilized a matrix transducer array in combination with the virtual source synthetic transmit aperture technique to achieve ultrafast spatial and temporal sampling in three dimensions. This enabled tracking of the shear wave at 3000 volumes per second and was able to visualize the volumetric propagation of the shear wave.

References

[1] T. Sugimoto, S. Ueha, K. Itoh, Tissue hardness measurement using the radiation force of focused ultrasound, Proc 1990 IEEE Ultrason. Symp. (1990) 1377–1380.

[2] C. Eckart, Vortices and streams caused by sound waves, Phys. Rev. 73 (1) (1948) 68–76.

[3] W.L.M. Nyborg, Acoustic streaming, in: W.P. Mason (Ed.), Physical Acoustics, Volume IIB, Academic Press, Inc., New York, 1965, pp. 265–331 (chapter 11).

[4] P.J. Westervelt, The theory of steady forces caused by sound waves, J. Acoust. Soc. Am. 23 (4) (1951) 312–315.

[5] R. Beyer, Radiation pressure-the history of a mislabeled tensor, J. Acoust. Soc. Am. 63 (4) (1978) 1025–1030.

[6] A.P. Sarvazyan, O.V. Rudenko, W.L. Nyborg, Biomedical applications of radiation force of ultrasound: historical roots and physical basis, Ultrasound Med. Biol. 36 (9) (2010) 1379–1394.

[7] A.H. Nayfeh, Perturbation Methods, John Wiley & Sons, New York, 2000.

[8] K.R. Nightingale, Acoustic radiation force impulse (ARFI) imaging: a review, Curr. Med. Imaging Rev. 7 (4) (2011) 328–339.

[9] K.R. Nightingale, M.L. Palmeri, R.W. Nightingale, G.E. Trahey, On the feasibility of remote palpation using acoustic radiation force, J. Acoust. Soc. Am. 110 (1) (2001) 625–634.

[10] K.R. Nightingale, M.S. Soo, R.W. Nightingale, G.E. Trahey, Acoustic radiation force impulse imaging: in vivo demonstration of clinical feasibility, Ultrasound Med. Biol. 28 (2) (2002) 227–235.

[11] K.R. Nightingale, P.J. Kornguth, W.F. Walker, B.A. McDermott, G.E. Trahey, A novel ultrasonic technique for differentiating cysts from solid lesions: preliminary results in the breast, Ultrasound Med. Biol. 21 (6) (1995) 745–751.

[12] G.F. Pinton, J.J. Dahl, G.E. Trahey, Rapid tracking of small displacements with ultrasound, IEEE Trans. Ultrason. Ferroelectr. Freq. Control 53 (6) (2005.) 1103–1117.

[13] C. Kasai, K. Namekawa, A. Koyano, R. Omoto, Real-time two-dimensional blood flow imaging using an autocorrelation technique, IEEE Trans. Ultrason. Ferroelectr. Freq. Control (3) (1985) 458–463.

[14] T. Loupas, J.T. Powers, An axial velocity estimator for ultrasound blood flow imaging, based on a full evaluation of the Doppler equation by means of a two-dimensional autocorrelation approach, IEEE Trans. Ultrason. Ferroelectr. Freq. Control 42 (4) (1998) 672−688.

[15] M.E. Anderson, Multi-dimensional velocity estimation with ultrasound using spatial quadrature, IEEE Trans. Ultrason. Ferroelectr. Freq. Control 45 (3) (1998) 852−861.

[16] J.A. Jensen, P. Munk, A new method for estimation of velocity vectors, IEEE Trans. Ultrason. Ferroelectr. Freq. Control 45 (3) (1998) 837−851.

[17] J.J. Dahl, G.F. Pinton, M.L. Palmeri, V. Agrawal, K.R. Nightingale, G.E. Trahey, A parallel tracking method for acoustic radiation force impulse imaging, IEEE Trans. Ultrason. Ferroelectr. Freq. Control 54 (2) (2007) 301−312.

[18] S.J. Hsu, R.R. Bouchard, D.M. Dumont, P.D. Wolf, G.E. Trahey, In vivo assessment of myocardial stiffness with acoustic radiation force impulse imaging, Ultrasound Med. Biol. 33 (11) (2007) 1706−1719.

[19] R.R. Bouchard, J.J. Dahl, S.J. Hsu, M.L. Palmeri, G.E. Trahey, Image quality, tissue heating, and frame-rate trade-offs in acoustic radiation force impulse imaging, IEEE Trans. Ultrason. Ferroelectr. Freq. Control 56 (1) (2009) 63−76.

[20] J.R. Doherty, D.M. Dumont, D. Hyun, J.J. Dahl, G.E. Trahey, Development and evaluation of pulse sequences for freehand ARFI imaging, Proc 2011 IEEE Int. Ultrason. Symp. (2011) 1281−1284.

[21] D.M. Dumont, J.R. Doherty, G.E. Trahey, Noninvasive assessment of wall-shear rate and vascular elasticity using combined ARFI/SWEI/spectral Doppler imaging system, Ultrason. Imaging 33 (3) (2011) 165−188.

[22] S.J. Hsu, R.R. Bouchard, D.M. Dumont, P.D. Wolf, G.E. Trahey, On the characterization of left ventricular function with acoustic radiation force impulse imaging, Proc. IEEE Ultrason. Symp. (2009) 1942−1945.

[23] J.R. Doherty, J.J. Dahl, G.E. Trahey, Harmonic tracking of acoustic radiation force-induced displacements, IEEE Trans. Ultrason. Ferroelectr. Freq. Control 60 (11) (2013) 2347−2358.

[24] J. J Dahl, N.M. Sheth, Reverberation clutter from subcutaneous tissue layers: simulation and in vivo demonstrations, Ultrasound Med. Biol. 40 (4) (2014) 714−726.

[25] G.F. Pinton, G.E. Trahey, J.J. Dahl, Sources of image degradation in fundamental and harmonic ultrasound imaging: a nonlinear fullwave simulation study, IEEE Trans. Ultrason. Ferroelectr. Freq. Control 58 (6) (2011) 1272−1283.

[26] A.I. El-Fallah, M.B. Plantec, K.W. Ferrara, Ultrasonic measurement of breast tissue motion and the implications for velocity estimation, Ultrasound Med. Biol. 23 (7) (1997) 1047−1057.

[27] B.J. Fahey, M.L. Palmeri, G.E. Trahey, The impact of physiological motion on tissue tracking during radiation force imaging, Ultrasound Med. Biol. 33 (7) (2007) 1149−1166.

[28] B.J. Fahey, S.J. Hsu, G.E. Trahey, A novel motion compensation algorithm for acoustic radiation force elastography, IEEE Trans. Ultrason. Ferroelectr. Freq. Control 55 (5) (2008) 1095−1111.

[29] C.M. Gallippi, K.R. Nightingale, G.E. Trahey, BSS-based filtering of physiological and ARFI-induced tissue and blood motion, Ultrasound Med. Biol. 29 (11) (2003) 1583−1592.

[30] D.M. Giannantonio, D.M. Dumont, G.E. Trahey, B.C. Byram, Comparison of physiological motion filters for in vivo cardiac ARFI, Ultrason. Imaging 33 (2) (2011) 89−108.

[31] A.C. Sharma, M.S. Soo, G.E. Trahey, K.R. Nightingale, Acoustic radiation force impulse imaging of in vivo breast masses, Proc. 2004 IEEE Ultrason. Symp. 1 (2004) 728–731.

[32] B.J. Fahey, R.C. Nelson, D.P. Bradway, S.J. Hsu, D.M. Dumont, G.E. Trahey, *In vivo* visualization of abdominal malignancies with acoustic radiation force elastography, Phys. Med. Biol. 53 (1) (2008) 279–293.

[33] W. Meng, G. Zhang, C. Wu, G. Wu, Y. Song, Z. Lu, Preliminary result of acoustic radiation force impulse (ARFI) ultrasound imaging of breast lesions, Ultrasound Med. Biol. 37 (9) (2011) 1436–1443.

[34] T. Shuang-Ming, Z. Ping, Q. Ying, C. Li-Rong, Z. Ping, L. Rui-Zhen, Usefulness of acoustic radiation force impulse imaging in the differential diagnosis of benign and malignant liver lesions, Acad. Radiol. 18 (7) (2011) 810–815.

[35] S.H. Cho, J.Y. Lee, J.K. Han, B.I. Choi, Acoustic radiation force impulse elastography for the evaluation of focal solid hepatic lesions: preliminary findings, Ultrasound Med. Biol. 36 (2) (2010) 202–208.

[36] G.E. Trahey, M.L. Palmeri, R.C. Bently, K.R. Nightingale, Acoustic radiation force impulse imaging of the mechanical properties of arteries: in vivo and ex vivo results, Ultrasound Med. Biol. 30 (9) (2004) 1163–1171.

[37] D.M. Dumont, R.H. Behler, T.C. Nichols, E.P. Merricks, C.M. Gallippi, ARFI imaging for noninvasive material characterization of atherosclerosis, Ultrasound Med. Biol. 32 (11) (2005.) 1703–1711.

[38] J.J. Dahl, D.M. Dumont, E.M. Miller, J.D. Allen, G.E. Trahey, Acoustic radiation force impulse imaging for noninvasive characterization of carotid artery atherosclerotic plaques: a feasibility study, Ultrasound Med. Biol. 35 (5) (2009) 707–716.

[39] L. Zhai, J. Madden, W.-C. Foo, M.L. Palmeri, V. Mouraviev, T.J. Polascik, K.R. Nightingale, Acoustic radiation force impulse imaging of human prostates ex vivo, Ultrasound Med. Biol. 36 (4) (2010) 576–588.

[40] L. Zhai, T.J. Polascik, W.-C. Foo, S. Rosenzweig, M.L. Palmeri, J. Madden, K.R. Nightingale, Acoustic radiation force impulse imaging of human prostates: initial *in vivo* demonstration, Ultrasound Med. Biol. 38 (1) (2012) 50–61.

[41] B.J. Fahey, K.R. Nightingale, S.A. McAleavey, M.L. Palmeri, P.D. Wolf, G.E. Trahey, Acoustic radiation force impulse imaging of myocardial radiofrequency ablation: initial in vivo results, IEEE Trans. Ultrason. Ferroelectr. Freq. Control 52 (4) (2005) 631–641.

[42] S.J. Hsu, B.J. Fahey, D.M. Dumont, P.D. Wolf, G.E. Trahey, Challenges and implementation of radiation-force imaging with an intracardiac ultrasound transducer, IEEE Trans. Ultrason. Ferroelectr. Freq. Control 54 (5) (2007) 996–1009.

[43] B.J. Fahey, R.C. Nelson, S.J. Hsu, D.P. Bradway, D.M. Dumont, G.E. Trahey, *In vivo* guidance and assessment of liver radio-frequency ablation with acoustic radiation force elastography, Ultrasound Med. Biol. 34 (10) (2008) 1590–1603.

[44] T.J. Czernuszewicz, J.W. Homeister, M.C. Caughey, M.A. Farber, J.J. Fulton, P.F. Ford, W.A. Marston, R. Vallabhaneni, T.C. Nichols, C.M. Gallippi, Noninvasive in vivo characterization of human carotid plaques with acoustic radiation force impulse ultrasound: comparison with histology after endarterectomy, Ultrasound Med. Biol. 41 (3) (2015) 685–697.

[45] M.L. Palmeri, T.J. Glass, Z.A. Miller, S.J. Rosenzweig, A. Buck, T.J. Ploscik, R.T. Gupta, A.F. Brown, J. Madden, K.R. Nightingale, Identifying clinically significant prostate cancers using 3-D *in vivo* acoustic radiation force impulse imaging

with whole-mount histology validation, Ultrasound Med. Biol. 42 (6) (2015.) 1251—1262.

[46] M.L. Palmeri, Z.A. Miller, T.J. Glass, K. Garcia-Reyes, R.T. Gupta, S.J. Rosenzweig, C. Kauffman, T.J. Polascik, A. Buck, E. Kulbacki, J. Madden, S.L. Lipman, N.C. Rouze, K.R. Nightingale, B-mode and acoustic radiation force impulse (ARFI) imaging of prostate zonal anatomy comparison with 3T T2-weighted MR imaging, Ultrason. Imaging 37 (1) (2015) 22—41.

[47] B.J. Fahey, M.L. Palmeri, G.E. Trahey, Frame rate considerations for real-time abdominal acoustic radiation force impulse imaging, Ultrason. Imaging 28 (4) (2005.) 193—210.

[48] S.A. Eyerly, S.J. Hsu, S.H. Agashe, G.E. Trahey, Y. Li, P.D. Wolf, An in vitro assessment of acoustic radiation force impulse imaging for visualizing cardiac radiofrequency ablation lesions, J. Cardiovasc. Electrophysiol. 21 (5) (2010) 557—563.

[49] S.A. Eyerly, T.D. Bahnson, J.I. Koontz, D.P. Bradway, D.M. Dumont, G.E. Trahey, P.D. Wolf, Intracardiac acoustic radiation force impulse imaging: a novel imaging method for intraprocedural evaluation of radiofrequency ablation lesions, Heart Rhythm 9 (11) (2012) 1855—1862.

[50] T.D. Bahnson, S.A. Eyerly, P.J. Hollender, J.R. Doherty, Y.-J. Kim, G.E. Trahey, P.D. Wolf, Feasibility of near real-time lesion assessment during radiofrequency catheter ablation in humans using acoustic radiation force impulse imaging, J. Cardiovasc. Electrophysiol. 25 (12) (2014) 1275—1283.

[51] S.J. Hsu, D.P. Bradway, R.R. Bouchard, P.J. Hollender, P.D. Wolf, G.E. Trahey, Parametric pressure-volume analysis and acoustic radiation force impulse imaging of left ventricular function, Proc. IEEE Ultrason. Symp. (2010) 698—701.

[52] S.J. Hsu, J.L. Hubert, P.D. Wolf, G.E. Trahey, Acoustic radiation force impulse imaging of mechanical stiffness propagation within myocardial tissue, Proc. IEEE Ultrason. Symp. (2007) 864—867.

[53] M. Fatemi, J.F. Greenleaf, Ultrasound-stimulated vibro-acoustic spectrography, Science 280 (3) (1998) 82—85.

[54] M. Fatemi, J.F. Greenleaf, Vibro-acoustography: an imaging modality based on ultrasound-stimulated acoustic emission, Proc. Natl. Acad. Sci. 96 (12) (1999) 6603—6608.

[55] S. Chen, M. Fatemi, R. Kinnick, J.F. Greenleaf, Comparison of stress field forming methods for vibro-acoustography, IEEE Trans. Ultrason. Ferroelectr. Freq. Control 51 (3) (2004) 313—321.

[56] M.W. Urban, C. Chalek, R.R. Kinnick, T.M. Kinter, B. Haider, J.F. Greenleaf, K.E. Thomenius, M. Fatemi, Implementation of vibro-acoustography on a clinical ultrasound system, IEEE Trans. Ultrason. Ferroelectr. Freq. Control 58 (6) (2011) 1169—1181.

[57] M. Mehrmohammadi, R.T. Fazzio, D.H. Whaley, S. Pruthi, R.R. Kinnick, M. Fatemi, Z. Alizad, Preliminary *in vivo* breast vibro-acoustography results with a quasi-2-D array transducer: a step forward toward clinical applications, Ultrasound Med. Biol. 40 (12) (2014) 2819—2829.

[58] M. Fatemi, L.E. Wold, A. Alizad, J.F. Greenleaf, Vibro-acoustic tissue mammography, IEEE Trans. Med. Imaging 21 (1) (2002) 1—8.

[59] A. Alizad, M. Fatemi, L.E. Wold, J.F. Greenleaf, Performance of vibro-acoustography in detecting microcalcifications in excised human breast tissue: a study of 74 tissue samples, IEEE Trans. Med. Imaging 23 (3) (2004) 307—312.

[60] F.G. Mitri, B.J. Davis, A. Alizad, J.F. Greenleaf, T.M. Wilson, L.A. Mynderse, M. Fatemi, Prostate cryotherapy monitoring using vibroacoustography: preliminary results of an ex vivo study and technical feasibility, IEEE Trans. Biomed. Eng. 55 (11) (2008) 2584–2592.

[61] F.G. Mitri, B.J. Davis, M.W. Urban, A. Alizad, J.F. Greenleaf, G.H. Lischer, T.M. Wilson, M. Fatemi, Vibro-acoustography imaging of permanent prostate brachytherapy seeds in an excised human prostate - preliminary results and technical feasibility, Ultrasonics 49 (3) (2009) 389–394.

[62] A. Alizad, M. Mehrmohammadi, F.G. Mitri, B.J. Davis, T.J. Sebo, L.A. Mynderse, R.R. Kinnick, J.F. Greenleaf, M. Fatemi, Application of vibro-acoustography in prostate tissue imaging, Med. Phys. 40 (2013) 022902, https://doi.org/10.1118/1.4773890.

[63] C. Pislaru, B. Kantor, R.R. Kinnick, J.L. Anderson, M.-C. Aubry, M.W. Urban, M. Fatemi, J.F. Greenleaf, In vivo vibroacoustography of large peripheral arteries, Investig. Radiol. 43 (4) (2008) 243–252.

[64] C. Pislaru, J.F. Greenleaf, B. Kantor, M. Fatemi, Atherosclerosis Disease Management (Chapter 21): Vibro-Acoustography of Arteries, Springer, New York, New York, 2011, pp. 675–698.

[65] H.A.S. Kamimura, L. Wang, A.A.O. Cameiro, R.R. Kinnick, K.-N. An, M. Fatemi, Vibroacoustography for the assessment of total hip arthroplasty, Clinics 68 (4) (2013) 463–468.

[66] E.E. Konofagou, K. Hynynen, Localized harmonic motion imaging: theory, simulations, and experiments, Ultrasound Med. Biol. 29 (10) (2003) 1405–1413.

[67] C. Maleke, E.E. Konofagou, Harmonic motion imaging for focused ultrasound (HMIFU): a fully integrated technique for sonication and monitoring of thermal ablation in tissues, Phys. Med. Biol. 53 (6) (2008) 1773–1793.

[68] C. Maleke, E.E. Konofagou, In vivo feasibility of real-time monitoring of focused ultrasound surgery (FUS) using harmonic motion imaging (HMI), IEEE Trans. Biomed. Eng. 57 (1) (2010) 7–11.

[69] L. Curiel, R. Chopra, K. Hynynen, Ultrasound Med. Biol. 35 (1) (2009) 65–78.

[70] M.R. Selzo, C.M. Gallippi, Viscoelastic response (VisR) imaging for assessment of viscoelasticity in voigt materials, IEEE Trans. Ultrason. Ferroelectr. Freq. Control 50 (12) (2013) 2488–2500.

[71] W.F. Walker, F.J. Fernandez, L.A. Negron, A method of imaging viscoelastic parameters with acoustic radiation force, Phys. Med. Biol. 45 (6) (2000) 1437–1447.

[72] F.W. Mauldin, M.A. Haider, E.G. Loboa, R.H. Behler, L.E. Euliss, T.W. Pfeiler, C.M. Gallippi, Monitored steady-state excitation and recovery (MSSER) radiation force imaging using viscoelastic models, IEEE Trans. Ultrason. Ferroelectr. Freq. Control 55 (7) (2008) 1597–1610.

[73] Food and Drug Administration - Center for Devices and Radiological Health, Information for Manufacturers Seeking Marketing Clearance of Diagnostic Ultrasound Systems and Transducers, Technical report, US Department of Health and Human Services, September 1997. Retrieved from: http://www.fda.gov/downloads/MedicalDevices/DeviceRegulationandGuidance/GuidanceDocuments/UCM070911.pdf.

[74] M.R. Selzo, C.J. Moore, M.M. Hossain, M.L. Palmeri, C.M. Gallippi, On the quantitative potential of viscoelastic response (VisR) ultrasound using the one- dimensional mass-spring-damper model, IEEE Trans. Ultrason. Ferroelectr. Freq. Control 63 (9) (2015.) 1276–1287.

[75] M.L. Palmeri, S.A. McAleavey, G.E. Trahey, K.R. Nightingale, Ultrasonic tracking of acoustic radiation force-induced displacements in homogeneous media, IEEE Trans. Ultrason. Ferroelectr. Freq. Control 53 (7) (2005.) 1300–1313.

[76] A.P. Sarvazyan, O.V. Rudenko, S.D. Swanson, J.B. Fowlkes, S.Y. Emelianov, Shear wave elasticity imaging: a new ultrasonic technology of medical diagnostics, Ultrasound Med. Biol. 24 (9) (1998) 1419–1435.

[77] J. Bercoff, M. Tanter, M. Fink, Supersonic shear imaging: a new technique for soft tissue elasticity mapping, IEEE Trans. Ultrason. Ferroelectr. Freq. Control 51 (4) (2004) 396–409.

[78] E. Bavu, J.-L. Gennisson, M. Couade adn, J. Vercoff, V. Mallet, M. Fink, A. Badel, A. Vallet-Pichard, B. Nalpas, M. Tanter, S. Pol, Noninvasive *in vivo* liver fibrosis evaluation using supersonic shear imaging: a clinical study on 113 Hepatitis C virus patients, Ultrasound Med. Biol. 37 (9) (2011) 1361–1373.

[79] J. Bercoff, M. Pernot, M. Tanter, M. Fink, Monitoring thermally-induced lesions with supersonic shear imaging, Ultrason. Imaging 26 (2) (2004) 71–84.

[80] F. Sebag, J. Vaillant-Lombard, J. Berbis, V. Griset, J.F. Henry, P. Petit, C. Oliver, Shear wave elastography: a new ultrasound imaging mode for the differential diagnosis of benign and malignant thyroid nodules, J. Clin. Endocrinol. Metab. 95 (12) (2010) 5281–5288.

[81] M. Muller, J.-L. Gennisson, T. Deffieux, M. Tanter, M. Fink, Quantitative viscoelasticity mapping of human liver using supersonic shear imaging: preliminary *in vivo* feasibility study, Ultrasound Med. Biol. 35 (2) (2009) 219–229.

[82] T. Deffieux, J.-L. Gennisson, L. Bousquet, M. Corouge, S. Cosconea, D. Amroun, S. Tripon, B. Terris, V. Mallet, P. Sogni, M. Tanter, S. Pol, Investigating liver stiffness and viscosity for fibrosis, steatosis and activity staging using shear wave elastography, J. Hepatol. 62 (2) (2015) 317–324.

[83] M. Muller, D. Ait-Belkacem, M. Hessabi, J.-L. Gennisson, G. Grange, F. Goffinet, E. Lecarpentier, D. Cabrol, M. Tanter, V. Tsatsaris, Assessment of the cervix in pregnant women using shear wave elastography: a feasibility study, Ultrasound Med. Biol. 41 (11) (2015) 2789–2797.

[84] M.L. Palmeri, M.H. Wang, J.J. Dahl, K.D. Frinkley, K.R. Nightingale, Quantifying hepatic shear modulus *in vivo* using acoustic radiation force, Ultrasound Med. Biol. 34 (4) (2008) 546–558.

[85] K.R. Nightingale, S.A. McAleavey, G.E. Trahey, Shear-wave generation using acoustic radiation force: *In vivo* and *ex vivo* results, Ultrasound Med. Biol. 29 (12) (2003) 1715–1723.

[86] M.L. Palmeri, D. Xu, L. Zhai, K.R. Nightingale, Acoustic radiation force based quantification of tissue shear modulus within the region of excitation, Proc. IUS 2008. IEEE Ultrason. Symp. (2008) 2009–2012.

[87] L. Sandrin, B. Fourquet, J.M. Hasquenoph, S. Yon, C. Fournier, F. Mal, C. Christidis, M. Ziol, B. Poulet, F. Kazemi, M. Beaugrand, R. Palau, Transient elastography: a new noninvasive method for assessment of hepatic fibrosis, Ultrasound Med. Biol. 29 (12) (2003) 1705–1713.

[88] K. Walsh, D.M. Dumont, M.L. Palmeri, B. Byram, On-axis radiation-force-based quantitative stiffness estimation with a bayesian displacement estimator, Proc. 2015 IEEE Int. Ultrason. Symp. (2015), https://doi.org/10.1109/ULTSYM.2015.0377.

[89]   M.L. Palmeri, M.H. Wang, N.C. Rouze, M.F. Abdelmalek, C.D. Guy, B. Moser, A.M. Diehl, K.R. Nightingale, Noninvasive evaluation of hepatic fibrosis using acoustic radiation force-based shear stiffness in patients with nonalcoholic fatty liver disease, J. Hepatol. 55 (3) (2011) 666—672.

[90]   S. Bota, F. Bob, I. Sporea, R. Şirli, A. Popescu, Factors that influence kidney shear wave speed assessed by acoustic radiation force impulse elastography in patients without kidney pathology, Ultrasound Med. Biol. 41 (1) (2015) 1—6.

[91]   T. Canas, T. Fontanilla, M. Miralles, A. Cacia, A. Malalana, E. Román, Normal values of spleen stiffness in healthy children assessed by acoustic radiation force impulse imaging (ARFI): comparison between two ultrasound transducers, Pediatr. Radiol. 45 (9) (2015) 1316—1322.

[92]   T. Fukuhara, E. Matsuda, Y. Endo, M. Takenobu, S. Izawa, K. Fujiwara, H. Ki- tano, Correlation between quantitative shear wave elastography and pathologic structures of thyroid lesions, Ultrasound Med. Biol. 41 (9) (2015) 2326—2332.

[93]   S.A. Eyerly, M. Vejdani-Jahromi, D.M. Dumont, G.E. Trahey, P.D. Wolf, The evolution of tissue stiffness at radiofrequency ablation sites during lesion formation and in the peri-ablation period, J. Cardiovasc. Electrophysiol. 26 (9) (2015) 1009—1018.

[94]   M. Vejdani-Jahromi, M. Nagle, G.E. Trahey, P.D. Wolf, Ultrasound shear wave elasticity imaging quantifies coronary perfusion pressure effect on cardiac compliance, IEEE Trans. Med. Imaging 34 (2) (2015) 465—473.

[95]   M. Vejdani-Jahromi, M. Nagle, Y. Jiang, G.E. Trahey, P.D. Wolf, A comparison of acoustic radiation force derived indices of cardiac function in the Langendorff perfused rabbit heart, IEEE Trans. Ultrason. Ferroelectr. Freq. Control 69 (9) (2015.) 1288—1295.

[96]   M. Tanter, J. Bercoff, A. Athanasiou, T. Deffieux, J.-L. Gennisson, G. Montaldo, M. Muller, A. Tardivon, M. Fink, Quantitative assessment of breast lesion viscoelasticity: initial clinical results using supersonic shear imaging, Ultrasound Med. Biol. 34 (9) (2008) 1373—1386.

[97]   F. Chamming's, M.-A. Le-Frere-Belda, H. Latorre-Ossa, V. Fitoussi, A. Redheuil, F. Assayag, L. Pidial, J.-L. Gennisson, M. Tanter, C.-A. Cuenod, L.S. Fournier, Supersonic shear wave elastography of response to anti-cancer therapy in a xenograft tumor model, Ultrasound Med. Biol. 42 (4) (2015.) 924—930.

[98]   M. Couade, M. Pernot, E. Messas, A. Bel, M. Ba, A. Hagege, M. Fink, M. Tanter, In vivo quantitative mapping of myocardial stiffening and transmural anisotropy during the cardiac cycle, IEEE Trans. Med. Imaging 30 (2) (2011) 295—305.

[99]   M. Pernot, M. Couade, P. Mateo, B. Crozatier, R. Fischmeister, M. Tanter, Realtime assessment of myocardial contractility using shear wave imaging, J. Am. Coll. Cardiol. 58 (1) (2011) 65—72.

[100] W. Kwiecinski, F. Bessiere, E.C. Colas, W.A. N'Djin, M. Tanter, C. Lafon, M. Pernot, Cardiac shear-wave elastography using a transesophageal transducer: application to the mapping of thermal lesions in ultrasound transesophageal cardiac ablation, Phys. Med. Biol. 60 (20) (2015) 7829—7846.

[101] B. Arnal, M. Pernot, M. Tanter, Monitoring of thermal therapy based on shear modulus changes: I. Shear wave thermometry, IEEE Trans. Ultrason. Ferroelectr. Freq. Control 58 (2) (2011) 369—378.

[102] B. Arnal, M. Pernot, M. Tanter, Monitoring of thermal therapy based on shear modulus changes: II. Shear wave imaging of thermal lesions, IEEE Trans. Ultrason. Ferroelectr. Freq. Control 58 (8) (2011) 1603—1611.

[103] P. Song, H. Zhao, A. Manduca, M.W. Urban, J.F. Greenleaf, S. Chen, Comb- push ul-trasound shear elastography (CUSE): a novel method for two-dimensional shear elas-ticity imaging of soft tissues, IEEE Trans. Med. Imaging 31 (9) (2012) 1821−1832.

[104] P. Song, M.W. Urban, A. Manduca, H. Zhao, J.F. Greenleaf, S. Chen, Comb-push ul-trasound shear elastography (CUSE) with various ultrasound push beams, IEEE Trans. Med. Imaging 32 (8) (2013) 1435−1447.

[105] G. Montaldo, M. Tanter, J. Bercoff, N. Benech, M. Fink, Coherent plane-wave com-pounding for very high frame rate ultrasonography and transient elastography, IEEE Trans. Ultrason. Ferroelectr. Freq. Control 56 (3) (2009) 489−506.

[106] T. Deffieux, J.-L. Gennisson, J. Bercoff, M. Tanter, On the effects of reflected waves in transient shear wave elastography, IEEE Trans. Ultrason. Ferroelectr. Freq. Control 58 (10) (2011) 2032−2035.

[107] A. Nabavizadeh, P. Song, S. Chen, J.F. Greenleaf, M.W. Urban, Multi-source and multi-directional shear wave generation with intersecting steered ultrasound push beams, IEEE Trans. Ultrason. Ferroelectr. Freq. Control 62 (4) (2015) 647−662.

[108] M. Denis, M. Mehrmohammadi, P. Song, D.D. Meixner, R.T. Fazzio, S. Pruthi, D.H. Whaley, S. Chen, M. Fatemi, Update on breast cancer detection using comb-push ultrasound shear elastography, IEEE Trans. Ultrason. Ferroelectr. Freq. Control 62 (9) (2015) 1644−1650.

[109] M. Mehrmohammadi, P. Song, D.D. Meixner, R.T. Fazzio, S. Chen, J.F. Greenleaf, M. Fatemi, A. Alizad, Comb-push ultrasound shear elastography (CUSE) for evalua-tion of thyroid nodules: preliminary *in vivo* results, IEEE Trans. Med. Imaging 34 (1) (2015) 97−106.

[110] J. Vappou, C. Maleke, E.E. Konofagou, Quantitative viscoelastic parameters measured by harmonic motion imaging, Phys. Med. Biol. 54 (11) (2009) 3579−3594.

[111] S. Catheline, J.-L. Gennisson, G. Delon, M. Fink, R. Sinkus, S. Abouelkaram, J. Culioli, Measurement of viscoelastic properties of homogeneous soft solid using transient elastography: an inverse problem approach, J. Acoust. Soc. Am. 116 (6) (2004) 3734−3741.

[112] I.Z. Nenadic, M.W. Urban, H. Zhao, W. Sanchez, P.E. Morgan, J.F. Greenleaf, S. Chen, Application of attenuation measuring ultrasound shearwave elastography in 8 post-transplant liver patients, Proc 2014 IEEE Int. Ultrason. Symp. (2014) 987−990.

[113] S. Chen, M. Fatemi, J.F. Greenleaf, Quantifying elasticity and viscosity from measure-ment of shear wave speed dispersion, J. Acoust. Soc. Am. 115 (6) (2004) 2781−2785.

[114] S. Chen, M.W. Urban, C. Pislaru adn, R. Kinnick, Y. Zheng, A. Yao, J.F. Greenleaf, Shearwave dispersion ultrasound vibrometry (SDUV) for measuring tissue elasticity and viscosity, IEEE Trans. Ultrason. Ferroelectr. Freq. Control 56 (1) (2009) 55−62.

[115] M.W. Urban, S. Chen, M. Fatemi, A review of shearwave dispersion ultrasound vibr-ometry (SDUV) and its applications, Curr. Med. Imaging Rev. 8 (1) (2012) 27−36.

[116] M.W. Urban, J.F. Greenleaf, Harmonic pulsed excitation and motion detection of a vibrating reflective target, J. Acoust. Soc. Am. 123 (1) (2008) 519−533.

[117] F.G. Mitri, M.W. Urban, M. Fatemi, J.F. Greenleaf, Shear wave dispersion ultrasonic vibrometry for measuring prostate shear stiffness and viscosity: an *in vitro* pilot study, IEEE Trans. Biomed. Eng. 58 (2) (2011) 235−242.

[118] C. Amador, M.W. Urban, S. Chen, J.F. Greenleaf, Shearwave dispersion ultrasound vibrometry (SDUV) on swine kidney, IEEE Trans. Ultrason. Ferroelectr. Freq. Control 58 (12) (2011) 2608−2619.

[119] J. Brum, M. Bernal, J.-L. Gennisson, M. Tanter, In vivo evaluation fo the elastic anisotropy of the human Achilles tendon using shear wave dispersion analysis, Phys. Med. Biol. 59 (3) (2014) 505–523.

[120] S. Chen, W. Sanchez, M.R. Callstrom, B. Gorman, J.T. Lewis, S.O. Sanderson, J.F. Greenleaf, H. Xie, Y. Shi, M. Pashley, V. Shamdasani, M. achman, S. Metz, Assessment of liver viscoelasticity by using shear waves induced by ultrasound radiation force, Radiology 266 (3) (2013) 964–970.

[121] I.Z. Nenadic, B. Qiang, M.W. Urban, L.H. de Araujo Vasconcelo, A. Nabavizadeh, A. Alizad, J.F. Greenleaf, M. Fatemi, Ultrasound bladder vibrometry method for measuring viscoelasticity of the bladder wall, Phys. Med. Biol. 58 (8) (2013) 2675–2695.

[122] P. Song, M.C. Macdonald, R.H. Behler, J.D. Lanning, M.H. Wang, M.W. Urban, A. Manduca, H. Zhao, M.R. Callstrom, A. Alizad, J.F. Greenleaf, S. Chen, Two-dimensional shear-wave elastography on conventional ultrasound scanners with time-aligned sequential tracking (TAST) and comb-push ultrasound shear elastography (CUSE), IEEE Trans. Ultrason. Ferroelectr. Freq. Control 62 (2) (2015) 290–302.

[123] S.A. McAleavey, M. Menon, J. Orszulak, Shear-modulus estimation by application of spatially-modulated impulsive acoustic radiation force, Ultrason. Imaging 29 (2) (2007) 87–104.

[124] P.J. Hollender, S.J. Rosenzweig, K.R. Nightingale, G.E. Trahey, Single and multiple-track-location shear wave and acoustic radiation force impulse imaging: matched comparison of contrast, contrast-to-noise ratio and resolution, Ultrasound Med. Biol. 41 (4) (2015) 1043–1057.

[125] S.A. McAleavey, E. Collins, J. Kelly, E. Elegbe, M. Menon, Validation of SMURF estimation of shear modulus in hydrogels, Ultrason. Imaging 31 (2) (2009) 131–150.

[126] M. Karaman, P.-C. Li, M. O'Donnell, Synthetic aperture imaging for small scale systems, IEEE Trans. Ultrason. Ferroelectr. Freq. Control 42 (3) (1995) 429–442.

[127] G.R. Lockwood, J.R. Talmand, S.S. Brunke, Real-time 3-D ultrasound imaging using sparse synthetic aperture beamforming, IEEE Trans. Ultrason. Ferroelectr. Freq. Control 45 (4) (1998) 980–988.

[128] M.-H. Bae, M.-K. Jeong, A study of synthetic-aperture imaging with virtual source elements in B-mode ultrasound imaging systems, IEEE Trans. Ultrason. Ferroelectr. Freq. Control 47 (6) (2000) 1510–1512.

[129] M. Correia, J. Provost, S. Chatelin, O. Villemain, M. Tanter, M. Pernot, Ultrafast harmonic coherent compound (UHCC) imaging for high frame rate echocardiography and shear-wave elastography, IEEE Trans. Ultrason. Ferroelectr. Freq. Control 63 (3) (2015.) 420–431.

[130] W.N. Lee, M. Pernot, M. Couade, E. Messas, P. Bruneval, A. Bel, A.A. Hagege, M. Fink, M. Tanter, Mapping myocardial fiber orientation using echocardiography-based shear wave imaging, IEEE Trans. Med. Imaging 31 (3) (2012) 554–562.

[131] D. Royer, E. Dieulesaint, Elastic Waves in Solids I. Free and Guided Propagation, Springer, Berlin, 2000.

[132] M.J.P. Musgrave, The propagation of elastic waves in crystals and other anisotropic media, Rep. Prog. Phys. 22 (1959) 74–96.

[133] M. Wang, B. Byram, M. Palmeri, N. Rouze, K. Nightingale, Imaging transverse isotropic properties of muscle by monitoring acoustic radiation force induced shear waves using a 2-D matrix ultrasound array, IEEE Trans. Med. Imaging 32 (9) (2013) 1671–1684.

[134] P.J. Hollender, S. Lipman, G.E. Trahey, O(STL-SWEI), Proc 2015 IEEE Int. Ultrason. Symp. (2015), https://doi.org/10.1109/ULTSYM.2015.0035.

[135] J.-L. Gennisson, J. Provost, T. Deffieux, C. Papadacci, M. Imbault, M. Pernot, M. Tanter, 4D ultrafast shear-wave imaging, IEEE Trans. Ultrason. Ferroelectr. Freq. Control 62 (6) (2015) 1059–1065.

Magnetic resonance elastography

6

Bogdan Dzyubak[1], Kevin J. Glaser[2]

[1]*Department of Medical Physics, Mayo Clinic, Rochester, MN, United States;* [2]*Medical Physics, Mayo Clinic, Rochester, MN, United States*

1. Introduction

Magnetic resonance elastography (MRE) is an MRI-based quantitative method for calculating tissue viscoelasticity. MRE was first introduced as a US FDA-cleared product for measuring liver stiffness in 2009. Its main clinical application is staging of hepatic fibrosis for which MRE has been shown to be the most effective method, having an accuracy, reproducibility, and success rate superior to ultrasound elastography [1—5] and far superior to other methods [6—9]. Biopsy, which has been a standard for diagnosing hepatic fibrosis for many years, is invasive and prone to variability due to sampling error and subjective interpretation [10,11]. While originally validated against biopsy, MRE overcomes these disadvantages. As of 2019, it has been adopted by over 1400 institutions across the world. Other applications of MRE are wide-ranging and include stiffness measurements in the breast, liver, brain, muscle, kidneys, and tumors [12].

Clinical liver MRE [13] is highly standardized, with the three major MRI vendors (GE, Siemens, and Philips) distributing the same hardware and very similar acquisitions and inversions as a package (General Electric MR-Touch and Siemens and Philips MR Elastography). In 2018, the Quantitative Imaging Biomarkers Alliance (QIBA), a standards organization, has published a consensus profile on ways of performing and quality controlling clinical liver MRE [14]. It uses an acoustic speaker, located outside the scanner room, to generate acoustic waves that are delivered to the patient via a set of air-filled tubes and a passive driver strapped to the chest. This setup delivers compressional (longitudinal) waves that undergo mode conversion to shear waves at tissue interfaces. In MRI, the images that are obtained are complex-valued images (i.e., signals with real and imaginary parts or magnitudes and phases) with magnitudes that show anatomic details and with phases that are often ignored for clinical purposes. In MRE, motion-sensitive magnetic-field gradients are added to standard magnetic resonance (MR) acquisitions to encode information about the tissue vibration or motion into the phase information of the MR images [15]]. Signal from the longitudinal waves is then filtered out, and the remaining shear wave information is used to calculate the absolute regional tissue stiffness. The use of mode conversion makes this system very versatile because shear waves attenuate quickly and this actuation approach is able to attain higher wave amplitudes in deeper tissues than other approaches. Additionally, because body interfaces act as secondary wave sources, the system is able to effectively image structures shielded by bone, such as the brain and the heart. Longitudinal waves are also able to pass through air and fluid, making it possible to perform MRE of the lungs and successful liver imaging of patients with ascites. This acoustic driver system is used clinically as well as for various research applications. Other drivers that employ electric motors, piezoelectric crystals, and, less commonly, ultrasound actuation have been developed, typically with the goal of generating higher frequency waves for imaging small or stiff structures [16].

Stiffness calculated by MRE is an absolute quantity and has been observed to agree well with dynamic mechanical testing in phantoms and ex vivo tissues [17,18] and with ultrasound in vivo [3,19]. The Young's modulus (the ratio of uniaxial stress to strain under small-amplitude deformations) reported in ultrasound elastography can be compared directly to MRE's shear modulus (the ratio of shear stress to shear strain) after dividing the Young's modulus by 3 for nearly incompressible materials, including most types of tissue. MRE uses continuous-wave actuation, which allows a specific vibration frequency to be used (e.g., 60-Hz waves are typically used for clinical MRE), rather than having a broad set of frequencies created by the transient actuation in ultrasound elastography. Although continuously driven waves can undergo interference due to reflections off of tissue boundaries, the attenuation coefficient in biological tissues is typically high enough that with an appropriately low-amplitude vibration, reflected waves are negligible. Finally, unlike most ultrasound elastography techniques, MRE delivers vibrations throughout entire organs and can measure whole volumes of tissue displacement information, including information about the full motion vector field rather than just the information about tissue motion in one direction. This leads to reduced operator dependence, as an imaging window does not need to be as carefully selected; however, care must still be taken to make sure the passive MRE driver is positioned close enough to the target structure to produce shear waves of adequate amplitude in the tissue of interest. The whole-volume vector-imaging capabilities of MRE allow it to use more sophisticated image processing methods to calculate the stiffness (i.e., inversion algorithms), which, although not used routinely for clinical imaging at this time, have been used to deal with complicated structures and to calculate a multitude of viscoelastic properties in research settings [20].

The potential clinical value of MRE-assessed tissue stiffness has been demonstrated in numerous studies of various organs, including the brain, lung, kidneys, spleen, heart, and muscle. It has been applied to both diffuse conditions (such as fibrosis, inflammation, and multiple sclerosis [MS]) and localized diseases (e.g., infarcts and tumors). While MRE is capable of measuring the absolute tissue stiffness, this can be challenging in small tumors or complex objects where high-frequency waves (which are subject to high attenuation) may be needed to increase the resolution and accuracy of the measurements. However, as long as the measurements are reproducible, even relative or effective stiffness values may be valuable for applications such as monitoring treatment response. Several studies have confirmed the reproducibility of hepatic MRE [21,22] and the agreement between MRE-assessed stiffness and the stiffness measured with dynamic mechanical analysis [17,18] and ultrasound elastography under controlled conditions. This chapter will provide an overview of several different MRE applications, including information about hardware configurations, inversion algorithms, quality control, and physiologic/clinical findings.

2. Acquisition

2.1 Generating and delivering mechanical waves

The most common driver system for producing MRE vibrations consists of an active driver (acoustic speaker) outside the scanner room and a passive driver (drum) connected to it via a series of plastic tubes (Fig. 6.1A). The passive driver may be made of plastic or a softer material and is secured to the patient's body near the organ being imaged. The compressed air inflates the passive driver generating longitudinal waves at the surface of the body, which then propagate into the body and are mode-converted to shear waves at interfaces between tissues of different acoustic impedances (organ boundaries). This type of driver system has been used to image

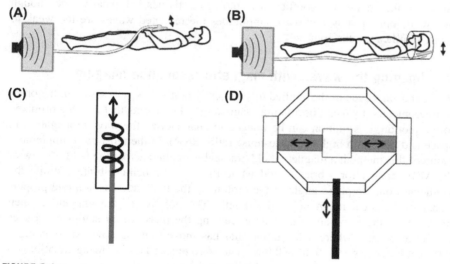

FIGURE 6.1

Types of magnetic resonance elastography drivers. (A,B) Applications of the acoustic driver system that delivers compressional waves from a speaker outside the scanner room to a passive drum driver connected to the chest or head of the patient. (C) Solenoid driver moving a rod or needle. (D) A piezoelectric driver system indirectly moving a rod with amplified amplitude. Drivers may deliver shear waves directly or longitudinal waves that are mode-converted to shear waves at tissue interfaces.

many organs, including the liver, kidneys, brain, heart, and lungs. Its diverse use stems from the ease and reproducibility of the setup due to the large driver size, the ability to produce high wave amplitudes in the frequency range of 10–100 Hz, and the low wave attenuation at these frequencies. However, at higher driving frequencies (>150 Hz) the amplitude of acoustic drivers decreases significantly. Piezoelectric and solenoid active drivers (Fig. 6.1B and C), which can push a rod or needle against the body, have better performance at higher frequencies, but the higher frequency waves they produce attenuate quickly in tissues. Owing to these factors, these drivers are most commonly used for imaging small or superficial structures such as tumors and muscle, as well as animal models.

Owing to the size of MRE drivers and the depth of penetration of the shear waves, MRE acquisitions are robust to modest variations in the patient setup and driver placement. In fact, during standard liver MRE examinations in which the driver is targeting the right side of the body, there can still be enough motion produced in the left side of the body to make it possible to measure the stiffness of the spleen. Nonetheless, it is still important to verify a good driver connection, coupling to the patient body, and centering of the driver over the organ to ensure the accuracy and reproducibility of the result. MRE uses continuous motion rather than transient impulses, which allows the vibration frequency to be set to a specific value. With continuous motion, wave reflections can cause interference patterns. However, the attenuation coefficient in biological tissues is typically high enough that with appropriately low wave amplitudes the reflected waves are too weak to cause significant interference.

2.2 Imaging the waves with magnetic resonance imaging

While the amplitude of the applied mechanical vibrations is very small, in the order of micrometers, by using oscillating magnetic field gradients that are synchronized to the motion, this motion can be imaged to characterize the wave propagation in space and time with high signal-to-noise ratio (SNR). In the most basic implementation of this method, a magnetic field gradient is applied across the field of view of the MRI scanner for a brief period while the tissue is being vibrated. While the gradient is on, the phase angle of the protons in the body accrues at a rate proportional to the local magnetic field strength. The polarity of the gradient is then reversed for the same amount of time, causing the phase of all stationary tissues to return to zero, whereas the tissue that has moved and is now experiencing a different local magnetic field will have a nonzero phase. The following equation describes the phase acquired during an MRE acquisition (i.e., phase of the MR signal at each voxel):

$$\varphi(r, \theta) = \frac{\gamma NT (G \cdot \xi_0)}{2} \cos(k \cdot r + \theta)$$

where γ is the gyromagnetic ratio (constant); T is the period of the vibration; N is the number of oscillation cycles of the gradient; G is the motion-encoding gradient

(MEG) amplitude; ξ is the amplitude of the particle motion; k is the stiffness-dependent wave number (2π/wavelength), which depends on the temporal frequency of the applied waves and the tissue stiffness; r is the spatial location vector; and θ is the relative phase shift of the particle motion with respect to the start of motion encoding [23]. Clinical MRI MEGs are single cycle and trapezoidal, which makes them sensitive to a broader range of frequencies. Owing to the broadband sensitivity of the MEGs, the period, duration, or frequency of the MEG may be different from the motion itself (e.g., to reduce the echo time) while still having substantial motion sensitivity. Note that the ability to encode motion is not related to the spatial resolution of MRI and allows micrometer-level motion to be easily encoded. To characterize the wave propagation in time, multiple images ("phase offsets") are acquired with different delays between the applied motion and the start of the MEGs. In general, three to eight phase offsets are acquired with the delays evenly spaced over one period of the motion. There are other processes through which MR images also accrue phase, such as inhomogeneous magnetic fields and off-resonance precession of protons, tissue vibration during the MRI gradients, and imperfections in the design of the MEGs themselves. Some of this additional phase information can be removed by performing two acquisitions with either alternating MEG polarities or the motion phase offsets 180 degrees apart, thus producing two images with the same static phase information but negative MRE motion-induced phase. Subtracting the two images reduces or removes the static phase information while doubling the MRE motion-induced signal, thus making the images easier to process when calculating the tissue stiffness. Fig. 6.2 illustrates the motion encoding performed in MRE.

In clinical liver MRE examinations, a small number of transverse (axial) images are acquired and the tissue motion (which in general has components in all three directions) is encoded along only one direction. In carefully controlled setups and large homogeneous organs, such as the liver, this is often sufficient to calculate the stiffness accurately and reliably. However, in more complex structures and

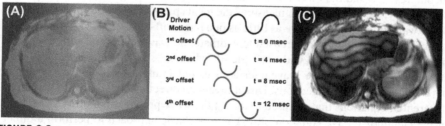

FIGURE 6.2

Imaging waves. (A) Magnetic gradient is applied across the target structure. Two opposite-sign gradients are applied sequentially so that stationary objects acquire zero net phase and moving objects acquire a phase proportional to displacement. (B) Motion encoding is applied in several passes with different offsets with respect to driver motion in order to image waves at different time points in the propagation cycle. (C) Example of encoded motion from a single offset.

deeper tissues the wave polarization may change and the wave propagation direction may not be just within the imaging plane. To avoid obtaining a biased stiffness when imaging these organs, an MRE acquisition with a more isotropic resolution in all three dimensions (3D) (e.g., using a 3D MR acquisition or a two dimensional [2D] acquisition with many thin slices) and motion encoding in all three directions should be used. The combination of 3D sampling of the spatial information, vector motion encoding, and 3D analysis of the images is often simply called 3D MRE and is the preferable technique to use for imaging small or complicated organs. The data sampling requirements (and thus the acquisition and processing time) for 3D vector MRE can be significantly more than those for 2D scalar MRE. Therefore small-volume imaging, acquisition acceleration (e.g., echo-planar and parallel imaging), and faster inversions are often performed to keep the acquisition and processing times reasonable.

3. Inversions

3.1 Overview of inversions and processing

The inversion algorithm used to invert the 2D MRE data and calculate the stiffness for the standard clinical hepatic MRE protocol is referred to as the multimodel direct inversion (MMDI) [24]. A multitude of other inversions have been used for various applications over the years, including phase-gradient, local frequency estimation (LFE), direct inversion, variational or integral, and finite-element-based inversions [25–27]. The general equations of motion for tissues, even those undergoing simple harmonic motion, can be very complex, as tissues can have heterogeneous, anisotropic, dispersive, viscoelastic, and nonlinear properties. Typically, to make the problem tractable and solvable in a reasonable amount of time, most inversions make assumptions such as that the tissue is isotropic, locally homogeneous, or incompressible and thus work best in large homogeneous organs, such as the liver. Very briefly, the LFE inversion measures the local spatial frequencies of the waves in MRE wave images and can use as little as one phase offset [28,29]. The local spatial frequency (wavelength) is then used to estimate the wave speed and stiffness of the tissue. When implemented with multiple filter bands, as is typically done, it is very stable with respect to noise, wave interference, and a range of stiffness values and is popular in emerging applications. Direct inversion directly solves one or a system of differential equations that are used to model the wave propagation, often a variation of the Helmholtz wave equation [30]. The most basic approach to perform a direct inversion is susceptible to produce numerically unstable results in noisy data, so smoothing the wave images, regularizing the inversion, and/or median filtering the results are often done as well. The MMDI [24], used for clinical MRE, uses local polynomial fits to smooth the data, stabilize the inversion, and assess the confidence in the calculated stiffness values and has become a robust tool for liver MRE. Finally, several different implementations of direct and iterative finite-element-based inversions have been developed over the years [20,27,31]. These methods can be more powerful than the other methods because they can handle complex factors such as boundary effects, poroelasticity, and heterogeneity and have been most commonly used for brain MRE [27,32,33]. A comparison for different inversion methods is shown in Fig. 6.3.

Before MRE phase images can be inverted to calculate the stiffness, they typically undergo some degree of preprocessing. Phase unwrapping is often required in MRE and other fields to remove discontinuities in the phase information ("phase wraps") that arise due to the ambiguity of only being able to define the phase angle of a complex-valued quantity within the range of $-\pi$ to π radian range of the sinusoidal spin precession. Directional filtering is often done to separate the measured wave field into multiple wave fields, each with waves propagating predominantly in one direction to reduce the effect of wave interference on the results [34]. Inversions can be applied to a single directionally filtered dataset to generate an estimate of the stiffness distribution in the tissue (elastograms), the directionally filtered wave images for filters oriented in different directions can be individually inverted and the resulting elastograms averaged together to produce the final elastograms, or the wave information from all the different directionally filtered datasets can be used simultaneously to solve a system of equations to produce a single elastogram [35]. Longitudinal (compressional) waves, which have long wavelengths and are not useful for typical in vivo tissue characterization, can be removed using either spatial frequency filtering or, in the case of data with 3D vector motion encoding, curl processing or the Helmholtz decomposition [36]. Some type of smoothing is also generally used either in the inversion itself (e.g., polynomial fitting in MMDI, intrinsic bandpass filtering in LFE, and adding low-pass filtering to the directional filters), on the input phase data (e.g., Gaussian or other low-pass filtering), or on the resulting elastogram (e.g., median filtering of direct inversion

	Resolution	Noise stability	Speed	Application	Reference
LFE^e	avg	↑	↑	Liver, tumors	28, 29
DI^e	↑	↓	↑	Brain, heart	26
MMDI	↓	↑	↑	Liver	24
PG^e	↑	↓	↑	Muscle	25
NLI	↑*	↑*	↓	Brain, heart	27, 32, 33
NN	avg*	↑	avg	Liver, Tumors	91

FIGURE 6.3

Comparison of properties and applications of the most common inversions: local frequency estimation (LFE), direct inversion (DI), multimodel direct inversion (MMDI), phase gradient (PG), nonlinear inversion (NLI), and neural network inversion (NNI). The performance marked with * depends on how heavily the network is regularized. ^e marks the inversions most commonly used in emerging applications.

elastograms) [25]. Processed phase images are commonly displayed instead of the raw phase data and are called "wave images." Finally, these wave images are processed by the inversion algorithm to calculate the elastograms with quantitative stiffness values at every pixel. The elastograms are often displayed in a continuous red, green, blue color scheme to intuitively communicate the difference between normal and abnormal tissues. For example, clinical liver MRE displays the stiffness elastograms using a range of 0—8 kPa such that healthy liver tissue stiffness values are in blue-green and fibrotic or cirrhotic liver tissues are yellow-red. Staging is performed by comparing calculating quantitative mean liver stiffness values from artifact-free regions-of-interest (ROIs) and comparing them against established thresholds. By contrast, in some emerging applications of MRE in complex tissues, which violate the assumptions of inversions, the stiffness information must be interpreted as semi-quantitative. Still, the values may be reproducible and diagnostically valuable.

3.2 Magnetic resonance elastographic outputs

The MRE acquisition produces magnitude images, which show anatomic information through typical MRI contrast mechanisms, and phase images, which contain the MRE motion information. The acquired phase images are typically the result of the phase-subtraction described earlier, which removes background artifacts. The images are typically phase-unwrapped, smoothed, and interpolated before being displayed to a user as the so-called "wave images." In the clinical 2D liver MRE processing pipeline, these images are often displayed using a red-blue map. From these data, the inversion algorithm produces the elastograms, containing quantitative stiffness information at every voxel and, usually, a wave-quality confidence metric. Fig. 6.4 shows the outputs of a typical MRE acquisition and inversion.

   Although the clinical application of liver MRE has become highly standardized with similar imaging parameters and the same inversion being used by all vendors, research applications have used a number of different techniques and reported a number of different tissue mechanical properties. Although many dynamic elastography publications, both MRE and ultrasound based, report a "stiffness" value, this word may refer to different parameters. The "stiffness" reported from clinical liver MRE examinations using the MMDI is the magnitude of the complex-valued shear modulus at the frequency of vibration (standardized to be 60 Hz):

$$G = \sqrt{G'^2 + G''^2}$$

where G' is the storage modulus (the real part of the complex shear modulus) describing the tissue's elastic and immediate response to motion and G'' is the loss modulus (the imaginary part of the complex shear modulus) describing the viscous properties characterized by a phase-delayed response to motion and attenuation. G, G', and G'' are in units of pascals (commonly reported in kilopascals, or kPa). MMDI, like many inversion algorithms, calculates the complex shear modulus initially, but then reports the magnitude as the stiffness. The quantities reported

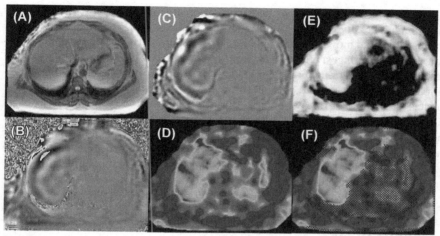

FIGURE 6.4

Magnetic resonance elastography outputs. (A) The acquired magnitude image, containing anatomic information, and the phase image (B), containing motion-encoding information. (C) A postprocessed wave image that underwent phase unwrapping and interpolation (smoothing). The wave image is often shown in color as seen in other figures. (D) The elastogram that contains quantitative stiffness values at every voxel and is calculated form the wave image. (E) A confidence map that indicates wave amplitude/smoothness. (F) The elastogram masked by the thresholded confidence map. (C—F) are results of the multimodel direct inversion. Other inversions typically have analogous outputs.

using ultrasound elastography are typically the Young's modulus (which is nearly equal to 3*G* for biological tissues because most tissues are nearly incompressible [i.e., Poisson's ratio ≈ 0.5]), or the shear wave speed. The LFE and phase-gradient inversions, on the other hand, assume that a tissue has no attenuation ($G'' = 0$) and use estimates of local shear wavelengths to calculate the storage modulus, which can be written as

$$G' = v^2\rho = \rho(\lambda f)^2$$

where v is the speed of the shear wave in the tissue, ρ is the tissue density (assumed to be about 1000 kg/m^3 with rare exceptions, such as the lungs), λ is the spatial frequency, and f is the frequency of applied motion.

The viscoelastic moduli increase with the frequency of vibration (a phenomenon called dispersion), so only stiffness values calculated at similar frequencies should be directly compared. By acquiring multifrequency data, a presumed mechanical model of the tissue (e.g., the Voigt model or a spring-pot power-law model) may be fit to the measured moduli or stiffness values to calculate frequency-independent tissue mechanical properties that may have diagnostic value for certain diseases. Both direct inversions and finite-element inversions can be modified to incorporate porosity or poroelasticity into the tissue model [31,37,38]. This model accounts for fluid movement through the extracellular matrix during the applied motion. It has the strongest effect at low frequencies (<10 Hz) and may be important in porous tissues such as the brain and kidneys. If MRE is performed under these conditions, taking porosity into account in the model may allow for more accurate estimates of tissue stiffness, or the porosity information itself may also be diagnostically valuable. Other quantities that have been reported in the literature include the complex shear modulus, loss tangent and phase of the complex shear modulus [39], shear wave attenuation [40], and frequency dispersion rate of these quantities [41]. The value of all of these parameters for specific applications continues to be investigated.

4. Applications

MRE has a multitude of applications, which illustrates the versatility of the technique. Liver MRE is FDA-cleared, highly standardized across all vendors, highly reproducible, and used by over 1000 institutions to diagnose and stage liver disease. Brain MRE has been working reliably in humans (although not yet cleared for commercialization) and has shown promising results in degenerative diseases, traumatic brain injury, treatment planning, and the study of aging. It is being advanced by several groups, using a variety of technical methods. On the hand, applications in tumors, skeletal muscle, heart, and lung face a variety of unique challenges but have, nonetheless, been developed to a level that shows initial promising results.

4.1 Liver

The liver MRE protocol used for fibrosis staging is highly standardized. It uses an acoustic driver system operating at 60 Hz with a passive drum driver placed over the liver, a 2D gradient echo MRE acquisition, and MMDI for processing. A typical examination obtains four slices using four 14-second breath holds, with four phase offsets per slice. In biopsy, only 1/50,000th of the liver is sampled, whereas a typical four-slice MRE examination samples between 30% and 70% of the hepatic tissue. Although 3D liver MRE is available [39,42], when the driver is placed over the liver and the imaging is performed in the middle of the liver, through-plane wave propagation is not a major problem and the simpler 2D MRE approach works well, with high reproducibility and low failure rate [4,43,44]. Liver stiffness is reported from an ROI selected by experienced readers, which avoids blood vessels, inversion hot spots, and partial volume effects. A confidence map, produced by the MMDI and having a threshold value of 0.95, is used to help guide the ROI selection. To help standardize this process, a fully automated ROI selection algorithm for reporting liver stiffness has also been developed [45]. It is used for patient examinations at the Mayo Clinic as well as for some research studies [46–49]. Examples of hepatic MRE examinations of a normal and a cirrhotic liver are shown in Fig. 6.5.

The average liver stiffness for the healthy population has been calculated to be 2.1–2.3 kPa at 60 Hz [50–52]. Sex, age, and body mass index do not appear to affect liver stiffness. The diagnostic threshold for separating healthy liver from stage 1 fibrosis is 2.93 kPa [53], although stiffness values above 2.5 kPa may indicate earlier inflammatory changes. Different studies have generated different stiffness thresholds for grading higher stages of fibrosis [3,4,53], although grading higher stage fibrosis is less diagnostically important than screening for the presence of fibrosis or following stiffness/fibrosis changes of a patient longitudinally during disease progression or treatment response. An example of stiffness thresholds for different stages of fibrosis is shown in Fig. 6.6. MRE elastograms for the liver are always displayed using a 0to 8-kPa colormap, standardized across all vendors, in which healthy livers appear blue-green and fibrotic livers are yellow-red. As liver MRE is standardized across vendors, the values obtained can be compared directly [54].

Liver MRE has been shown to have the highest accuracy (with respect to biopsy) and highest reproducibility of staging hepatic fibrosis, as well as a comparatively

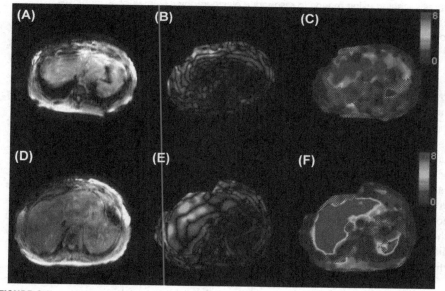

FIGURE 6.5

Comparison of magnetic resonance elastographic images from patients with normal (A–C) and cirrhotic (D–F) livers. (A, D) Transverse magnitude (anatomic) images of the abdomen. (B, E) Wave images showing the 60-Hz wave propagation in the liver and the abdomen. (C, F) Elastograms calculated using the multimodel direct algorithm, with quantitative stiffness values reported at every pixel in kilopascals. The checkerboard pattern indicates the low-confidence area to be excluded due to poor wave propagation.

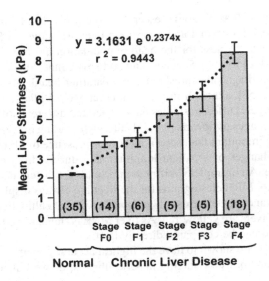

FIGURE 6.6

Fibrosis staging using magnetic resonance elastography. Liver stiffness increases progressively with fibrosis stage in many chronic liver diseases. The separation between normal liver tissue, mild fibrosis, and severe fibrosis can be performed effectively.

Reprinted from M. Yin, J.A. Talwalkar, K.J. Glaser et al. Assessment of hepatic fibrosis with magnetic resonance elastography. Clin. Gastroenterol. Hepatol. 5 (2007):1207–13.

low failure rate of 6%. The coefficient of variation of the hepatic stiffness measurement with MRE is approximately 10% [24,44,55], and the QIBA consensus recommends that a stiffness change of 19% or more be considered a true physiologic change [14]. Hepatic inflammation [46] and increased portal venous blood pressure, caused by meal consumption [56] or disease, have been observed to affect liver stiffness. For these reasons, longitudinal stiffness changes of at least ±0.5 kPa are typically considered significant. MRE has been used to stage fibrosis caused by a wide range of chronic liver diseases including hepatitis B [6], hepatitis C [51], nonalcoholic steatohepatitis (NASH) [57], and primary sclerosing cholangitis [48]. It is also showing promise for monitoring liver transplant acceptance [58] and for indirectly diagnosing pericarditis [49] wherein cardiac disease causes blood-pressure-induced changes in liver stiffness.

Recently, liver MRE research has focused on using 3D to improve reproducibility, calculating additional parameters, and combining multiparametric measurements to increase diagnostic value. One study has demonstrated the ability to predict NASH using a combination of stiffness and damping ratio calculated with 3D MRE acquired at two motion frequencies (40 and 60 Hz) and proton density fat fraction calculated from MRI (e.g., IDEAL or mDixon). This parallels the standard biopsy-based evaluation, which evaluates fibrosis, inflammation, and fat deposition in the liver. The three parameters were fit into a linear regression model to calculate a predicted NASH score [59]. A similar study also demonstrated the ability to predict nonalcoholic fatty liver disease in mice using single-frequency two-parameter MRE (shear stiffness, $|G*|$, and loss modulus, $|G''|$) in combination with fat fraction [60].

4.2 Brain

The most common approach for performing brain MRE has involved either vibrating the whole head by placing one or more small drum or pillow drivers under the head or vibrating the cradle in which the head is resting. Shear waves are then generated by the skull itself, so the shielding by the skull is not a problem in brain MRE as in other types of elastography (Fig. 6.7). The regional stiffness distribution of the brain has been investigated by multiple groups either using gray matter-white matter segmentation or atlas-based registration of the MRE magnitude images [61,62] (Table 6.1). The differences in the measured values are largely attributable to the use of different inversion models and vibration frequencies. Stiffness differences with gender and a progressive decrease with age have been observed, so subjects should be matched for any comparative studies [62,63].

One application of brain MRE is for the early diagnosis and progression monitoring of diffuse diseases such as MS, Alzheimer disease (AD), normal pressure hydrocephalus (NPH), and autoimmune encephalomyelitis (AE). Studies of patients with moderately advanced MS have observed a significant global reduction in brain stiffness (-20%) with respect to normal subjects [64]. If remyelination occurred restoring a normal state, the stiffness was also found to return to normal [68]. In AD, the brain stiffness in both humans [69] and animal mouse models [70] was found to be reduced with respect to normal. Similarly, in patients with NPH [71] and in a mouse model with AE [72], brain stiffness was found to be lower than normal in the affected brain, but the stiffness increased toward the normal value after

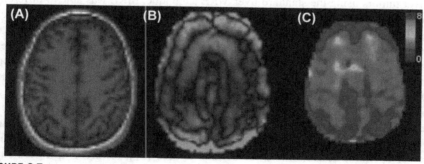

FIGURE 6.7

Demonstration of brain magnetic resonance elastography performed in a normal subject using 60-Hz vibrations. (A) The transverse magnitude (anatomic) image. (B) A wave image showing shear waves penetrating the entire brain. (C) The elastogram calculated from the curl of the wave field using a direct inversion algorithm.

Table 6.1 Summary of stiffnesses for different regions of the brain reported in current studies. The studies use different frequencies and inversions which contributes to the difference in observed numbers.

No. of subjects	Frequency (Hz)	Stiffness (kPa)	Quantity calculated	Reference
38	25–62.5	WB: 3.3	$\|G\|$	64
10	60	WB: 2.99, frontal: 3.15, occipital: 3.21, parietal: 2.87, temporal: 3.17 Deep GM/WM: 3.41 CB: 2.38	$\|G\|$	65
8	80	WM: $2.41 + i1.21$	$G' + iG''$	66
5	90	GM: $3.1 + i2.5$, WM: $2.7 + i2.5$	$G' + iG''$	67

CB, *cerebellum;* GM, *gray matter,* WB, *whole brain;* WM, *white matter* "i" *indicates imaginary part obtained from the complex-valued inversion.*

treatment. Degenerative diseases impairing motor function have also been observed to reduce regional brain stiffness, with Parkinson disease affecting the frontal and mesencephalic regions [73] and cerebral palsy reducing stiffness but increasing the damping ratio of gray matter in children [74]. More studies are required to investigate whether MRE-based diagnosis is superior to, or adds value when combined with, existing diagnostic methods.

4.3 Tumors

Tumor MRE has been performed with standard whole-organ imaging protocols as well as dedicated protocols, particularly for animal studies. In a preliminary study of liver tumors using the standard liver MRE technique, MRE was able to differentiate benign tumors from malignant tumors with 100% accuracy [52], while another study found that MRE had superior accuracy to diffusion-weighted MRI [75]. In the breast, MRE studies in the 65—100 Hz range also showed significant differences in several viscoelastic tissue parameters between normal tissue and benign and malignant tumors, including the storage modulus, loss modulus, attenuation, and power-law exponent. Of these, the storage modulus generally allowed for the best separation [76,77]. When compared to contrast-enhanced MRI, which has a good sensitivity for detecting breast tumors, MRE had the same accuracy, while the combination of the two techniques had improved specificity, raising the area under the receiver operating characteristic curve value to 0.96 [77]. As precompression can affect tissue stiffness, an indirect driving method, such as placing a narrow acoustic driver on the sternum [78], may be preferable for superficial tumors and tissues. In the brain, a standard brain MRE setup has also been used to aid in surgical planning for removing brain tumors. Soft meningiomas (Fig. 6.8A), for example, are easily

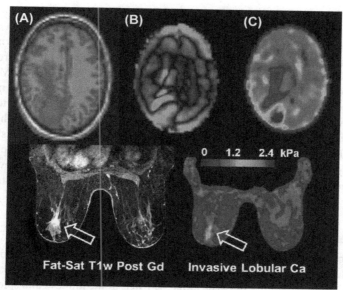

FIGURE 6.8

Tumor magnetic resonance elastography (MRE). (A–C) MRE was used to calculate the stiffness of a meningioma for surgical planning. The red area (round white outline, in print) indicates the tumor is significantly stiffer than the surrounding tissues, while the purple area (dark-grey, in print) corresponds to a softer region of edema. (D,E) MRE was used to calculate the stiffness of a breast cancer. The breast indicated with arrows cancer was found to be stiffer than the surrounding fibroglandular tissue.

(A–C) Courtesy of Drs. Arvin Arani and John Houston. (D,E) Courtesy of Dr. Jun Chen.

resected from the brain via suctioning, whereas stiff meningiomas require a lengthy operation. Brain MRE allows tumor consistency to be assessed in advance and has been found valuable by brain surgeons to predict the time and tools/experience required for the procedure [79]. MRE has also been used to evaluate schwannomas in the brain [80]—the study found that shear stiffness significantly decreased with tumor grade. Additionally, the IDH1 gene mutation, associated with improved outcome, led to a significantly reduced stiffness compared with the wild type. The authors point to the possibility of using MRE to predict genetics to improve patient stratification; however, multiple parameters will be needed to do so in parallel with predicting the tumor grade.

The vibration frequencies used in these MRE studies are typically in the range of 50–100 Hz, while many of the tumors were 1–4 cm in size. As a result, the calculated stiffness of the tumors was likely affected by the tumor geometry and size. To improve the quantitative characterization of small tumors and early diagnosis, and to more accurately track tumor stiffness changes over time, much higher frequencies need to be used. A dedicated piezoelectric or inertial driver is generally required to deliver waves of sufficient amplitude at these frequencies; however, only superficial structures can be imaged this way because of the high attenuation. A study of implanted brain tumors in mice with vibrations delivered at 1000 Hz via a bite-bar transducer was able to segregate several different types of tumors from each other and from normal tissue. Two other studies evaluated the ability of MRE performed at 800 [81] and 1000 Hz [82] to detect chemotherapy response in mice and found a significantly different response in the treatment group days after the start of treatment, which was significantly earlier than the response detected with volume changes or diffusion-based MRI. The images of the mouse tumor are shown in Fig. 6.8B. The ability to predict treatment response early has significant implications for reducing side effects to patients and for performing clinical trials. However, additional studies are needed to evaluate the ability of MRE to predict the final treatment outcome.

4.4 Other organs

MRE has been applied to a wide variety of other organs each with their own challenges, solutions, and applications, including the kidneys, spleen, pancreas, lungs, and muscles. The feasibility of performing 3D MRE of the kidneys and spleen using 30- to 90-Hz vibrations has been demonstrated in healthy volunteers. Furthermore, it was found that the stiffness of the spleen and kidneys changes significantly with changes in blood pressure in the aorta and portal vein, respectively, which may be useful diagnostically or as a way of removing blood pressure as a confounder from the liver MRE, or another examination. Preliminary work has also shown the possibility of performing lung MRE, which is a highly challenging application because of the oxygen-related MRI signal loss and low tissue density. Special spin-echo sequences with short echo times [83] were developed for this application, as they experience significantly less signal loss. Breath holds are used to minimize motion artifacts, and a proton-weighted imaging sequence is used to calculate tissue density, which varies during the respiratory cycle (from about 0.08 to 0.5 g/mL) and is necessary for an accurate inversion. The stiffness of healthy lungs in several subjects using 50-Hz vibrations was measured to vary between 0.95 kPa at full expiration and 1.5 kPa at full inspiration [84].

The mechanical properties of skeletal and cardiac muscles are important physiologically and in the progression, diagnosis, and treatment of many diseases. The stiffness of muscles is known to change substantially during passive stretching and active contraction. Furthermore, muscular atrophy, necrosis, neuromuscular diseases, and scarring are all known to result in changes in stiffness [85] and loss of function. Preliminary MRE studies have focused mainly on establishing the normal stiffness range for various skeletal muscles and obtaining preliminary information about muscles affected by specific diseases. Muscle is a markedly anisotropic tissue with greater stiffness along the fibers than across them [86]. Several different image acquisition and analysis methods have been developed for muscle MRE. These techniques include acquiring single 2D image planes aligned with the muscle fibers and extracting one-dimensional profiles of the wave propagation along the fibers from which to estimate the shear wavelength, wave speed, and stiffness; applying 2D isotropic tissue inversions to 2D MRE data; and using knowledge of the fiber orientations in 3D MRE acquisitions to perform inversions based on anisotropic equations of wave propagation [87]. Cardiac MRE is especially challenging because of the dynamic motion of the heart, its changing mechanical properties during the cardiac cycle, and the thin, anisotropic nature of the cardiac muscle. Preliminary studies in normal volunteers and patients have demonstrated that waves at mechanical frequencies ranging from 25 to 220 Hz can be introduced into the heart and imaged with breath-held, cardiac-gated gradient-recalled echo and spin-echo echo planar imaging acquisitions, and new imaging and image processing techniques are constantly being explored to improve the quality and accuracy of these methods.

5. Artifacts and quality control

The most extensive study of issues related to MRE quality or success has been a retrospective review of 1377 clinical liver MRE cases by Yin et al. [43]. The failure rate was found to be fairly low at 5.6%, with the primary cause of failure being attributed to high iron content (and thus low SNR) in the livers of some patients. Since then, specialized spin-echo-based acquisitions with short echo times have been adopted from lung MRE and can be used to image many of these patients effectively [47]. The remaining failures observed in this study were mostly attributed to poor breath holds causing image artifacts (a problem that can be partly addressed by repeating the acquisition and coaching the patient about the proper breath-holding technique) and to the passive driver being disconnected from the active driver or not attached being securely to the patient (issues that can be addressed by carefully checking the setup of the driver). Liver MRE can be performed at 1.5T or 3T, with the latter not only providing higher SNR but also being more susceptible to SNR loss due to iron overload. For this reason, spin-echo-based MRE sequences are preferred at 3T. For example, the GE implementation of MRE (called MR-Touch) uses a gradient echo acquisition, with higher SNR for the same acquisition time as well as higher iron sensitivity, at 1.5T but a spin-echo acquisition at 3T.

MRE failure due to low SNR is detectable by an uncharacteristically dark MRE anatomic magnitude image and noisy phase images. Such SNR problems typically can only be addressed by performing the acquisition again using a spin-echo MRE sequence and/or reducing the echo time. Although these changes may change the anatomic contrast in the magnitude images, this does not negatively impact the inversion as long as the wave images have sufficient SNR for analysis. If the passive driver becomes disconnected (and thus no motion is being produced), this is identifiable via phase/wave images that are very flat and smooth (neither much noise nor waves). All driver connections and the coupling of the driver to the patient should be checked and the acquisition repeated. For large or obese patients, the driver amplitude may be too low for the waves to penetrate all the way into the tissue of interest, in which case the driver amplitude may need to be increased and the acquisition repeated. For small patients, if the phase images have too much phase wrap (which can cause inversion artifacts) or if the magnitude images have dark bands near the driver (causing low SNR due to intravoxel phase dispersion), the amplitude should be reduced. If the wave amplitude is high at the body wall but low in the target organ, the cause may be due to poor driver coupling (in which case the driver should be adjusted and secured more firmly) or the significant wave attenuation at that vibration frequency (in which case the vibration frequency may have to be reduced, in a research application). As MRE inversions are biased by noise, noisy images or low-amplitude wave images cannot be used to calculate the quantitative stiffness. Some inversion algorithms, such as the MMDI, provide estimates of the wave quality that indicates areas with low SNR and can be used to guide ROI selection. Illustrations of SNR-, driver-, and breath-hold-related liver MRE failures are shown in Fig. 6.9.

Partial volume effects should also be avoided whenever possible when analyzing MRE images. Most MRE inversion processing uses multivoxel kernels to calculate the elastogram (at least 3 pixels in each dimension if little or no filtering is used, but as many as 11 pixels in the case of the liver MMDI algorithm). Therefore the effective point spread function or resolution of the processing can be much larger than that in the original acquired images, which increases the impact of partial volume effects on the elastograms. When creating MRE ROIs from which to report the final tissue stiffness, the ROIs should be eroded from any detectable organ edges and inclusions (e.g., blood vessels) by the extent of the inversion processing kernel. As most inversions assume a degree of local tissue homogeneity, very localized areas of increased or decreased stiffness are likely artifactual and should also be avoided. These may occur either due to unresolvable wave patterns (such as wave interference near boundaries or too much phase wrap) or due to the presence of a tumor, vessel, or another anatomic structure that is not easily identifiable in the MRE magnitude images (which are not optimized for anatomic contrast). In 2D MRE, out-of-plane wave propagation may cause a bias that increases the apparent wavelength, and so the calculated stiffness, of organs. The driver should be carefully placed over the center of the organ, whenever possible, especially along the direction of motion encoding, to minimize this effect. Reliable detection of through-plane wave propagation from a completed 2D MRE examination is very challenging. However, a monotonic stiffness change across slices may indicate this issue. Finally, as MRE uses continuous rather than transient vibrations, standing wave patterns may develop when the attenuation is low or the organ is small compared with the wavelengths and incident waves that interfere with reflected waves. Inversions that include some form of directional filtering are generally robust to such effects.

In clinical liver MRE, trained technicians review all magnitude, phase, and elasticity images to select an ROI that avoids extraneous tissues, partial volume effects, and sharp features in the phase and elasticity images but contains as much liver tissue as possible. The MMDI confidence map with a 95% confidence threshold is used to guide the ROI selection as well as to mask out noisy areas from the final ROI. In

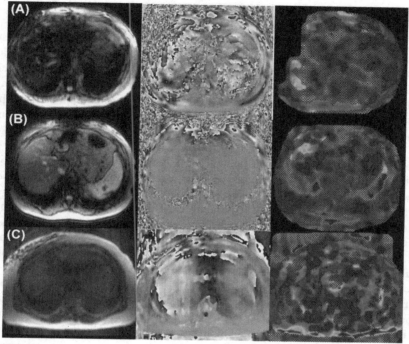

FIGURE 6.9

Examples of magnetic resonance elastographic (MRE) examination failures. (A) Iron overload: Significant iron accumulation in the liver resulted in low magnetic resonance (MR) signal and noisy MRE phase/wave images, which produced an unreliable elastogram (as indicated by most of the liver being masked out by the multimodel direct inversion confidence mask in the elastogram). (B) Poor driver setup: The driver was placed medially too far away from the liver and was not secured firmly to the abdomen. Although some motion was produced in the immediate vicinity of the driver, there was very little motion in the liver itself. So even though the liver had good MR signal, the elastogram has low confidence. (C) Poor breath hold: The patient was not able to hold breath during the acquisition causing significant ghosting artifacts throughout the images that blurred and distorted the MRE phase/wave information. Because of blurring, the apparent signal-to-noise ratio is improved and some portions of the liver have high confidence; however, in reality, none of the wave information in the liver was satisfactory. Failed breath hold caused magnetic resonance imaging ghosting, making the tissue contribute to more than one voxel and blurring the phase image.

the brain, measurements using partial volume exclusion [65], low-SNR area exclusion (e.g., based on octahedral shear strain SNR [88]), and registration to a brain atlas [61,65] to report stiffness have been used. For each new MRE application, the criteria for data quality control must be reevaluated and established through repeatability and reproducibility studies.

6. Summary and conclusions

MRE is a powerful method that allows for the measurement of a number of tissue mechanical properties in a wide variety of organs. It has several important advantages, including the ability to calculate these mechanical quantities quantitatively, the ability to image shielded and deeply situated structures, and the ability to image whole organs and encode motion in three planes. More established applications (specifically liver and brain MRE) have become largely operator independent, with only the hardware setup needing to be verified. However, the stiffness reporting step (i.e., ROI selection) is still often performed by human readers. The reproducibility of these measurements is fairly high (within 10%), however, and the established liver and brain applications are moving toward automated and semiautomated analyses.

MRE of the liver is well established, highly standardized, and available at approximately 1400 institutions. It is being used to stage the degree of fibrosis and monitor treatment response. Hepatic fibrosis can be caused by many diseases and conditions including hepatitis, fatty liver disease, pharmacologic toxicity, and cholangitis. Studies are underway to determine if different diagnostic thresholds should be used for different fibrosis causes and to see if MRE can be used as an effective screening method for these diseases. As the disease cause may impact the relationship between disease severity and stiffness change from normal, and because there are confounding factors that affect tissue stiffness (e.g., blood pressure and inflammation), MRE results should always be interpreted in the context of patient history.

Brain MRE, although not yet FDA approved, also uses fairly established procedures and has a high success rate. Preliminary work has established the normal stiffness of various regions of the brain as well as normal differences between the genders and changes with age. Preliminary utility has also been shown for both diffuse diseases, such as MS, AD, and hydrocephalus, and for focal lesions. In other organs the methods are far more variable and the procedures for image acquisition, inversion, and stiffness measurement are still being developed. Significant differences between diseased and normal tissues, and between benign and malignant tumors, have been observed in many preliminary studies. However, additional validation studies and a comparison of the diagnostic effectiveness of MRE alone and that of MRE in combination with existing diagnostic techniques is a topic for future work.

Recent research has also investigated some radically different approaches to MRE. Rather than calculating a mechanical property of a single tissue, Yin et al. [89] have proposed a technique for measuring adhesion between tissues called the slip interface. The technique relies on the intravoxel phase dispersion artifact that occurs in MRE when motion causes phases of spins within a voxel to dephase, leading to signal dropout. The authors normalize the data to unit magnitude, enhance the effect by reducing resolution to make larger effective voxels, and quantify the amount of dropout. The technique has been evaluated for determining tumor adhesion to aid in preoperative planning and for measuring the skull-to-brain adhesion with the eventual goal of predicting traumatic brain injury [90]. A promising novel approach to stiffness inversion has also been recently proposed—the authors optimized a neural network to calculate elastograms by training on simulated phase data with small 1-cm inclusions and noise. The network showed higher stiffness accuracy than the direct inversion used by the same group for brain MRE, particularly under increasing noise conditions. In an in vivo study with 88 liver examinations, the neural network also showed a high correlation with the clinical MMDI and a comparable agreement with biopsy for predicting the fibrosis stage [91].

Many tissue mechanical parameters, such as the storage modulus, porosity, loss modulus, power-law exponent, anisotropy, and attenuation can be calculated from MRE examinations, particularly using 3D and multifrequency acquisitions. Currently, the storage modulus and the magnitude of the complex shear modulus appear to yield the best predictive power for most applications and many studies report one of these values. The use of a single parameter simplifies the use of MRE information in the clinical environment. However, the use of additional parameters may have a significant clinical value in differentiating tumors (which have highly differing extracellular matrix and cell properties) or for differentiating fibrotic, inflammatory, and hydrostatic tissue changes. MRE has proven to be a versatile, powerful, accurate, and reproducible tool that can deliver important and diverse information about the mechanical properties of a wide variety of organs and diseases as well as information that will allow us to improve disease diagnosis and refine new treatment methods.

References

[1] E. Tsochatzis, K. Gurusamy, S. Ntaoula, E. Cholongitas, B. Davidson, A. Burroughs, Elastography for the diagnosis of severity of fibrosis in chronic liver disease: a meta-analysis of diagnostic accuracy, J. Hepatol. 54 (4) (2011) 650–659.

[2] J. Nierhoff, A. Chavez Ortiz, E. Herrmann, S. Zeuzem, M. Friedrich-Rust, The efficiency of acoustic radiation force impulse imaging for the staging of liver fibrosis: a meta-analysis, Eur. Radiol. 23 (11) (2013) 3040–3053.

[3] S. Ichikawa, U. Motosugi, H. Morisaka, et al., Comparison of the diagnostic accuracies of magnetic resonance elastography and transient elastography for hepatic fibrosis, Magn. Reson. Imaging 33 (1) (2015) 26–30.

[4] L. Huwart, C. Sempoux, E. Vicaut, et al., Magnetic resonance elastography for the noninvasive staging of liver fibrosis, Gastroenterology 135 (1) (2008) 32–40.

[5] J. Yoon, J. Lee, I. Joo, et al., Hepatic fibrosis: prospective comparison of MR elastography and US shear-wave elastography for evaluation, Radiology 273 (3) (2014) 772–782.

[6] S. Ichikawa, U. Motosugi, H. Morisaka, et al., Validity and reliability of magnetic resonance elastography for staging hepatic fibrosis in patients with chronic hepatitis B, Magn. Reson. Med. Sci. 14 (3) (2015) 211–221.

[7] J. Cui, B. Ang, W. Haufe, et al., Comparative diagnostic accuracy of magnetic resonance elastography vs. eight clinical prediction rules for non-invasive diagnosis of

advanced fibrosis in biopsy-proven non-alcoholic fatty liver disease: a prospective study, Aliment. Pharmacol. Ther. 41 (12) (2015) 1271—1280.

[8] B. Taouli, A. Tolia, M. Losada, et al., Diffusion-weighted MRI for quantification of liver fibrosis: preliminary experience, Am. J. Roentgenol. 189 (2007) 799—806.

[9] G. Xiao, S. Zhu, X. Xiao, L. Yan, J. Yang, G. Wu, Comparison of laboratory tests, ultrasound, or magnetic resonance elastography to detect fibrosis in patients with nonalcoholic fatty liver disease: A meta-analysis, Hepatology 66 (5) (2017) 1486—1501.

[10] V. Ratziu, F. Charlotte, A. Heurtier, et al., Sampling variability of liver biopsy in nonalcoholic fatty liver disease, Gastroenterology 128 (7) (2005) 1898—1906.

[11] A. Regev, M. Berho, L.J. Jeffers, et al., Sampling error and intraobserver variation in liver biopsy in patients with chronic HCV infection, Am. J. Gastroenterol. 97 (10) (2002) 2614—2618.

[12] K.J. Glaser, A. Manduca, R.L. Ehman, Review of MR elastography applications and recent development, J. Magn. Reson. Imaging 36 (2012) 757—774.

[13] S. Venkatesh, J. Talwalkar, When and how to use magnetic resonance elastography for patients with liver disease in clinical practice, Am. J. Gastroenterol. 113 (2018) 913—926.

[14] QIBA MR Elastography of the Liver Biomarker Committee. Magnetic Resonance Elastography of the Liver. Version Profile Stage: Consensus. QIBA, May 2, 2018.

[15] R. Muthupillai, R.L. Ehman, Magnetic resonance elastography, Nat. Med. 2 (5) (1996) 601—603.

[16] Z.T.H. Tse, H. Janssen, A. Hamed, M. Ristic, I. Young, M. Lamperth, Magnetic resonance elastography hardware design: a survey, Proc. Inst. Mech. Eng. H 223 (4) (2009) 497—514.

[17] S.I. Ringleb, Q. Chen, D.S. Lake, A. Manduca, R.L. Ehman, K.N. An, Quantitative shear wave magnetic resonance elastography: comparison to a dynamic shear material test, Magn. Reson. Med. 53 (2005) 1197—1201.

[18] S. Arunachalam, P. Rossman, A. Arani, et al., Quantitative 3D magnetic resonance elastography: comparison with dynamic mechanical analysis, Magn. Reson. Med. 77 (3) (2017) 1184—1192.

[19] P. Song, D. Mellema, S. Sheedy, et al., Performance of 2-D ultrasound shear wave elastography in liver fibrosis detection using magnetic resonance elastography as the reference standard: a pilot study, J. Ultrasound Med. 35 (2) (2016) 401—412.

[20] D. Fovargue, D. Nordsletten, R. Sinkus, Stiffness reconstruction methods for MR elastography, NMR Biomed. (2018) [E-pub ahead of print].

[21] Y. Shi, Q. Guo, F. Xia, J. Sun, Y. Gao, Short- and midterm repeatability of magnetic resonance elastography in healthy volunteers at 3.0 T, Magn. Reson. Imaging 32 (2014) 665—670.

[22] Y.J. Lee, J.M. Lee, J.E. Lee, et al., MR elastography for noninvasive assessment of hepatic fibrosis: reproducibility of the examination and reproducibility and repeatability of the liver stiffness value measurement, J. Magn. Reson. Imaging 39 (2014) 326—331.

[23] R. Muthupillai, P.J. Rossman, D.J. Lomas, J.F. Greenleaf, S.J. Riederer, R.L. Ehman, Magnetic resonance imaging of transverse acoustic strain waves, Magn. Reson. Med. 36 (2) (1996) 266—274.

[24] B. Dzyubak, K. Glaser, M. Yin, et al., Automated liver stiffness measurements with magnetic resonance elastography, J. Magn. Reson. Imaging 38 (2013) 371—379.

[25] A. Manduca, T.E. Oliphant, M.A. Dresner, et al., Magnetic resonance elastography: non-invasive mapping of tissue elasticity, Med. Image Anal. 5 (4) (2001) 237—254.

[26] M.D.J. McGarry, E.E.W. Van Houten, A.J. Pattison, J.B. Weaver, K.D. Paulsen, Comparison of iterative and direct inversion MR elastography algorithms, Mech. Biol. Syst. Mater. 2 (2011) 49–56.

[27] M. Honarvar, R. Rohling, S. Salcudean, A comparison of direct and iterative finite element inversion techniques in dynamic elastography, Phys. Med. Biol. 61 (8) (2016) 3026–3048.

[28] A. Manduca, R. Muthupillai, P.J. Rossman, J.F. Greenleaf, R.L. Ehman, Image processing for magnetic resonance elastography, in: M.H. Loew, K.M. Hanson (Eds.), SPIE's International Symposium Medical Imaging 1996. Newport Beach, SPIE - The International Society for Optical Engineering, California, USA, 1996, pp. 616–623.

[29] H. Knutsson, C.F. Westin, G. Granlund, Local multiscale frequency and bandwidth estimation, Proc. IEEE Int. Conf. Image Proc. (1994) 36–40.

[30] T.E. Oliphant, A. Manduca, R.L. Ehman, J.F. Greenleaf, Complex-valued stiffness reconstruction for magnetic resonance elastography by algebraic inversion of the differential equation, Magn. Reson. Med. 45 (2001) 299–310.

[31] M. McGarry, C. Johnson, B. Sutton, et al., Suitability of poroelastic and viscoelastic mechanical models for high and low frequency MR elastography, Med. Phys. 42 (2) (2015) 947–957.

[32] E.E.W. Van Houten, M.I. Miga, J.B. Weaver, F.E. Kennedy, K.D. Paulsen, Three-dimensional subzoned-based reconstruction algorithm for MR elastography, Magn. Reson. Med. 45 (2001) 827–837.

[33] M. McGarry, C.L. Johnson, B.P. Sutton, et al., Including spatial information in nonlinear inversion MR elastography using soft prior regularization, IEEE Trans. Med. Imaging 32 (10) (2013) 1901–1909.

[34] A. Manduca, D.S. Lake, S.A. Kruse, R.L. Ehman, Spatio-temporal directional filtering for improved inversion of MR elastography images, Med. Image Anal. 7 (2003) 465–473.

[35] A. Romano, M. Scheel, S. Hirsch, J. Braun, I. Sack, In vivo waveguide elastography of white matter tracts in the human brain, Magn. Reson. Med. 68 (2012) 1410–1422.

[36] R. Sinkus, M. Tanter, T. Xydeas, S. Catheline, J. Bercoff, M. Fink, Viscoelastic shear properties of in vivo breast lesions measured by MR elastography, Magn. Reson. Imaging 23 (2005) 159–165.

[37] S. Hirsch, J. Guo, R. Reiter, et al., Towards compression-sensitive magnetic resonance elastography of the liver: sensitivity of harmonic volumetric strain to portal hypertension, J. Magn. Reson. Imaging 39 (2014) 298–306.

[38] S. Hirsch, F. Beyer, J. Guo, et al., Compression-sensitive magnetic resonance elastography, Phys. Med. Biol. 58 (2013) 5287–5299.

[39] S. Hirsch, J. Guo, R. Reiter, et al., MR elastography of the liver and the spleen using a piezoelectric driver, single-shot wave-field acquisition, and multifrequency dual parameter reconstruction, Magn. Reson. Med. 71 (2014) 267–277.

[40] Z.J. Domire, M.B. McCullough, Q. Chen, K.N. An, Wave attenuation as a measure of muscle quality as measured by magnetic resonance elastography: initial results, J. Biomech. 42 (2009) 537–540.

[41] L. Juge, A. Petiet, S. Lambert, et al., Microvasculature alters the dispersion properties of shear waves–a multi-frequency MR elastography study, NMR Biomed. 28 (12) (2015) 1763–1771.

[42] G.I. Nedredal, M. Yin, T. McKenzie, et al., Portal hypertension correlates with splenic stiffness as measured with MR elastography, J. Magn. Reson. Imaging 34 (2011) 79−87.

[43] M. Yin, K. Glaser, J. Talwalkar, J. Chen, A. Manduca, R. Ehman, Hepatic MR elastography: clinical performance in a series of 1377 consecutive examinations, Radiology 278 (1) (2016) 114−124.

[44] N.J. Shire, M. Yin, J. Chen, et al., Test-retest repeatability of MR elastography for noninvasive liver fibrosis assessment in Hepatitis C, J. Magn. Reson. Imaging 34 (2011) 947−955.

[45] B. Dzyubak, S. Venkatesh, A. Manduca, J. Glaser, R. Ehman, Automated liver elasticity calculation for MR elastography, J. Magn. Reson. Imaging 43 (5) (2016) 1055−1063.

[46] Y. Shi, Q. Guo, F. Xia, et al., MR elastography for the assessment of hepatic fibrosis in patients with chronic Hepatitis B infection: does histologic necroinflammation influence the measurement of hepatic stiffness? Radiology 273 (1) (2014) 88−98.

[47] Y. Mariappan, B. Dzyubak, K. Glaser, et al., Application of modified spin-echo based sequences for hepatic MR elastography: evaluation, comparison to the conventional gradient echo sequence and preliminary clinical experience, Radiology 282 (2) (2016) 390−398.

[48] J. Eaton, B. Dzyubak, S. Venkatesh, et al., Performance of magnetic resonance elastography in primary sclerosing cholangitis (submitted to), Clin. Gastroenterol. Pathol. (2015).

[49] E. Fenstad, B. Dzyubak, J. Oh, et al., Evaluation of liver stiffness with magnetic resonance elastography in patients with constrictive pericarditis: preliminary findings, J. Magn. Reson. Imaging 44 (1) (2016) 81−88.

[50] H. Lee, J. Lee, J. Han, B. Choi, MR elastography of healthy liver parenchyma: normal value and reliability of the liver stiffness value measurement, J. Magn. Reson. Imaging 38 (2013) 1215−1223.

[51] M. Batheja, H. Vargas, A.M. Silva, et al., Magnetic resonance elastography (MRE) in assessing hepatic fibrosis: performance in a cohort of patients with histological data, Abdom. Imag. 40 (4) (2015) 760−765.

[52] S.K. Venkatesh, M. Yin, J.F. Glockner, et al., MR elastography of liver tumors: preliminary results, Am. J. Roentgenol. 190 (2008) 1534−1540.

[53] M. Yin, J.A. Talwalkar, K.J. Glaser, et al., Assessment of hepatic fibrosis with magnetic resonance elastography, Clin. Gastroenterol. Hepatol. 5 (2007) 1207−1213.

[54] S. Serai, M. Yin, H. Wang, R. Ehman, D. Podberesky, Cross-vendor validation of liver magnetic resonance elastography, Abdom. Imag. 40 (4) (2015) 789−794.

[55] U. Motosugi, T. Ichikawa, K. Sano, et al., Magnetic resonance elastography of the liver: preliminary results and estimation of inter-rater reliability, Jpn. J. Radiol. 28 (2010) 623−627.

[56] M. Yin, J.A. Talwalkar, S.K. Venkatesh, R.L. Ehman, MR Elastography of Dynamic Postprandial Hepatic Stiffness Augmentation in Chronic Liver Disease, International Society for Magnetic Resonance in Medicine, Honolulu, Hawaii, 2009, p. 110.

[57] J. Chen, J.A. Talwalkar, M. Yin, K.J. Glaser, S.O. Sanderson, R.L. Ehman, Early detection of nonalcoholic steatohepatitis in patients with nonalcoholic fatty liver disease by using MR elastography, Radiology 259 (2011) 749−756.

[58] S. Crespo, M. Bridges, R. Nakhleh, A. McPhail, S. Pungpapong, A.P. Keaveny, Noninvasive assessment of liver fibrosis using magnetic resonance elastography in liver transplant recipients with hepatitis C, Clin. Transplant. 27 (2013) 652−658.

[59] A. Allen, V. Shah, T. Therneau, et al., The role of three-dimensional magnetic resonance elastography in the diagnosis of nonalcoholic steatohepatitis in obese patients undergoing bariatric surgery, Hepatology 0 (0) (2019).

[60] Z. Yin, M.C. Murphy, J. Li, et al., Prediction of nonalcoholic fatty liver disease (NAFLD) activity score(NAS) with multiparametric hepatic magnetic resonance imaging and elastography, Eur. Radiol. (2019) [E-pub ahead of print].

[61] C.L. Johnson, M.D.J. McGarry, A.A. Gharibans, et al., Local mechanical properties of white matter structures in the human brain, Neuroimage 79 (2013) 145–152.

[62] A. Arani, M. Murphy, K. Glaser, et al., Measuring the effects of aging and sex on regional brain stiffness with MR elastography in healthy older adults, Neuroimage 111 (2015) 59–64.

[63] L. Hiscox, C. Johnson, M. McGarry, et al., High-resolution magnetic resonance elastography reveals differences in subcortical gray matter viscoelasticity between young and healthy older adults, Neurobiol. Aging 65 (158–67) (2018).

[64] K.J. Streitberger, I. Sack, D. Krefting, et al., Brain viscoelasticity alteration in chronic-progressive multiple sclerosis, PLoS One 7 (1) (2012) e29888.

[65] M.C. Murphy, J. Huston, C.R. Jack, et al., Measuring the characteristic topography of brain stiffness with magnetic resonance elastography, PLoS One 8 (12) (2013) e81668.

[66] J. Zhang, M.A. Green, R. Sinkus, L.E. Bilston, Viscoelastic properties of human cerebellum using magnetic resonance elastography, J. Biomech. 44 (2011) 1909–1913.

[67] M.A. Green, L.E. Bilston, R. Sinkus, *In vivo* brain viscoelastic properties measured by magnetic resonance elastography, NMR Biomed. 21 (7) (2008) 755–764.

[68] K. Schregel, E. Wuerfel nee Tysiak, P. Garteiser, et al., Demyelination reduces brain parenchymal stiffness quantified in vivo by magnetic resonance elastography, Proc. Natl. Acad. Sci. 109 (2012) 6650–6655.

[69] M.C. Murphy, J. Huston III, C.R. Jack Jr., et al., Decreased Brain Stiffness in Alzheimer's Disease Determined by Magnetic Resonance Elastography, International Society for Magnetic Resonance in Medicine, Montreal, Canada, 2011.

[70] M.C. Murphy, G.L. Curran, K.J. Glaser, et al., Magnetic resonance elastography of the brain in a mouse model of Alzheimer's disease: initial results, Magn. Reson. Imaging 30 (2012) 535–539.

[71] F.B. Freimann, K.J. Streitberger, D. Klatt, et al., Alteration of brain viscoelasticity after shunt treatment in normal pressure hydrocephalus, Neuroradiology 54 (2012) 189–196.

[72] K. Riek, J.M. Millward, I. Hamann, et al., Magnetic resonance elastography reveals altered brain viscoelasticity in experimental autoimmune encephalomyelitis, Neuroimage: Clin. 1 (2012) 81–90.

[73] A. Lipp, C. Skowronek, A. Fehlner, K. Streitberger, J. Braun, I. Sack, Progressive supranuclear palsy and idiopathic Parkinson's disease are associated with local reduction of in vivo brain viscoelasticity, Eur. Radiol. 28 (8) (2018) 3347–3354.

[74] C. Chaze, G. McIlvain, D. Smith, et al., Altered brain tissue viscoelasticity in pediatric cerebral palsy measured by magnetic resonance elastography, Neuroimage Clin. (2019) [E-Pub ahead of print].

[75] T. Hennedige, J. Hallinan, F. Leung, et al., Comparison of magnetic resonance elastography and diffusion-weighted imaging for differentiating benign and malignant liver lesions, Eur. Radiol. 26 (2) (2015) 398–406.

[76] R. Sinkus, K. Siegmann, T. Xydeas, M. Tanter, C. Claussen, M. Fink, MR elastography of breast lesions: understanding the solid/liquid duality can improve the specificity of contrast-enhanced MR mammography, Magn. Reson. Med. 58 (2007) 1135–1144.

[77] K. Siegmann, T. Xydeas, R. Sinkus, B. Kraemer, U. Vogel, C. Claussen, Diagnostic value of MR elastography in addition to contrast-enhanced MR imaging of the breast -initial clinical results, Eur. Radiol. 20 (2010) 318–325.

[78] J. Chen, K.J. Glaser, E.G. Stinson, J.L. Kugel, R.L. Ehman, Non-Contact Driver System for MR Elastography of the Breast, International Society for Magnetic Resonance in Medicine, Montreal, Canada, 2011.

[79] M.C. Murphy, J. Huston III, K.J. Glaser, et al., Preoperative Assessment on Meningioma Stiffness by MR Elastography, International Society for Magnetic Resonance in Medicine, Melbourne, Australia, 2012, p. 3183.

[80] K. Pepin, K. McGee, A. Arani, et al., Magnetic resonance elastography analysis of glioma stiffness and IDH1 mutation status, Am. J. Neuroradiol. 39 (1) (2018) 31–36.

[81] K.M. Pepin, J. Chen, K.J. Glaser, et al., MR elastography derived shear stiffness - a new imaging biomarker for the assessment of early tumor response to chemotherapy, Magn. Reson. Med. 71 (2014) 1834–1840.

[82] J. Li, Y. Jamin, J.K.R. Boult, et al., Tumour biomechanical response to the vascular disrupting agent ZD6126 in vivo assessed by magnetic resonance elastography, Br. J. Canc. 110 (2014) 1727–1732.

[83] Y.K. Mariappan, K.J. Glaser, R.D. Hubmayr, A. Manduca, R.L. Ehman, K.P. McGee, MR elastography of human lung parenchyma: technical development, theoretical modeling and in vivo validation, J. Magn. Reson. Imaging 33 (2011) 1351–1361.

[84] Y. Mariappan, K. Glaser, D. Levin, et al., Estimation of the absolute shear stiffness of human lung parenchyma using 1H spin echo, echo planar MR elastography, J. Magn. Reson. Imaging 40 (2014) 1230–1237.

[85] J.R. Basford, T.R. Jenkyn, K.N. An, R.L. Ehman, G. Heers, K.R. Kaufman, Evaluation of healthy and diseased muscle with magnetic resonance elastography, Arch. Phys. Med. Rehabil. 83 (2002) 1530–1536.

[86] S. Papazoglou, J. Rump, J. Braun, I. Sack, Shear wave group velocity inversion in MR elastography of human skeletal muscle, Magn. Reson. Med. 56 (2006) 489–497.

[87] A.J. Romano, P.B. Abraham, S.I. Ringleb, P.J. Rossman, J.A. Bucaro, R.L. Ehman, Determination of Anisotropic Velocity Profiles in Muscle Using Wave-Guide Constrained Magnetic Resonance Elastography, International Society for Magnetic Resonance in Medicine, Seattle, WA, U.S.A., 2006, p. 1725.

[88] M. McGarry, E. Van Houten, P. Perrinez, A. Pattison, J. Weaver, K. Paulsen, An octahedral shear strain-based measure of SNR for 3D MR elastography, Phys. Med. Biol. 56 (2011) N153–N164.

[89] Z. Yin, K. Glaser, A. Manduca, et al., Slip interface imaging predicts tumor-brain adhesion in vestibular schwannomas 277 (2) (2015) 507–517, 2015.

[90] Z. Yin, Y. Sui, J. Trzasko, et al., In vivo characterization of 3D skull and brain motion during dynamic head vibration using magnetic resonance elastography, Magn. Reson. Med. 80 (6) (2018) 2573–2585.

[91] M.,A.M. Murphy, J.,K.J.G. Trzasko, J.R.L.E. Huston III, Artificial neural networks for stiffness estimation in magnetic resonance elastography, Magn. Reson. Med. 80 (1) (2018) 351–360.

Reconstructive elastography

7

Marvin M. Doyley

Department of Electrical and Computer Engineering, University of Rochester,
Rochester, NY, United States

1. Introduction

Elastography was developed in the late 1980s to early 1990s to improve ultrasonic imaging [1—4]. The success of ultrasonic elastography has also inspired investigators to develop analogues based on magnetic resonance imaging [5—8], and optical coherence tomography [9—11].

Elastography's goal is to produce images that faithfully represent the mechanical properties of the underlying tissue structures. This goal is realized only when the internal stress and strain distributions are known. Conventional imaging modalities provide information about the spatial variations of induced tissue strain; however, none of them can visualize stress distribution. In this chapter, we will demonstrate that considering elastography within the framework of solving an inverse problem circumvents this difficulty. A requirement for solving the inverse elastography problem (i.e., computing tissue mechanical properties from the measured strain or displacements) is having a theoretical model that accurately describes the mechanical behavior of the underlying tissue using knowledge of the external boundary conditions and the tissue properties. This model provides the framework for computing mechanical properties using either a direct or iterative inverse reconstruction techniques. Failure to consider elastography outside the framework of solving an inverse problem will at best provide inexact estimates of the underlying tissue properties for very limited cases.

In this chapter, we provide a very concise survey of reconstruction approaches, focusing on ultrasonic techniques with a brief reference to approaches based on magnetic resonance imaging. Specifically, we will describe (a) the general principles of quasi-static, harmonic, and transient elastography—the most popular approaches to elastography—and (b) the underlying equations of motion that govern each approach. For a comprehensive survey of modulus reconstruction, the readers are encouraged to consult [12].

Tissue Elasticity Imaging. https://doi.org/10.1016/B978-0-12-809661-1.00007-8

2. Solving the forward elasticity problem

Before discussing the inverse problem, we must consider the forward problem—predicting the mechanical response of a material using knowledge of biomechanical properties and external boundary conditions. More specifically, solving a system of partial differential equations [13,14]:

$$\nabla \cdot \mu \nabla \mathbf{u} + \nabla(\lambda + \mu)\nabla \cdot \mathbf{u} = \rho \frac{\partial^2 \mathbf{u}}{\partial t^2},$$ 7.1

where ρ is the density of the material, $\mathbf{u}$ is the displacement vector, λ and μ are Lamé constants, and t is time. For quasi-static deformations, Eq. (7.6) reduces to:

$$\nabla \cdot \mu \nabla \mathbf{u} + \nabla(\lambda + \mu)\nabla \cdot \mathbf{u} = 0$$ 7.2

For harmonic deformations, the time-independent (steady-state) equations in the frequency domain give [8,15]:

$$\nabla \cdot \mu \nabla \mathbf{u} + \nabla(\lambda + \mu)\nabla \cdot \mathbf{u} = \rho \omega^2 \mathbf{u},$$ 7.3

where ω is the angular frequency of the sinusoidal excitation. For transient deformations, the shear-wave velocity, c_2, is given by:

$$c_2 = \sqrt{\frac{\mu}{\rho}},$$ 7.4

where ρ is the density of the tissue. The governing equations for quasi-static, harmonic, and transient elastographic imaging methods can be solved analytically [16−19] or numerically [15,20−27].

3. Solving the inverse elastography problem

3.1 Quasi-static elastography

Quasi-static elastography visualizes the strain induced within tissue using either an external or internal source. A small motion is induced within the tissue (typically on the order of 1%−2% of the axial dimension) with a quasi-static mechanical source. The axial component of the internal tissue displacement is measured by performing cross-correlation analysis on pre- and postdeformed radio-frequency (RF) echo frames [3,4,28], and strain is estimated by spatially differentiating the axial displacements. In quasi-static elastography, soft tissues are typically viewed as a series of one-dimensional springs that are arranged in a simple fashion. For this simple mechanical model, the measured axial strain (ε_{xx}) is related to the internal stress (σ_{xx}) as follows (Hooke's Law):

$$\sigma_{xx} = E\varepsilon_{xx}$$ 7.5

No method can measure the internal stress distribution in vivo; consequently, the internal stress distribution is assumed to be constant (i.e., $\sigma \approx 1$). An approximate estimate of Young's modulus is computed from the reciprocal of the measured strain. The disadvantage of computing modulus elastograms in this manner is that neither stress decay nor stress concentration are accounted for; consequently, quasi-static elastograms typically contain target-hardening artifacts [22,25].

Despite this limitation, several groups have obtained good elastograms in applications where accurate quantification of Young's modulus is not essential.

Direct and iterative inversion schemes have been developed to make quasi-static elastograms more quantitative. These techniques compute the Young's, or shear, modulus from the measured displacement or strain using the forward elasticity model that was previously described. Direct-inversion schemes use a linear system of equations that are derived by rearranging the PDEs that describe the forward elastography problem [6,18,29].

$$(\partial_{yy} - \partial_{xx})(\varepsilon_{xy}\mu) + \partial_{xy}(\varepsilon_{xy}\mu) = 0 \qquad 7.6$$

Eq. (7.2) contains high-order derivatives that amplify measurement noise, which compromises the quality of the ensuing modulus elastograms (reconstructed). Iterative inversion techniques [30,31] overcome this issue by considering the inverse problem as a parameter-optimization task, where the goal is to find the Young's modulus that minimizes the error between measured displacement or strain fields, and those computed by solving the forward elastography problem. The matrix solution at the $(k+1)$ iteration has the general form:

$$\mu^{k+1} = \Delta\mu^k + \left[J(\mu^k)^T J(\mu^k) + \rho I\right]^{-1} J(\mu^k)^T (u_m - u\{\mu^k\}), \qquad 7.7$$

where $\Delta\mu^k$ is a vector of shear modulus updates at all coordinates in the reconstruction field and J is the Jacobian, or sensitivity, matrix. The Hessian matrix, $\left[J(\mu^k)^T J(\mu^k)\right]$, is ill-conditioned. Therefore, to stabilize performance in the presence of measurement noise, the matrix is regularized using one of three variational methods: the Tikhanov [31], the Marquardt [32], or the total variational method [33,34]. Fig. 7.1 shows an example of modulus elastograms computed with the iterative inversion approach. For further details on implementation, tradeoffs, etc., the readers can consult Section 4.1 of [12].

3.2 Dynamic elastography based on local frequency estimation

In dynamic elastography [1,2,5,21,35], low frequency acoustic waves (typically < 1 kHz) are transmitted within the tissue using a sinusoidal mechanical source. The phase and amplitude of the propagating waves are visualized using either color Doppler imaging [21,35,36]— or phase-contrast MRI [5,7,8]. Lerner and Parker were pioneers in elasticity imaging and termed their method *sonoelastography*.

With an assumption that shear waves propagate with plane-wavefronts, an approximate estimate of the local shear modulus (μ) may be computed from local estimates of the wavelength using Eq. (7.4).

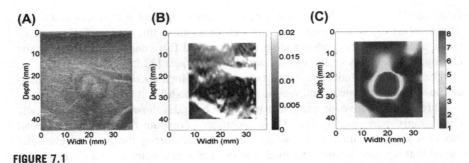

FIGURE 7.1

Sonogram (A), strain elastogram (B), and modulus elastogram (C) of RF ex vivo ablated bovine liver.

Courtesy of Drs. T. J. Hall, T. Varghese, and J. Jiang (University of Wisconsin–Madison).

In a homogeneous tissue, shear modulus can be estimated from local estimates of instantaneous frequency [37,38]. Although shear modulus estimated using this approach is insensitive to measurement noise, the spatial resolution of the ensuing modulus elastograms is limited. A further weakness of the approach is that the plane-wave approximation breaks down in complex organs such as the breast and brain, when waves reflected from internal tissue boundaries interfere constructively and destructively. Fig. 7.2 shows a representative example of an elastogram obtained from a health volunteer by solving the inverse harmonic elastography problem.

3.3 Transient elastography based on arrival time estimation

A major limitation of dynamic elastography is that shear waves attenuate rapidly as they propagate through soft tissues, which limits the depth of penetration. The transient approach to elastography overcomes this limitation by using the acoustic radiation force of an ultrasound transducer to perturb tissue locally [39–42]. 2D transient elastography uses an ultrasound scanner with an ultra-high frame rate (i.e., 10,000 fps) to track the propagation of shear waves (**see** Fig. 7.3). As in harmonic elastography, local estimates of shear modulus are estimated from local estimates of wavelength. However, the reflections of shear waves at internal tissue boundaries make it difficult to measure shear-wave velocity—this limitation can be overcome by computing wave speeds directly from the arrival times as discussed in Ref. [43]. For further details on implementation, tradeoffs, etc., the readers can consult Section 4.2 of [12].

4. Advanced reconstruction methods

The vast majority of work that have been done in solving the inverse elasticity problem assume the underlying tissue exhibit linear elastic isotropic mechanical

(A)   **(B)**

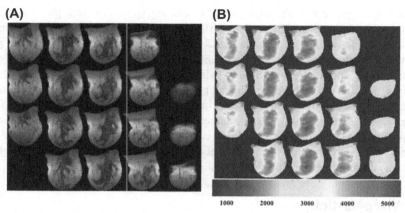

1000   2000   3000   4000   5000

FIGURE 7.2

Montage of MR magnitude images (A) and shear modulus elastograms (B) recovered from a healthy volunteer using the subzone inversion scheme.

Courtesy of Drs. J. B. Weaver and K. D. Paulsen, Dartmouth College, Thayer School of Engineering. The resolution of the elastograms was sufficiently high to visualize fibroglandular tissue from the adipose tissue.

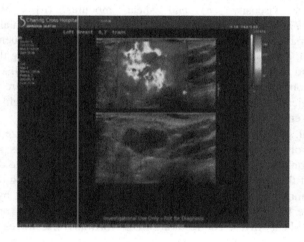

FIGURE 7.3

Transient shear wave images of a breast with pathology confirmed invasive ductal carcinoma. The maximum diameter on the longitudinal axis on B-mode was 17 mm, whereas both elastographic techniques indicated a larger footprint of the cancer.

Courtesy of Dr. W. Svensson, Imperial College, London.

behavior. Although doing so simplifies the inverse problem (i.e., requiring only two parameters to describe the mechanical behavior of the underlying tissues), doing introduces errors and limits the scope of elastography. It is well known that soft tissues display several biomechanical properties, including viscosity and nonlinearity, which may improve the diagnostic value of elastography. For example, clinicians could use mechanical nonlinearity to differentiate between benign and malignant breast tumors [44]. Furthermore, there has been evidence that other mechanical parameters, namely viscosity [45,46] and anisotropy [46] could also differentiate between benign and malignant tissues—similar claims have also been made for Young's modulus [46]. These mechanical parameters discriminate between different tissue types, but they may provide value in other clinical areas, including brain imaging [47,48], distinguishing the mechanical properties of active and passive muscle groups [49–51], characterizing blood clots [52], and diagnosing edema [53]. Several groups are developing inverse reconstruction techniques to visualize other mechanical parameters such as viscoelasticity and nonlinearity.

4.1 Viscoelasticity

Soft tissues display viscoelastic creep, viscoelastic relaxation, and hysteresis. Fluid-like and elastic mechanical behavior, where stress and strain are both dependent on time. Researchers have used rheological models (Maxwell, Kelvin-Voigt, etc.) to measure the viscoelastic properties of normal and diseased liver tissue taken from patients with grade 3 and 4 liver fibrosis [49]. They observed that fibrotic liver tissue had a higher viscosity (η) and shear modulus ($\mu_1 = 2.91 \pm 0.84$ and $\mu_2 = 4.83 \pm 1.77$) than normal liver tissue. Their results revealed that although liver tissue is dispersive, it appeared as nondispersive between the frequency range of 25–50 Hz [54], computed the shear modulus and viscosity by fitting the measured speed of sound and attenuation equation to Voigt's and Maxwell's rheological models. They observed that the recovered shear modulus values were independent of the rheological model employed, but viscosity values were highly dependent on the models employed.

Sinkus et al. [55] developed a direct-inversion scheme to visualize the mechanical properties of viscoelastic materials, in which a curl operation was performed on the time-harmonic displacement field $\mathbf{u}(\mathbf{x},t) = \mathbf{u}(\mathbf{x})e^{i\omega t}$ to remove the displacement contribution of the compressional wave. They derived the governing equation that describes the motion incurred in an isotropic, viscoelastic medium by computing the curl of the PDEs that describe the motion incurred by both transverse and compressional shear waves. The resulting PDEs for transverse waves are given in compact form by:

$$\rho \partial_t^2 \mathbf{u} = \mu \nabla^2 \mathbf{u} + \eta \partial_t \nabla^2 \mathbf{u} \qquad \qquad 7.8$$

Sinkus et al. [55] developed methods to compute μ and η directly from Eq. (7.4). Using simulated data, they demonstrated that they could accurately compute shear modulus. With more realistic displacements accurate estimate of both viscosity and shear modulus was obtained only for soft tissues (<8 kPa). Nevertheless, patient studies revealed that the shear modulus values of malignant breast tumors were noticeably higher than those of benign fibroadenomas, but there was no significant difference observed in the viscosity of the tumor types, a result that would appear to contradict those reported in Ref. [45].

4.2 Nonlinearity

When soft tissues deform by a small amount (an infinitesimal deformation), their geometry in the undeformed and deformed states is similar, and thus the deformation is characterized using engineering strain. The reference configuration must be defined to characterize finite deformation, which is the geometry of the tissue under investigation in either the deformed or undeformed state. The Green-Lagrangian strain is defined as:

$$\varepsilon_{ij} = \frac{1}{2}\left[\frac{\partial u_i}{\partial x_j} + \frac{\partial u_j}{\partial x_i} + \frac{\partial u_k}{\partial x_i}\frac{\partial u_k}{\partial x_j}\right] \qquad 7.9$$

Some materials exhibit nonlinear material properties that are typically described using a strain energy density function. Among the strain energy functions proposed in the literature, the most widely used for modeling tissues are the Neo-Hookean hyperelastic model, and the Neo-Hookean model with an exponential term. Oberai et al. [56] used a different model, the Veronda-Westman strain energy density function, to describe the finite displacement of a hyperelastic solid that is undergoing finite deformation.

Preliminary clinical evaluation of their approach suggests that nonlinear elasticity imaging may be used to differentiate between benign and malignant tumors.

5. Discussion

Although developing elastography within the framework of solving an inverse problem should provide more accurate estimates of the mechanical parameters of human tissues than the simple approaches described in Section 2 of this paper, several concerns remain to be resolved before model-based elastography could become the prevailing approach to quasi-static, harmonic, and transient elastography. These concerns include (1) developing practical techniques to transform ill-posed problems into a well-posed one; (2) minimizing model-data mismatch; and (3) developing better test procedures to evaluate and optimize the performance of advanced reconstruction methods. Using a priori information to constrain the inverse problem will transform model-based elastography from an ill-posed inverse problem to a well-posed one [57—59]; however, care should be exercised since doing this could introduce errors in the reconstruction processing. Researchers have demonstrated that neural network can be used to construct constitutive relation from elastographic data; however, successful utilization of this technique will also require some knowledge of the forces on the boundary [60—62]. More effort should be devoted to developing anthropomorphic elasticity phantoms [63,64] since this will provide better insight into how different inverse reconstruction method will perform in the clinical setting.

References

[1] R.M. Lerner, K.J. Parker, Sonoelasticity images, ultrasonic tissue charactertization and echographic imaging, in: J. Thijssen (Ed.), 7th European Communities Workshop, Nijmegen, The Netherlands, 1987.

[2] R.M. Lerner, K.J. Parker, J. Holen, R. Gramiak, R.C. Waag, Sono-elasticity: medical elasticity images derived from ultrasound singals in mechanically vibrated targets, Acoust. Imaging 16 (1988) 317−327.

[3] J. Ophir, I. Cespedes, H. Ponnekanti, Y. Yazdi, X. Li, Elastography: a quantitative method for imaging the elasticity of biological tissues, Ultrason. Imaging 13 (1991) 111−134.

[4] M. O'Donnell, A.R. Skovoroda, B.M. Shapo, S.Y. Emelianov, Internal displacement and strain imaging using ultrasonic speckle tracking, IEEE Trans. Ultrason. Ferroelectr. Freq. Control 41 (3) (1994) 314−325.

[5] R. Muthupillai, D.J. Lomas, P.J. Rossman, J.F. Greenleaf, A. Manduca, R.L. Ehman, Magnetic-resonance elastography by direct visualization of propagating acoustic strain waves, Science 269 (1995) 1854−1857.

[6] J. Bishop, A. Samani, J. Sciarretta, D. Plewes, Two-dimensional MR elastography with linear inversion reconstruction: methodology and noise analysis 45 (2000) 2081−2091.

[7] J.B. Weaver, E.E. Van Houten, M.I. Miga, F.E. Kennedy, K.D. Paulsen, Magnetic resonance elastography using 3D gradient echo measurements of steady-state motion, Med. Phys. 28 (2001) 1620−1628.

[8] R. Sinkus, J. Lorenzen, D. Schrader, M. Lorenzen, M. Dargatz, D. Holz, High-resolution tensor MR elastography for breast tumour detection, Phys. Med. Biol. 45 (2000) 1649−1664.

[9] A.S. Khalil, R.C. Chan, A.H. Chau, B.E. Bouma, M.R.K. Mofrad, Tissue elasticity estimation with optical coherence elastography: toward mechanical characterization of in vivo soft tissue, Ann. Biomed. Eng. 33 (2005) 1631−1639.

[10] S.J. Kirkpatrick, R.K. Wang, D.D. Duncan, OCT-based elastography for large and small deformations, Opt. Express 14 (2006) 11585−11597.

[11] H.J. Ko, W. Tan, R. Stack, S.A. Boppart, Optical coherence elastography of engineered and developing tissue, Tissue Eng. 12 (2006) 63−73.

[12] M.M. Doyley, Model-based elastography: a survey of approaches to the inverse elasticity problem, Phys. Med. Biol. 57 (2012) R35−R73.

[13] S.P. Timoshenko, J.N. Goodier, Theory of Elasticity, McGraw-Hill, Singapore, 1970.

[14] Y.C. Fung, Biomechanics: Mechanical Properties of Living Tissue, Springer, New York, 1981.

[15] E.E. Van Houten, M.I. Miga, J.B. Weaver, F.E. Kennedy, K.D. Paulsen, Three-dimensional subzone-based reconstruction algorithm for MR elastography, Magn. Reson. Med. 45 (2001) 827−837.

[16] M. Bilgen, M. Insana, Elastostatics of a Spherical Inclusion in Homogeneous Biological Media, vol. 43, 1998, pp. 1−20.

[17] A. Love, The stress produced in a semi-infinite solid by pressure on part of the boundary, Philos. Trans. R. Soc. London (1929) 377–420.

[18] C. Sumi, A. Suzuki, K. Nakayama, Estimation of shear modulus distribution in soft-tissue from strain distribution, IEEE Trans. Biomed. Eng. 42 (1995) 193–202.

[19] F. Kallel, M. Bertrand, J. Ophir, Fundamental limitations on the contrast-transfer efficiency in elastography: an analytic study, Ultrasound Med. Biol. 22 (1996) 463–470.

[20] J. McLaughlin, D. Renzi, Shear wave speed recovery in transient elastography and supersonic imaging using propagating fronts, Inverse Probl. 22 (2006) 681–706.

[21] K.J. Parker, S.R. Huang, R.A. Musulin, R.M. Lerner, Tissue-response to mechanical vibrations for sonoelasticity imaging, Ultrasound Med. Biol. 16 (1990) 241–246.

[22] H. Ponnekanti, J. Ophir, I. Cespedes, Ultrasonic-imaging of the stress-distribution in elastic media due to an external compressor, Ultrasound Med. Biol. 20 (1994) 27–33.

[23] A. Samani, J. Bishop, D.B. Plewes, A constrained modulus reconstruction technique for breast cancer assessment, IEEE Trans. Med. Imaging 20 (2001) 877–885.

[24] M.I. Miga, A new approach to elastography using mutual information and finite elements, Phys. Med. Biol. 48 (2003) 467–480.

[25] E. Konofagou, P. Dutta, J. Ophir, I. Cespedes, Reduction of stress nonuniformities by apodization of compressor displacement in elastography, Ultrasound Med. Biol. 22 (1996) 1229–1236.

[26] T.J. Hall, M. Bilgen, M.F. Insana, T.A. Krouskop, Phantom materials for elastography, IEEE Trans. Ultrason. Ferroelectr. Freq. Control 44 (1997) 1355–1365.

[27] J.C. Brigham, W. Aquino, F.G. Mitri, J. Greenleaf, M. Fatemi, Inverse estimation of viscoelastic material properties for solids immersed in fluids using vibroacoustic techniques, J. Appl. Phys. 101 (2007), 023509-0 − 09-14.

[28] J.C. Bamber, N.L. Bush, Freehand Elasticity Imaging Using Speckle Decorrelation Rate, Plenum Press, New York, 1995.

[29] A.R. Skovoroda, S.R. Aglyamov, On reconstruction of elastic properties of soft biological tissues exposed to low-frequencies, Biofizika 40 (1995) 1329–1334.

[30] M.M. Doyley, J.C. Bamber, T. Shiina, M.O. Leach, Reconstruction of elasticity modulus distribution from envelope detected B-mode data, in: M. Levy, S.C. Scneider, B.R. McAvoy (Eds.), Proc IEEE Ultrason Symp, 1996, pp. 1611–1614.

[31] F. Kallel, M. Bertrand, Tissue elasticity reconstruction using linear perturbation method, IEEE Trans. Med. Imaging 15 (1996) 299–313.

[32] M.M. Doyley, P.M. Meaney, J.C. Bamber, Evaluation of an iterative reconstruction method for quantitative elastography, Phys. Med. Biol. 45 (2000) 1521–1540.

[33] M.S. Richards, P.E. Barbone, A.A. Oberai, Quantitative three-dimensional elasticity imaging from quasi-static deformation: a phantom study, Phys. Med. Biol. 54 (2009) 757–779.

[34] J. Jiang, T. Varghese, C.L. Brace, E.L. Madsen, T.J. Hall, S. Bharat, M.A. Hobson, J.A. Zagzebski, F.T. Lee Jr., Young's modulus reconstruction for radio-frequency ablation electrode-induced displacement fields: a feasibility study, IEEE Trans. Med. Imaging 28 (2009) 1325–1334.

[35] Y. Yamakoshi, J. Sato, T. Sato, Ultrasonic imaging of internal vibration of soft tissue under forced vibration, IEEE Trans. Ultrason. Ferroelectr. Freq. Control 17 (1990) 45–53.

[36] R.M. Lerner, S.R. Huang, K.J. Parker, "Sonoelasticity" images derived from ultrasound signals in mechanically vibrated tissues, Ultrasound Med. Biol. 16 (1990) 231–239.

[37] A. Manduca, T.E. Oliphant, M.A. Dresner, J.L. Mahowald, S.A. Kruse, E. Amromin, J.P. Felmlee, J.F. Greenleaf, R.L. Ehman, Magnetic resonance elastography: non-invasive mapping of tissue elasticity, Med Imaging Anal 5 (2001) 237–254.

[38] Z. Wu, K. Hoyt, D.J. Rubens, K.J. Parker, Sonoelastographic imaging of interference patterns for estimation of shear velocity distribution in biomaterials, J. Acoust. Soc. Am. 120 (2006) 535–545.

[39] A.P. Sarvazyan, O.V. Rudenko, S.D. Swanson, J.B. Fowlkes, S.Y. Emelianov, Shear wave elasticity imaging: a new ultrasonic technology of medical diagnostics, Ultrasound Med. Biol. 24 (1998) 1419–1435.

[40] S. McAleavey, E. Collins, J. Kelly, E. Elegbe, M. Menon, Validation of SMURF estimation of shear modulus in hydrogels, Ultrason. Imaging 31 (2009) 131–150.

[41] J. Bercoff, S. Chaffai, M. Tanter, L. Sandrin, S. Catheline, M. Fink, J. Gennisson, M. Meunier, In vivo breast tumor detection using transient elastography, Ultrasound Med. Biol. 29 (2003) 1387–1396.

[42] K. Nightingale, S. McAleavey, G. Trahey, Shear-wave generation using acoustic radiation force: in vivo and ex vivo results, Ultrasound Med. Biol. 29 (2003) 1715–1723.

[43] L. Ji, J.R. McLaughlin, D. Renzi, J.R. Yoon, Interior elastodynamics inverse problems: shear wave speed reconstruction in transient elastography, Inverse Probl. 19 (2003) S1–S29.

[44] T.A. Krouskop, T.M. Wheeler, F. Kallel, B.S. Garra, T. Hall, Elastic moduli of breast and prostate tissues under compression, Ultrason. Imaging 20 (1998) 260–274.

[45] Y.P. Qiu, M. Sridhar, J.K. Tsou, K.K. Lindfors, M.F. Insana, Ultrasonic viscoelasticity imaging of nonpalpable breast tumors: preliminary results, Acad. Radiol. 15 (2008) 1526–1533.

[46] R. Sinkus, M. Tanter, T. Xydeas, S. Catheline, J. Bercoff, M. Fink, Viscoelastic shear properties of in vivo breast lesions measured by MR elastography, Magn. Reson. Imaging 23 (2005) 159–165.

[47] U. Hamhaber, D. Klatt, S. Papazoglou, M. Hollmann, J. Stadler, I. Sack, J. Bernarding, J. Braun, In vivo magnetic resonance elastography of human brain at 7 T and 1.5 T, J. Magn. Reson. Imaging 32 (2010) 577–583.

[48] I. Sack, B. Beierbach, J. Wuerfel, D. Klatt, U. Hamhaber, S. Papazoglou, P. Martus, J. Braun, The impact of aging and gender on brain viscoelasticity, Neuroimage 46 (2009) 652–657.

[49] P. Asbach, D. Klatt, U. Hamhaber, J. Braun, R. Somasundaram, B. Hamm, I. Sack, Assessment of liver viscoelasticity using multifrequency MR elastography, Magn. Reson. Med. 60 (2008) 373–379.

[50] K. Hoyt, B. Castaneda, K.J. Parker, Two-dimensional sonoelastographic shear velocity imaging, Ultrasound Med. Biol. 34 (2008) 276–288.

[51] P.R. Perrinez, F.E. Kennedy, E.E.W. Van Houten, J.B. Weaver, K.D. Paulsen, Modeling of soft poroelastic tissue in time-harmonic MR elastography, IEEE Trans. Biomed. Eng. 56 (2009) 598–608.

[52] C. Schmitt, G. Soulez, R.L. Maurice, M.F. Giroux, G. Cloutier, Noninvasive vascular elastography: toward a complementary characterization tool of atherosclerosis in carotid arteries, Ultrasound Med. Biol. 33 (2007) 1841–1858.

[53] R. Righetti, B.S. Garra, L.M. Mobbs, C.M. Kraemer-Chant, J. Ophir, T.A. Krouskop, The feasibility of using poroelastographic techniques for distinguishing between normal and lymphedematous tissues in vivo, Phys. Med. Biol. 52 (2007) 6525–6541.

[54] S. Catheline, J. Gennisson, G. Delon, M. Fink, R. Sinkus, S. Abouelkaram, J. Culioli, Measurement of viscoelastic properties of homogeneous soft solid using transient elastography: an inverse problem approach, J. Acoust. Soc. Am. 116 (2004) 3734–3741.

[55] R. Sinkus, M. Tanter, S. Catheline, J. Lorenzen, C. Kuhl, E. Sondermann, M. Fink, Imaging anisotropic and viscous properties of breast tissue by magnetic resonance-elastography, Magn. Reson. Med. 53 (2005) 372–387.

[56] A.A. Oberai, N.H. Gokhale, S. Goenezen, P.E. Barbone, T.J. Hall, A.M. Sommer, J.F. Jiang, Linear and nonlinear elasticity imaging of soft tissue in vivo: demonstration of feasibility, Phys. Med. Biol. 54 (2009) 1191–1207.

[57] M.M. Doyley, S. Srinivasan, E. Dimidenko, N. Soni, J. Ophir, Enhancing the performance of model-based elastography by incorporating additional a priori information in the modulus image reconstruction process, Phys. Med. Biol. 51 (2006) 95–112.

[58] R. Baldewsing, F. Mastik, J. Schaar, P. Serruys, A. van der Steen, Young's modulus reconstruction of vulnerable atherosclerotic plaque components using deformable curves, Ultrasound Med. Biol. 32 (2006) 201–210.

[59] S. Le Floc'h, J. Ohayon, P. Tracqui, G. Finet, A.M. Gharib, R.L. Maurice, G. Cloutier, R.I. Pettigrew, Vulnerable atherosclerotic plaque elasticity reconstruction based on a segmentation-driven optimization procedure using strain measurements: theoretical framework, IEEE Trans. Med. Imaging 28 (2009) 1126–1137.

[60] C. Hoerig, J. Ghaboussi, M.F. Insana, Data-driven elasticity imaging using cartesian neural network constitutive models and the autoprogressive method, IEEE Trans. Med. Imaging 38 (2019) 1150–1160.

[61] C. Hoerig, J. Ghaboussi, M.F. Insana, An information-based machine learning approach to elasticity imaging, Biomech. Model. Mechanobiol. 16 (2017) 805–822.

[62] K.T. Rich, C.L. Hoerig, M.B. Rao, T.D. Mast, Relations between acoustic cavitation and skin resistance during intermediate- and high-frequency sonophoresis, J. Control. Release 194 (2014) 266–277.

[63] M.A. Hobson, E.L. Madsen, G.R. Frank, J.F. Jiang, H.R. Shi, T.J. Hall, T. Varghese, Anthropomorphic phantoms for assessment of strain imaging methods involving saline-infused sonohysterography, Ultrasound Med. Biol. 34 (2008) 1622–1637.

[64] E.L. Madsen, M.A. Hobson, G.R. Frank, H. Shi, J. Jiang, T.J. Hall, T. Varghese, M.M. Doyley, J.B. Weaver, Anthropomorphic breast phantoms for testing elastography systems, Ultrasound Med. Biol. 32 (2006) 857–874.

Lateral and shear strain imaging for ultrasound elastography

Tomy Varghese

Department of Medical Physics University of Wisconsin School of Medicine and Public Health
University of Wisconsin—Madison, Madison, WI, United States

1. Introduction

Commercial ultrasonographic systems that incorporate elastographic imaging software providing maps of tissue stiffness have proliferated over the past decade, with approaches that include both strain and shear wave imaging approaches. However, most commercial implementations of strain imaging include information on only the axial strain tensor component. The axial displacement vector and strain tensor denote the most commonly described and ubiquitous ultrasound strain images. This is primarily due to the manner in which backscattered ultrasound data is acquired, as the most accurate, high-resolution (both spatial and temporal), and precise strain estimation is possible only along the axial or beam propagation direction.

Elastography fundamentally is a three-dimensional (3D) imaging problem (four-dimensional if time is also incorporated as a variable), with three normal displacement vectors and corresponding normal strain tensors to be estimated. In addition to the normal strain tensors, the associated shear strain components are of interest. We present expressions for the normal strain tensors [1] obtained from a gradient of the corresponding displacement estimates as shown in Eq. (8.1):

$$\varepsilon_{zz} = \frac{\partial d_z}{\partial z}, \quad \varepsilon_{xx} = \frac{\partial d_x}{\partial x}, \text{ and } \varepsilon_{yy} = \frac{\partial d_y}{\partial y} \qquad 8.1$$

where ε_{zz}, ε_{xx}, and ε_{yy} denote the three normal strain tensor components, namely, axial strain, lateral strain, and elevational strain, along the z, x, and y directions, respectively. Displacement vectors are denoted by d_z, d_x, and d_y along the z, x, and y directions, respectively (see Fig. 8.1). Axial strain is computed from the gradient of the axial displacement, generally estimated along the axial or beam propagation direction. Lateral strain is estimated from the gradient of the displacement along the lateral direction (defined as perpendicular to the beam direction) and elevational strain from the gradient perpendicular to both beam propagation and scan plane direction. An exception to the above occurs with affine-based strain estimators that do not use a derivative operation [2,3].

Tissue Elasticity Imaging. https://doi.org/10.1016/B978-0-12-809661-1.00008-X

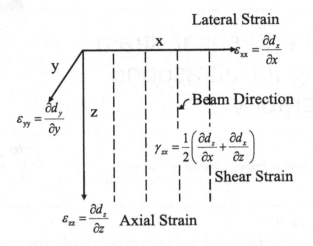

FIGURE 8.1

The axial, lateral, and shear strains within the *zx* imaging plane.

The full shear strain tensor in the "*zx*" imaging or scan plane includes both an axial shear and lateral shear strain tensor component [1] as defined in Eq. (8.2):

$$\text{Shear strain } \gamma_{zx} = \frac{1}{2}\left(\frac{\partial d_z}{\partial x} + \frac{\partial d_x}{\partial z}\right) \qquad 8.2$$

where z and x represent the axial and lateral directions, respectively; $\partial d_z/\partial x$ represents the axial shear strain tensor; and $\partial d_x/\partial z$ represents the lateral shear strain tensor (Fig. 8.1). The elevational component is ignored here for data acquired for a conventional two-dimensional (2D) ultrasound scan planes. Elevational components have to be considered while solving for the 3D problem, which will become feasible with the advent of 2D transducer arrays for 3D imaging. The general 3D strain distribution or engineering strain matrix [1] can therefore be written as follows:

$$\text{Strain matrix} = \begin{bmatrix} \varepsilon_{zz} & \gamma_{zx} & \gamma_{zy} \\ \gamma_{xz} & \varepsilon_{xx} & \gamma_{xy} \\ \gamma_{yz} & \gamma_{yx} & \varepsilon_{yy} \end{bmatrix} \qquad 8.3$$

Ideally one should be able to obtain images of the three normal strain tensors and the corresponding six shear strain tensor components. In this chapter, we will discuss the imaging of the additional normal strain tensor and shear strain components that can be estimated from 2D datasets. We will not focus on results reported in the flow imaging literature on the estimation of blood flow in multiple dimensions or purely displacement mapping without strain tensor estimations. We will instead focus the discussion on approaches that utilize backscattered radio frequency (RF) echo signals. Note that most of the approaches can also be performed using envelope or B-mode signals albeit at significantly reduced accuracy, resolution, and precision.

In addition to the uniaxial deformation utilized with the one-dimensional (1D) or 2D deformation tracking and estimation approaches described earlier, investigators have evaluated approaches such as the use of lateral shear deformations to enhance the visualization of the shear strain distribution around masses [4].

2. Classification of lateral and shear strain estimation methods

Classification of approaches that estimate the lateral displacement vector and subsequently the lateral strain tensor in this chapter is based on the dimensions of the kernel utilized for tracking and matching pre-and postdeformation RF echo signals. We can therefore classify these methods into 1D and 2D kernel- based approaches to lateral and shear strain estimation. Note that extension of these approaches to 3D kernels for 3D displacement vector and strain estimation would be straightforward when 3D datasets become available with 2D transducer arrays.

3. One-dimensional deformation tracking and estimation

The axial displacement vector and the axial strain tensor were initially estimated using 1D cross-correlation-based techniques [5]. The basic premise was to match a small rectangular windowed segment of RF echo signals from a predeformation frame by searching for a similar segment on the postdeformation frame of data following an applied deformation [5]. As the estimation was performed along the beam propagation direction, on RF data typically sampled at or above the Nyquist frequency, estimation of the axial deformation provided the highest quality displacement and strain estimates [6]. However, due to the 1D estimation of the time delays between the pre- and postcompression data, lateral deformation leading to lateral displacements and strain could not be accurately estimated. Additionally, due to the lack of phase information to track in the lateral and elevational directions, most previous studies were based on strain estimated along the axial or the beam propagation direction [5], ignoring displacement and strain components in the lateral and elevational directions [7,8], respectively. Researchers, however, have presented several techniques to estimate the lateral displacement and strain with 1D processing, which are described in the following sub-sections.

3.1 Incompressibility assumption

The simplest approach utilized was to harness the assumption of tissue incompressibility under plane strain conditions to obtain high-resolution lateral strain estimates from the estimated axial displacement and strain [9]. Assuming plane strain conditions, the axial and lateral strains are related by the following relationship, which describes the deformation in the lateral direction with respect to the corresponding axial deformation:

$$\nu = -\frac{\varepsilon_{xx}}{\varepsilon_{zz}} \qquad\qquad 8.4$$

where the term ν denotes Poisson's ratio, which is approximately equal to 0.495 for incompressible biological tissues. Under the assumption of incompressible tissue and plane strain conditions, the lateral strain map can be directly estimated from the estimated axial strain distribution. The disadvantage of this approach is that no additional new information is present in the lateral strain, which is now a scaled replica of the corresponding axial strain.

3.2 Weighted interpolation and recorrelation approach

An approach to improve lateral strain estimation using a weighted interpolation approach between neighboring RF A-lines was presented by Konofagou ad Ophir [10]. An iterative algorithm was then utilized for the estimation of both axial and lateral displacements followed by an iterative correction of the underlying lateral and axial signal decorrelation, which is then followed by reestimation of the axial and lateral displacements to finally converge onto an optimal axial and lateral displacement pair. An illustrative figure from the paper is shown in Fig. 8.2, demonstrating the iterative reconstruction of the axial and lateral strain distribution. They also reported on the use of the corrected axial and lateral strain images to obtain a map of Poisson's ratio using Eq. (8.4), under a plane strain assumption. Shear strain images for the corresponding imaging plane can also be obtained using Eq. (8.2). A possible disadvantage of this approach is the computational load, which may not be a significant issue with the current computers with a large memory and/or graphics processing unit (GPU)-based computing.

3.3 Perpendicular insonification using dual transducers

Another straightforward approach described in the literature involves the use of dual transducers aligned orthogonally to estimate both the axial and lateral displacements following an applied deformation [11]. As RF data acquired orthogonally are along the beam propagation direction, high—resolution axial and lateral displacements can be estimated; however, data acquisition would require specialized hardware to operate and synchronize data acquisition on both transducers. However, this approach is limited in its applicability because both transducers have to be ultrasonically coupled to the tissue being deformed, and along the same imaging plane. In addition, errors from the possible speed of sound variations due to changes in the thickness of the different tissue layers can introduce refraction or beam deflection errors, as illustrated in Fig. 8.3. Similar errors are also associated with beam-steered data acquisitions described later in this chapter.

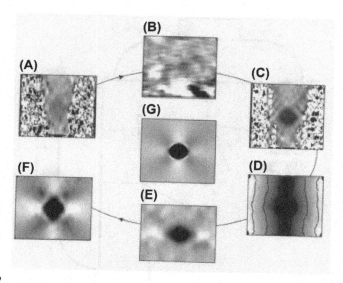

FIGURE 8.2

Weighted interpolation and iterative method to estimate axial and lateral strain tensors in a simulated inhomogeneous single inclusion phantom for a 3% applied deformation. (A,B) The first axial and lateral strain image estimates without correction; (C) the second iteration used to estimate the axial strain image after global temporal stretching, with (D) the corresponding lateral displacement image; and (E) the second corrected lateral strain image. (F) The axial strain images obtained after the third iteration with global stretching and lateral correction, with (G) the ideal axial strain image.

Reproduced with permission from Elsevier Publishing Company through the Copyright Clearance Center. Originally Figure 8 in E. Konofagou, J. Ophir, A new elastographic method for estimation and imaging of lateral displacements, lateral strains, corrected axial strains and Poisson's ratios in tissues, Ultrasound Med. Biol. 24 (1998):1183–1199.

3.4 Altered point spread functions and synthetic aperture system

In this method the authors propose to introduce nonaxial oscillations or a modulation of their point spread function by utilizing specialized beamformers and focusing to improve lateral motion estimation [12,13].

Several other approaches to increase the accuracy of lateral tracking by interpolation techniques have been proposed [10,14]. Synthetic lateral phase information has also been utilized to improve lateral strain estimation [15]. Synthetic aperture systems have been utilized to estimate axial and lateral strains, by increasing the A-line density [16].

3.5 Angular beam-steered data acquisition approach

An approach to estimate both the axial and lateral shear strains within a 2D imaging plane was proposed by Techavipoo et al. [17]. They utilized a phased array

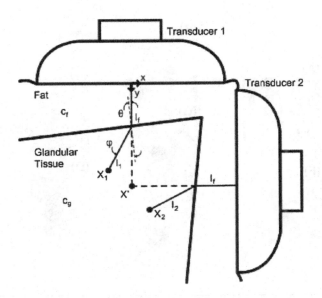

FIGURE 8.3

A limitation that is associated with the use of dual transducers includes the possible variations in the speed of sound between the two paths. If the two transducers are imaging the same point X′ in the figure, due to speed of sound changes, there might be beam deflection and the transducers may end up scanning the points X_1 and X_2.

Reproduced with permission from the Elsevier Publishing Company through the Copyright Clearance Center. Originally Figure 10 in J.M. Abeysekera, R. Zahiri Azar, O. Goksel, R. Rohling, S.E. Salcudean, Analysis of 2-D motion tracking in ultrasound with dual transducers, Ultrasonics 52 (2012):156–168.

transducer for data acquisition that was translated over a tissue surface to obtain angular RF data over ±45 degrees along a 2D scan plane. RF data at specified beam-steered angles from the translated phased array dataset were collated into pre- and postdeformation pairs and processed as a set of beam-steered datasets to estimate local angular displacements along the beam propagation direction [17]. Unique axial and lateral displacement estimates are from the angular displacements using a least squares solution.

This approach was extended to linear array transducers using electronic beam-steering by Rao et al. [18], eliminating registration errors incurred in the translation of a phased array transducer. Mechanical translation of the phased array transducer can lead to motion errors in transducer positioning, which is absent when the transducer beams are steered electronically with linear array transducers. RF frames in a phased array geometry acquired using this approach had to be reorganized to generate RF frames at specific beam-steered angles. Beam-steered data acquisition on a linear array transducer is, however, limited to the ±15-degree range owing to possible grating lobes and increased noise artifacts. Utilization of beam-steered data also improves the elastographic signal-to-noise and contrast-to-noise ratios in the resulting strain tensor images. Beam-steered approaches, in addition, are not pitch limited like some of the previously described approaches that utilize altered transmit pulses.

Normal and shear strain tensors were initially estimated under the assumption that the angular displacement noise was independent and identically distributed [17]. Angular displacement noise was incorporated into the least squares estimation approach using a cross-correlation matrix of the displacement noise artifacts [19], thereby eliminating any assumptions on the angular displacement noise properties. In addition, this approach was utilized to perform axial and lateral strain estimation without the use of the incompressibility assumption [20].

4. Two-dimensional deformation tracking and estimation

Algorithms that utilized a 2D kernel for deformation tracking using block-matching were described as early as 1998 for elasticity imaging [21], which were adapted from approaches for 2D Doppler velocity estimations [22]. However, many of these approaches were initially focused only on axial displacement and strain estimation but were later expanded using 2D normalized cross-correlation approaches to the estimation of lateral displacements and strain [23−30].

4.1 Subsample displacement estimation

Accurate and unbiased 2D subsample or subpixel estimation of the displacement is essential for the displacement vector and strain tensor estimation. Simultaneous 2D subsample estimation of the displacement vector is recommended over 1D interpolation approaches to improve the accuracy and precision of the unbiased strain tensor estimation [26,31−33].

4.2 Direct axial and lateral displacement estimation using two-dimensional kernels

Many investigators have reported on the use of 2D kernels for axial and lateral displacement and strain estimation, which has been summarized by Lopata et al. [23]. Lopata et al. report that a coarse to fine approach [24] that utilizes envelope data initially and RF data in the final stage, with 2D parabolic interpolation to obtain subsample displacement estimates, provided the best results for axial and lateral displacement and strain estimation [25]. A multilevel coarse-to-fine hierarchic approach utilizing nonoverlapping data kernels for deformation tracking and strain estimation using a 2D cross-correlation kernel [34] has demonstrated accurate tracking of both axial and lateral displacements and strain. Several processing strategies such as dynamic frame skip, Bayesian regularization [35], and subsample interpolation [31] were incorporated into this algorithm, enabling estimation of both axial and lateral displacements and strain with signal-to-noise ratios in the same order of magnitude for specific ranges of deformation [34,36]. A GPU-based implementation of this algorithm has been reported [36].

4.3 Coupled axial and lateral displacement estimation using two-dimensional kernels

Coupled displacement estimation has also been utilized to further improve the accuracy and precision of lateral displacement and strain estimation [26–29]. Jiang and Hall initially estimate strain using 2D block-matching followed by appropriate scaling of the postdeformation kernel, assuming local tissue incompressibility. The correlation function around the peak is then used to determine an isocontour center to obtain the lateral displacement and strain. Brusseau et al. [27] utilized a direction-dependent model of tissue motion, along with constrained maximization, to estimate the axial and lateral strains. Multidimensional spline-based motion estimation [28] or minimization of cost functions [29] have also been proposed to obtain the axial and lateral displacements.

4.4 Two-dimensional processing of angular beam-steered data acquisitions

Processing of beam-steered RF data using 2D kernels [37] enables the use of improved deformation tracking with smaller kernel dimensions, enabling high spatial resolution and high signal-to-noise ratio strain images from multiple insonifications [38,39]. With beam-steered data, the 2D kernels are parallelogram shaped as opposed to rectangular kernels [38]. In addition, use of 2D kernels provides additional orthogonal displacement vectors (along and perpendicular to the beam propagation direction) that can be utilized to further improve the reliability of strain estimation utilizing the least squares approach used with 1D processing [17].

4.5 Model-based elastography

Lateral strain estimation methods have also been developed in the context of solving the inverse problem without information regarding the 3D stress tensor [16], or those using a Lagrangian speckle model [40–42] or affine model based approaches [2,43]. Model-based approaches may not require the use of derivatives to estimate the strain tensors.

4.6 Minimization and regularization

Regularization algorithms have been utilized to enhance the displacement tracking accuracy to obtain improved lateral strain estimation [35,44–47]. Regularization approaches utilized include Bayesian approaches [35], constraints such as tissue incompressibility [44], displacement variances [45,46], and geometric regularization [47]. Analytic minimization of a regularized cost function has also been utilized to perform lateral strain estimation [29].

4.7 Plane wave compounding and directional beamforming

Newer beamforming strategies such as plane wave compounding [48–50] and transverse oscillations [51,52] have also been utilized to improve lateral strain estimation. Ultrasonographic systems that utilize plane wave imaging, transverse oscillations, and directional beamforming approaches are becoming more widespread [51,53]. Systems that utilize plane wave imaging for B-mode, Doppler, and strain imaging provide significantly improved frame rates that can be utilized to improve imaging performance. Plane wave compounding has the potential to provide high-precision lateral strain estimation [48–50]. Coherent plane wave compounding approaches have also been used for lateral strain estimation using the incompressibility assumption for vascular applications [54]. Coherent compounding of diverging waves has been utilized for cardiac strain imaging [55].

5. Clinical applications of lateral and shear strain estimation

In the current practice of ultrasound elastography, because of the widespread availability of only 2D scan planes, only the axial and lateral strain components within this 2D imaging plane have been estimated and described in the literature. For cardiac applications, imaging of orthogonal components has translated to radial and circumferential strain imaging within the 2D imaging plane.

Lateral and shear strain estimation has found clinical applications in the classification of breast cancers into malignant or benign classes based on the mobility of the mass [57], poroelastography [58], and the characterization of instability in carotid plaque due to the deformation of softer plaque tissue over a cardiac cycle [40,56,59–61]. Shear strain imaging provides information on the bonding of the mass boundary with the surrounding background tissue. The resulting strain distribution at the interface provides the ability to differentiate between benign and malignant masses [62,63]. However, it has to be noted that this approach is not limited to cardiovascular or breast cancer imaging applications. An example of the axial, lateral, and shear strain tensors obtained from the carotid artery of a patient is illustrated in Fig. 8.4. The data analysis for this patient was performed using a hierarchic multilevel approach that utilizes axial and lateral displacements computed using a 2D cross-correlation kernel [34]. In Fig. 8.4, positive and negative strains (denoted by the colorbar) are associated with stretching and compression of plaque over a cardiac cycle, with the variation of the strain distribution inside the plaque possibly due to plaque heterogeneity. Clinical utility of axial, lateral, and shear strain indices derived from axial, lateral, and shear strain tensors includes the relationship associating high strain indices to poorer cognitive scores [64,65]. In asymptomatic

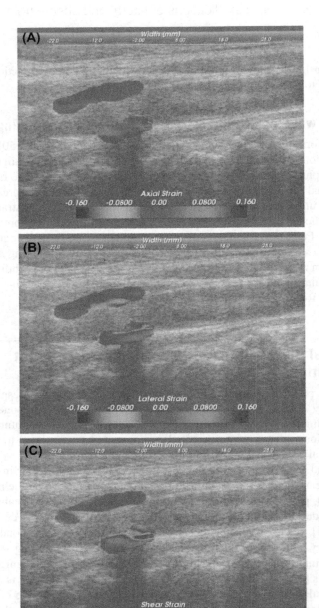

FIGURE 8.4

(A) Axial, (B) lateral, and (C) shear strain images in the segmented region overlaid on the B-mode image.

Reproduced with permission from the Elsevier Publishing Company through Copyright Clearance Center. Originally Figure 2 in X. Wang, D.C. Jackson, C.C. Mitchell, T. Varghese, S.M. Wilbrand, B.G. Rocque, et al., Classification of symptomatic and asymptomatic patients with and without cognitive decline using non-invasive carotid plaque based strain indices as biomarkers, Ultrasound Med. Biol. 42 (2016):909–918.

patients, these indices are related to the initial decline in executive function, whereas in symptomatic patients, the indices are also correlated with other cognitive domains such as language, visuospatial reasoning, and memory function [66]. We also observed increased correlation of lateral strain indices to composite cognitive scores in symptomatic patients, which is hypothesized to be due to the increased lateral strain with possible fissures, when compared with intact plaque in asymptomatic patients [56].

Researchers have also utilized imaging of only the axial shear component (lateral gradient of the axial displacement estimates) of the full shear strain distribution, as lateral displacement and strain estimation have poorer spatial resolution when compared with the corresponding axial displacement and strain estimation, to characterize the infiltration of breast tumors to normal background tissue at their interface [62,63]. As many approaches only estimate axial displacement and strain, axial shear has been used to provide the supplementary bonding information. An example of axial shear strain images comparing benign and malignant breast masses is shown in Fig. 8.5. The normalized axial shear strain area (normalized to the lesion size, applied strain, and lesion strain contrast) described in Ref. [67], and abbreviated as NASSA [68], is outlined in purple in Fig. 8.5 by an observer for both masses. NASSA values are significantly larger for malignant breast masses than for benign masses, as shown in Fig. 8.5 [62,63]. The NASSA feature has been utilized in classifiers to differentiate between benign and malignant masses [62,69], demonstrating significantly better performance than the size ratio or the strain stiffness contrast previously utilized for differentiation [62].

Poroelastography, that is, the imaging of the lateral to axial strain ratio under stress relaxation, is another parameter that has been utilized to characterize edematous soft tissue [58]. Poroelastic tissue has been characterized as a solid matrix

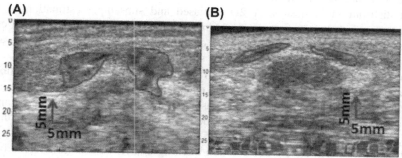

(A)   **(B)**

FIGURE 8.5

Axial shear strain images overlaid in color on B-mode images of (A) a cancer and (B) a fibroadenoma.

Reproduced with permission from the Elsevier Publishing Company through Copyright Clearance Center. Originally Figure 4 in A.K. Thittai, J.M. Yamal, J. Ophir, Small breast lesion classification performance using the normalized axial-shear strain area feature, Ultrasound Med. Biol. 39 (2013):543–548.

containing fluid that is displaced under the application of a constant stress or deformation. The lateral-to-axial strain ratio of this tissue matrix decreases exponentially with time with the time constant dependent on the modulus of the matrix, permeability, and tissue dimensions along the direction of fluid movement [58].

5.1 Radial and circumferential strain

Radial and circumferential strain has been utilized for cardiac and cardiovascular applications instead of the axial and lateral strain tensor components [59,70—73]. Approaches to convert axial and lateral strain tensors to their corresponding polar coordinates have also been reported [71]. For cardiac strain imaging of the left ventricle a majority of the algorithms are based on 2D and 3D speckle tracking of B-mode echo signals to obtain longitudinal, radial, and circumferential strains [74]. However, several groups have reported on strain tensor estimation for cardiac applications using RF echo signals, including estimation of radial and circumferential strain tensors [70—73]. Cross-sectional imaging of blood vessels is another area where radial and circumferential strain has been utilized [59]. Intravascular elastography is based on specialized transducers and imaging geometries that feature radial and circumferential strain tensor estimations [59,75—77].

6. Conclusion

Commercial implementations of strain imaging only focus on the estimation and imaging of the axial component of the displacement vector and strain tensors. However, lateral displacement vector and strain estimation, along with concomitant estimation of the shear strain component, within the 2D imaging plane can be obtained. Lateral strain estimation is essential to obtain an accurate depiction of the strain distribution in tissues. A 2D unbiased and subsample estimation of the displacement vector is necessary to obtain accurate and precise lateral strain estimates. Improvements in the accuracy of 2D subsample estimation have also enhanced the quality of lateral and shear strain estimation [34,36].

With the advent of 2D transducer arrays, future estimation of the elevational displacement vector and strain tensor will eventually become feasible, enabling estimation of all normal and shear strain tensors. However, independent estimation of the normal strain tensors has to be performed without the use of simplifying assumptions, such as incompressibility, to obtain uncorrelated information.

Although 3D kernels for deformation tracking and strain estimation have been reported in the literature for data acquired using 1D or 1.5 D transducer arrays, high-resolution elevational displacements and strain have not been estimated. We anticipate that the lateral and shear strain imaging approaches described previously will be extended to elevational strain and the corresponding shear strain imaging when full 2D transducer arrays with larger than the current 32×32 aperture become available.

3D estimation of the displacement vector field and the corresponding normal and shear strain will enable quantification and thereby enhance the reproducibility of strain imaging. Young's modulus reconstruction approaches will produce accurate and unique solutions in the presence of more accurate and reliable displacement and strain tensor information [78]. In a similar manner, noninvasive temperature estimation techniques using ultrasonic techniques will more accurately estimate the resulting tissue expansion or contraction in three dimensions.

References

[1] A.S. Saada, Elasticity: Theory and Applications, Pergamon Press, New York, 1974.
[2] X. Pan, K. Liu, J. Shao, J. Gao, L. Huang, J. Bai, et al., Performance comparison of rigid and affine models for motion estimation using ultrasound radio-frequency signals, IEEE Trans. Ultrason. Ferroelectr. Freq. Control 62 (2015) 1928–1943.
[3] M.H. RoyCardinal, M.H.G. Heusinkveld, Z. Qin, R.G.P. Lopata, C. Naim, G. Soulez, et al., Carotid artery plaque vulnerability assessment using noninvasive ultrasound elastography: validation with MRI, Am. J. Roentgenol. 209 (2017) 142–151.
[4] M. Rao, T. Varghese, E.L. Madsen, Shear strain imaging using shear deformations, Med. Phys. 35 (2008) 412–423.
[5] J. Ophir, I. Cespedes, H. Ponnekanti, Y. Yazdi, X. Li, Elastography: a quantitative method for imaging the elasticity of biological tissues, Ultrason. Imaging 13 (1991) 111–134.
[6] T. Varghese, J. Ophir, A theoretical framework for performance characterization of elastography: the strain filter, IEEE Trans. Ultrason. Ferroelectr. Freq. Control 44 (1997) 164–172.
[7] J. Ophir, F. Kallel, T. Varghese, M. Bertrand, I. Cespedes, H. Ponnekanti, Elastography: a systems approach, Int. J. Imaging Syst. Technol. 8 (1997) 89–103.
[8] T. Varghese, Quasi-Static ultrasound elastography, Ultrasound Clin. 4 (2009) 323–338.
[9] M.A. Lubinski, S.Y. Emelianov, K.R. Raghavan, A.E. Yagle, A.R. Skovoroda, M. O'Donnell, Lateral displacement estimation using tissue incompressibility, IEEE Trans. Ultrason. Ferroelectr. Freq. Control 43 (1996) 247–256.
[10] E. Konofagou, J. Ophir, A new elastographic method for estimation and imaging of lateral displacements, lateral strains, corrected axial strains and Poisson's ratios in tissues, Ultrasound Med. Biol. 24 (1998) 1183–1199.
[11] J.M. Abeysekera, R. Zahiri Azar, O. Goksel, R. Rohling, S.E. Salcudean, Analysis of 2-D motion tracking in ultrasound with dual transducers, Ultrasonics 52 (2012) 156–168.
[12] M.E. Anderson, Multi-dimensional velocity estimation with ultrasound using spatial quadrature, IEEE Trans. Ultrason. Ferroelectr. Freq. Control 45 (1998) 852–861.
[13] J.A. Jensen, P. Munk, A new method for estimation of velocity vectors, IEEE Trans. Ultrason. Ferroelectr. Freq. Control 45 (1998) 837–851.
[14] H. Chen, T. Varghese, Multilevel hybrid 2D strain imaging algorithm for ultrasound sector/phased arrays, Med. Phys. 36 (2009) 2098–2106.
[15] X. Chen, M.J. Zohdy, S.Y. Emelianov, M. O'Donnell, Lateral speckle tracking using synthetic lateral phase, IEEE Trans. Ultrason. Ferroelectr. Freq. Control 51 (2004) 540–550.

[16] S. Korukonda, M.M. Doyley, Estimating axial and lateral strain using a synthetic aperture elastographic imaging system, Ultrasound Med. Biol. 37 (2011) 1893−1908.

[17] U. Techavipoo, Q. Chen, T. Varghese, J. Zagzebski, Estimation of displacement vectors and strain tensors in elastography using angular insonifications, IEEE Trans. Med. Imaging 23 (2004) 1479−1489.

[18] M. Rao, Q. Chen, H. Shi, T. Varghese, E.L. Madsen, J.A. Zagzebski, et al., Normal and shear strain estimation using beam steering on linear-array transducers, Ultrasound Med. Biol. 33 (2007) 57−66.

[19] H. Chen, T. Varghese, Noise analysis and improvement of displacement vector estimation from angular displacements, Med. Phys. 35 (2008) 2007−2017.

[20] M. Rao, T. Varghese, Spatial angular compounding for elastography without the incompressibility assumption, Ultrason. Imaging 27 (2005) 256−270.

[21] F. Yeung, S.F. Levinson, K.J. Parker, Multilevel and motion model-based ultrasonic speckle tracking algorithms, Ultrasound Med. Biol. 24 (1998) 427−442.

[22] L.N. Bohs, B.H. Friemel, B.A. McDermott, G.E. Trahey, A real time system for quantifying and displaying two-dimensional velocities using ultrasound, Ultrasound Med. Biol. 19 (1993) 751−761.

[23] R.G. Lopata, M.M. Nillesen, H.H. Hansen, I.H. Gerrits, J.M. Thijssen, C.L. de Korte, Performance evaluation of methods for two-dimensional displacement and strain estimation using ultrasound radio frequency data, Ultrasound Med. Biol. 35 (2009) 796−812.

[24] C. Pellot-Barakat, F. Frouin, M.F. Insana, A. Herment, Ultrasound elastography based on multiscale estimations of regularized displacement fields, IEEE Trans. Med. Imaging 23 (2004) 153−163.

[25] H. Shi, T. Varghese, Two-dimensional multi-level strain estimation for discontinuous tissue, Phys. Med. Biol. 52 (2007) 389−401.

[26] J. Jiang, T.J. Hall, A coupled subsample displacement estimation method for ultrasound-based strain elastography, Phys. Med. Biol. 60 (2015) 8347−8364.

[27] E. Brusseau, J. Kybic, J.F. Deprez, O. Basset, 2-D locally regularized tissue strain estimation from radio-frequency ultrasound images: theoretical developments and results on experimental data, IEEE Trans. Med. Imaging 27 (2008) 145−160.

[28] F. Viola, R.L. Coe, K. Owen, D.A. Guenther, W.F. Walker, MUlti-Dimensional Spline-Based Estimator (MUSE) for motion estimation: algorithm development and initial results, Ann. Biomed. Eng. 36 (2008) 1942−1960.

[29] H. Rivaz, E.M. Boctor, M.A. Choti, G.D. Hager, Real-time regularized ultrasound elastography, IEEE Trans. Med. Imaging 30 (2011) 928−945.

[30] H. Li, Y. Guo, W.N. Lee, Systematic performance evaluation of a cross-correlation-based ultrasound strain imaging method, Ultrasound Med. Biol. 42 (2016) 2436−2456.

[31] M.M. McCormick, T. Varghese, An approach to unbiased subsample interpolation for motion tracking, Ultrason. Imaging 35 (2013) 76−87.

[32] C. Sumi, Fine elasticity imaging utilizing the iterative RF-echo phase matching method, IEEE Trans. Ultrason. Ferroelectr. Freq. Control 46 (1999) 158−166.

[33] E.S. Ebbini, Phase-coupled two-dimensional speckle tracking algorithm, IEEE Trans. Ultrason. Ferroelectr. Freq. Control 53 (2006) 972−990.

[34] M. McCormick, T. Varghese, X. Wang, C.C. Mitchell, M.A. Kliewer, R.J. Dempsey, Methods for robust in vivo strain estimation in the carotid artery, Phys. Med. Biol. 57 (2012) 7329−7353.

[35] M. McCormick, N. Rubert, T. Varghese, Bayesian regularization applied to ultrasound strain imaging, IEEE Trans. Biomed. Eng. 58 (2011) 1612–1620.

[36] N.H. Meshram, T. Varghese, GPU accelerated multilevel Lagrangian carotid strain imaging, IEEE Trans. Ultrason. Ferroelectr. Freq. Control 65 (2018) 1370–1379.

[37] P. Verma, M.M. Doyley, Revisiting the Cramér Rao lower bound for elastography: predicting the performance of axial, lateral and polar strain elastograms, Ultrasound Med. Biol. 43 (2017) 1780–1796.

[38] H. Xu, T. Varghese, Normal and shear strain imaging using 2D deformation tracking on beam steered linear array datasets, Med. Phys. 40 (2013) 012902.

[39] H.H.G. Hansen, R.G.P. Lopata, C.L. de Korte, Noninvasive carotid strain imaging using angular compounding at large beam steered angles: validation in vessel phantoms, IEEE Trans. Med. Imaging 28 (2009) 872–880.

[40] E. Mercure, G. Cloutier, C. Schmitt, R.L. Maurice, Performance evaluation of different implementations of the Lagrangian speckle model estimator for non-invasive vascular ultrasound elastography, Med. Phys. 35 (2008) 3116–3126.

[41] R.L. Maurice, M. Bertrand, Speckle-motion artifact under tissue shearing, IEEE Trans. Ultrason. Ferroelectr. Freq. Control 46 (1999) 584–594.

[42] W.N. Lee, Z. Qian, C.L. Tosti, T.R. Brown, D.N. Metaxas, E.E. Konofagou, Preliminary validation of angle-independent myocardial elastography using MR tagging in a clinical setting, Ultrasound Med. Biol. 34 (2008) 1980–1997.

[43] C. Carvalho, P. Slagmolen, S. Bogaerts, L. Scheys, J. D'hooge, K. Peers, et al., 3D tendon strain estimation using high-frequency volumetric ultrasound images: a feasibility study, Ultrason. Imaging 40 (2) (2018) 67–83.

[44] L. Guo, Y. Xu, Z. Xu, J. Jiang, A PDE-based regularization algorithm toward reducing speckle tracking noise: a feasibility study for ultrasound breast elastography, Ultrason. Imaging 37 (2015) 277–293.

[45] C. Sumi, K. Sato, Regularization for ultrasonic measurements of tissue displacement vector and strain tensor, IEEE Trans. Ultrason. Ferroelectr. Freq. Control 55 (2008) 787–799.

[46] C. Sumi, T. Itoh, Spatially variant regularization of lateral displacement measurement using variance, Ultrasonics 49 (2009) 459–465.

[47] F. Kremer, H.F. Choi, S. Langeland, E. D'Agostino, P. Claus, J. D'hooge, Geometric regularization for 2-D myocardial strain quantification in mice: an in-silico study, Ultrasound Med. Biol. 36 (2010) 1157–1168.

[48] R. Nayak, S. Huntzicker, J. Ohayon, N. Carson, V. Dogra, G. Schifitto, et al., Principal strain vascular elastography: simulation and preliminary clinical evaluation, Ultrasound Med. Biol. 43 (2017) 682–699.

[49] H.H. Hansen, A.E. Saris, N.R. Vaka, M.M. Nillesen, C.L. de Korte, Ultrafast vascular strain compounding using plane wave transmission, J. Biomech. 47 (2014) 815–823.

[50] Q. He, L. Tong, L. Huang, J. Liu, Y. Chen, J. Luo, Performance optimization of lateral displacement estimation with spatial angular compounding, Ultrasonics 73 (2017) 9–21.

[51] B. Heyde, N. Bottenus, J. D'hooge, G.E. Trahey, Evaluation of the transverse oscillation technique for cardiac phased array imaging: a theoretical study, IEEE Trans. Ultrason. Ferroelectr. Freq. Control 64 (2017) 320–334.

[52] H. Liebgott, A. Basarab, P. Gueth, D. Friboulet, P. Delachartre, Transverse oscillations for tissue motion estimation, Ultrasonics 50 (2010) 548–555.

[53] J. Jensen, C.A. Villagomez-Hoyos, M.B. Stuart, C. Ewertsen, M.B. Nielsen, J.A. Jensen, Fast plane wave 2-D vector flow imaging using transverse oscillation and directional beamforming, IEEE Trans. Ultrason. Ferroelectr. Freq. Control 64 (2017) 1050−1062.

[54] J. Porée, D. Garcia, B. Chayer, J. Ohayon, G. Cloutier, Noninvasive vascular elastography with plane strain incompressibility assumption using ultrafast coherent compound plane wave imaging, IEEE Trans. Med. Imaging 34 (2015) 2618−2631.

[55] J. Grondin, V. Sayseng, E.E. Konofagou, Cardiac strain imaging with coherent compounding of diverging waves, IEEE Trans. Ultrason. Ferroelectr. Freq. Control 64 (2017) 1212−1222.

[56] X. Wang, D.C. Jackson, C.C. Mitchell, T. Varghese, S.M. Wilbrand, B.G. Rocque, et al., Classification of symptomatic and asymptomatic patients with and without cognitive decline using non-invasive carotid plaque based strain indices as biomarkers, Ultrasound Med. Biol. 42 (2016) 909−918.

[57] E.E. Konofagou, T. Harrigan, J. Ophir, Shear strain estimation and lesion mobility assessment in elastography, Ultrasonics 38 (2000) 400−404.

[58] R. Righetti, B.S. Garra, L.M. Mobbs, C.M. Kraemer-Chant, J. Ophir, T.A. Krouskop, The feasibility of using poroelastographic techniques for distinguishing between normal and lymphedematous tissues in vivo, Phys. Med. Biol. 52 (2007) 6525−6541.

[59] H. Ribbers, R.G. Lopata, S. Holewijn, G. Pasterkamp, J.D. Blankensteijn, C.L. de Korte, Noninvasive two-dimensional strain imaging of arteries: validation in phantoms and preliminary experience in carotid arteries in vivo, Ultrasound Med. Biol. 33 (2007) 530−540.

[60] H. Shi, C.C. Mitchell, M. McCormick, M.A. Kliewer, R.J. Dempsey, T. Varghese, Preliminary in vivo atherosclerotic carotid plaque characterization using the accumulated axial strain and relative lateral shift strain indices, Phys. Med. Biol. 53 (2008) 6377−6394.

[61] C. Schmitt, G. Soulez, R.L. Maurice, M.F. Giroux, G. Cloutier, Noninvasive vascular elastography: toward a complementary characterization tool of atherosclerosis in carotid arteries, Ultrasound Med. Biol. 33 (2007) 1841−1858.

[62] H. Xu, T. Varghese, J. Jiang, J.A. Zagzebski, In vivo classification of breast masses using features derived from axial-strain and axial-shear images, Ultrason. Imaging 34 (2012) 222−236.

[63] A. Thitaikumar, L.M. Mobbs, C.M. Kraemer-Chant, B.S. Garra, J. Ophir, Breast tumor classification using axial shear strain elastography: a feasibility study, Phys. Med. Biol. 53 (2008) 4809−4823.

[64] R.J. Dempsey, T. Varghese, D.C. Jackson, X. Wang, N. Meshram, C.C. Mitchell, et al., Ultrasound plaque strain determines carotid atherosclerotic plaque instability and cognition in asymptomatic patients with significant stenosis, J. Neurosurg. 128 (2018) 111−119.

[65] N.H. Meshram, T. Varghese, C.C. Mitchell, D.C. Jackson, S.M. Wilbrand, B.P. Hermann, et al., Quantification of carotid artery plaque stability with multiple region of interest based ultrasound strain indices and relationship with cognition, Phys. Med. Biol. 62 (2017) 6341−6360.

[66] N.H. Meshram, D. Jackson, T. Varghese, C.C. Mitchell, S.M. Wilbrand, R.J. Dempsey, et al., A cross-sectional investigation of cognition and ultrasound-based vascular strain indices, Arch. Clin. Neuropsychol. (2019), https://doi.org/10.1093/arclin/acz006 [Epub ahead of print].

[67] A. Thitaikumar, T.A. Krouskop, B.S. Garra, J. Ophir, Visualization of bonding at an inclusion boundary using axial-shear strain elastography: a feasibility study, Phys. Med. Biol. 52 (2007) 2615–2633.

[68] H. Xu, M. Rao, T. Varghese, A. Sommer, S. Baker, T.J. Hall, et al., Axial-shear strain imaging for differentiating benign and malignant breast masses, Ultrasound Med. Biol. 36 (2010) 1813–1824.

[69] A.K. Thittai, J.M. Yamal, J. Ophir, Small breast lesion classification performance using the normalized axial-shear strain area feature, Ultrasound Med. Biol. 39 (2013) 543–548.

[70] R.G. Lopata, M.M. Nillesen, J.M. Thijssen, L. Kapusta, C.L. de Korte, Three-dimensional cardiac strain imaging in healthy children using RF-data, Ultrasound Med. Biol. 37 (2011) 1399–1408.

[71] C. Ma, X. Wang, T. Varghese, Segmental analysis of cardiac short axes views using Lagrangian radial and circumferential strain, Ultrason. Imaging 38 (2016) 363–383.

[72] S.J. Okrasinski, B. Ramachandran, E.E. Konofagou, Assessment of myocardial elastography performance in phantoms under combined physiologic motion configurations with preliminary in vivo feasibility, Phys. Med. Biol. 57 (2012) 5633–5650.

[73] W.N. Lee, C.M. Ingrassia, S.D. Fung-Kee-Fung, K.D. Costa, J.W. Holmes, E.E. Konofagou, Theoretical quality assessment of myocardial elastography with in vivo validation, IEEE Trans. Ultrason. Ferroelectr. Freq. Control 54 (2007) 2233–2245.

[74] M. Alessandrini, B. Heyde, S. Queiros, S. Cygan, M. Zontak, O. Somphone, et al., Detailed evaluation of five 3D speckle tracking algorithms using synthetic echocardiographic recordings, IEEE Trans. Med. Imaging 35 (8) (2016) 1915–1926.

[75] M.S. Richards, M.M. Doyley, Non-rigid image registration based strain estimator for intravascular ultrasound elastography, Ultrasound Med. Biol. 39 (2013) 515–533.

[76] Y. Liang, H. Zhu, M.H. Friedman, Measurement of the 3D arterial wall strain tensor using intravascular B-mode ultrasound images: a feasibility study, Phys. Med. Biol. 55 (2010) 6377–6394.

[77] C.L. de Korte, A.F. van der Steen, Intravascular ultrasound elastography: an overview, Ultrasonics 40 (2002) 859–865.

[78] M.S. Richards, P.E. Barbone, A.A. Oberai, Quantitative three-dimensional elasticity imaging from quasi-static deformation: a phantom study, Phys. Med. Biol. 54 (2009) 757–779.

Optical elastography on the microscale

9

Philip Wijesinghe[1,2], Brendan F. Kennedy[2,3], David D. Sampson[1,4]

[1]*Optical+Biomedical Engineering Laboratory, Department of Electrical, Electronic and Computer Engineering, The University of Western Australia, Perth, WA, Australia;* [2]*BRITElab, Harry Perkins Institute of Medical Research, QEII Medical Centre, Nedlands, WA, Australia;* [3]*Department of Electrical, Electronic and Computer Engineering, School of Engineering, The University of Western Australia, Perth, WA, Australia;* [4]*University of Surrey, Surrey, United Kingdom*

1. Introduction

Optical elastography describes a suite of techniques that use optical methods to form images of the mechanical properties of tissues. Properties such as elasticity and viscosity are of vital importance in tissue form and function and can serve as indicators or predictors of disease [1−5]. Optical elastography has been researched for the past 20 years, with over 500 papers[1] published to date, ~75% of which were published in the past 5 years. The recent rapid increase in momentum is credited to the maturation of optoelectronic and photonics technology, the provision of sufficient computing power, and related developments in biomechanical imaging in the fields of cell mechanics and medicine [6−8].

Optical elastography offers a window into tissue mechanics on the microscale between that of whole organs, as probed by magnetic resonance (MR) and ultrasound (US) elastography, and that of cells, as probed by microscopic techniques, such as atomic force microscopy (AFM) and traction force microscopy. With demonstrated resolutions in the range 2−100 μm, optical elastography presents new opportunities for the detection and understanding of diseases, such as cancer [9,10], eye diseases [11−13], and vascular diseases [14−17], as well as insight into the underlying cellular biomechanics [7,18−20]. The goal of optical elastography research is broadly twofold: to distinguish tissue features using mechanical contrast, often toward the goal of detecting signatures of disease in medicine, and to accurately measure an intrinsic mechanical property, often used to inform on the biological mechanisms of disease formation.

In this chapter, we describe the main technologies and methods of optical elastography within the context of their application and touch briefly upon the history and provide our perspectives on the likely future directions of the field. Although

---

[1] Web of Science: topic optical elastography.

Tissue Elasticity Imaging. https://doi.org/10.1016/B978-0-12-809661-1.00009-1

a plethora of optical elastography techniques have been demonstrated, we focus mainly on two techniques that have become prominent over the past 5 years: optical coherence elastography (OCE) and Brillouin microscopy. We briefly describe other techniques, including laser speckle imaging (LSI), photoacoustic elastography, and US-modulated optical tomography. We limit this chapter to *tissue elasticity imaging*, i.e., techniques involving the imaging of solid tissue mechanics. Given our scope, we do not discuss flow measurement, rheology, and point-measurement techniques. For a more comprehensive treatment of optical measurement techniques in mechanics, we direct the reader to Refs. [7,17,21].

1.1 Brief history of optical elastography

For the purposes of this chapter, we consider optical elastography to be the mapping of mechanical properties of tissues into an image. Optical elastography has been a relatively recent development, but the origins of using optics to observe mechanical behavior date back over 300 years [22]. In 1665, in his seminal *Micrographia*, Robert Hooke communicated his observations of the microscopic world though a microscope. Independently, Hooke was concurrently theorizing on the law of elasticity in springs (which we now know as Hooke's law), stimulating Newton's later discoveries of the laws of motion [23], which form the cornerstone of contemporary continuum mechanics. It was between 1676 and 1723 that Antonie van Leeuwenhoek sent close to 190 letters to The Royal Society describing his first observations of microorganisms and cells, noting, in particular, their proclivity for extreme motion [22,24] (i.e., noting that cells are dynamic and not simply a static scaffold of life). Centuries later, in much of the early work in cellular physiology conducted in the first half of the 20th century, mechanics would feature prominently. In studying the protoplasm (contents of a cell) [25], which was a prominent field of research in the late 1800s and early 1900s, the earliest widely quantified mechanical property of the cell was viscosity, which was postulated to have a profound effect on the dynamics of microorganisms and cells seen under a microscope [26]. Photographic and stroboscopic analysis in the 1950s allowed motion to be studied in more detail, elucidating biological mechanisms such as ciliary motion [27]. Stroboscopic analysis was used to complement bulk rheometer characterization of the viscoelasticity of tissues [28]. The 1950s also saw the emergence of microscopy-based micromanipulation techniques, such as the cell elastimeter [29], which was later renamed micropipette aspiration.

Early methods were held back by the inability to readily record or to digitize the signal, making subsequent manipulation to extract deeper insights largely unfeasible. The digital revolution in computing that started in the late 1950s allowed the field to progress from mostly qualitative observation to more readily quantifiable mechanical measurement. Furthermore, the digital revolution helped initiate the rise of medical imaging in the 1970s, with the development and adoption of X-ray computed tomography, MR imaging, and ultrasonography. MR- and US-based elastography were proposed thereafter [30—32]. The resolution scale of these techniques, however, did not reach that achievable with optical methods, and in contrast to the optics-based techniques, which looked into the cellular and biophysical makeup of life, MR- and US-based elastography were mainly motivated to noninvasively diagnose and characterize a disease. In fact, it is in the context of these techniques that the term *elastography* was first coined [30–32]. Such elastography was, and is, primarily motivated to exploit disease-altered mechanical properties of tissues to enhance image contrast and, thereby, diagnosis. It is often compared to *manual palpation*—the common practice in which a physician manually feels the stiffness of a patient's tissues to help inform diagnosis. These techniques have been closely connected with gross-scale biomechanics research, such as the seminal works surveyed by Fung since 1969 [5] and, additionally, share some connection to industrial materials testing techniques, such as uniaxial testing and sonic resonance [33].

The emergence of optical elastography was, perhaps, stimulated by the phenomenon of *laser speckle*—a granularity observed in all coherent imaging techniques arising from the superposition of waves scattered from subwavelength microstructures [34]. Speckle was found to provide intrinsic information about the movement and dynamics of tissues. Around the 1970s, laser speckle fluctuation was rigorously related to the dynamics of particle motion for the first time [17] and was also used to measure vibration of nonbiological samples [35].

By the late 1980s, laser speckle was used to record biological activity [36] and blood flow [37]. By the late 1990s, a myriad of optical techniques for mechanics had been proposed [21]. The demonstration of OCE by Schmitt in 1998 [38] (described in detail in Section 2) first shifted the focus away from predominantly rheology (the study of the flow of matter in liquids and solids, e.g., viscosity) to elasticity imaging. The work of Schmitt, which represented a major milestone in optical elastography, was inspired by the early works in US elastography [30]. In fact, the phenomenon of speckle, via the scattering of acoustic waves, is also present in ultrasonographic imaging. Because of the close similarity of the underlying physics, much of the early work in US elastography has influenced current optical elastography research. Similarly to US elastography, initial demonstrations employed compression loading [38,39], wherein a tissue or material is compressed, and the relative deformation is used to infer elasticity. Stiffer features would deform less, whereas softer features deform more. Beyond this initial work, a broad range of tissue loading methods and analyses of deformation would be explored, and in keeping with the earlier success of US elastography, methods employing acoustic excitation, such as acoustic radiation force (ARF) [40,41], and methods employing the propagation of mechanical waves, such as surface acoustic and shear wave elasticity imaging [42,43], were developed for optical elastography. We will see much of this reflected in Section 2. A major turning point for OCE came with the development of the phase-sensitive method in Fourier-domain OCT for detecting much smaller displacements than had been hitherto possible [44], which we detail in Section 2.2. The higher spatial resolution of optical elastography has further separated it from ultrasonography both in the loading methods used and in the applications being targeted.

We briefly recapitulate the origins of the field and the scope of this chapter. Optical elastography is part of the broader generalization of *using optics to measure mechanics*. The roots of the latter can be largely traced to (1) techniques originating from, and building upon, the first observations of cell and microorganism dynamics, such as AFM and traction force microscopy, and (2) techniques inspired by the physics of speckle, materials testing, and the precedent of MR and US elastography. Although their origins are different, these techniques are merging in applications. This is owing to the recent developments in optical elastography, detailed in this chapter, in which new methods are approaching a mechanical resolution suitable for cellular-scale research.

1.2 Optical elastography: a matter of scale

The relationship between a biological tissue, a disease, the associated mechanical properties and forces (critical in understanding the disease [45]), and the imaging tool fundamentally depends on the *length scale* of observation, which, in turn, determines the necessary spatial and temporal resolution [46]. As the fundamental unit of life, the molecular, biochemical, and mechanical processes that take place within and around a cell determine its form and function. Cells are hierarchically organized into tissues, which come together to form organs that perform specialized physiologic functions. Although much has been achieved on the molecular scale, a remaining challenge in the postgenomic era is to relate dysfunctions in cellular-scale molecular and biochemical processes to organ-scale symptoms of disease [2]—spanning what we might call the 'mesoscale'. Over the past 50 years, the role of biomechanics has been explored across the entire range of this mesoscale [1,2,5,47], laying the foundations of contemporary optical elastography research.

The scale on which mechanics is probed by an elastographic tool often dictates the approach to be taken to detect and understand disease and establishes a niche of suitable applications. For instance, MR and US elastographic methods, with a resolution of ~ 1 mm and ~ 250 μm, respectively, predominantly rely on measuring systemic changes in the mechanics of organs and their immediate surroundings; on this scale, they are particularly suited to the detection and characterization of pathologic conditions, such as cancer and liver diseases, at fairly advanced stages [48–50]. At the other extreme, AFM (resolution of ~ 100 nm) is employed to understand underlying mechanisms, wherein microscopic forces and the mechanics of the cell's microenvironment drive cellular processes and influence the earliest stages of disease [51].

Optical elastography, having demonstrated a resolution in the range $\sim 2-100$ μm, is ideally placed to provide mechanical information on the scale bridging that of cells and organs [52]. This niche has seen optical elastography techniques link to, or mirror, approaches from either side of this range. At the longer end of the range, optical elastography has followed the precedent of MR and US elastography in their attempt to aid in the diagnosis, assessment, and monitoring of diseases, such as cancer [9,10] and eye diseases [11–13]. Although optical elastography offers greater spatial resolution and, thus, potentially greater sensitivity and earlier detection of disease, it is hindered by the limited optical penetration depth ($\sim$1–2 mm) and field of view (typically $\sim$10–15 mm). Thus current challenges lie in increasing the field of view and contrast, developing probe-based methods for deep tissue imaging, and finding applications for which these specifications are sufficient.

At the shorter end of the resolution range, optical elastography aims to translate the understanding and methods of cellular-scale biomechanics to multicellular structures and tissues, in situ and in three dimensions (3D) [6,7]. The challenge, in this case, is the improvement of the spatial resolution and sensitivity with which mechanical properties can be determined, along with the accurate quantification of intrinsic mechanical properties and forces, which necessarily requires computational methods [53].

A key consideration in elastographic research is that the spatial resolution of the mechanical property measurement (mechanical resolution) in these techniques is distinct to the spatial resolution of the optical image (optical resolution). Optical resolution, under ideal conditions, is largely dependent on the resolution of the imaging system and the optical properties of the sample, and sets a limit on mechanical resolution, whereas the mechanical resolution is additionally dependent on the structure and mechanical properties of the tissue, as well as on the model used to reconstruct these mechanical properties from raw measurements of, for example, displacement or velocity; it is typically at least a few times coarser than the optical resolution. The link between mechanical resolution and mechanical properties stems from the continuity and incompressibility of tissues, meaning that, within a local region the response to mechanical loading is also dependent on the response of tissue features in the surrounding regions.

Critical to accurately estimating the mechanical properties of tissues are the model employed and the interplay between it and the loading method. By necessity of speed and robustness, mechanical models employed in elastography are typically derived from some inherent assumption of homogeneity or uniformity of tissue mechanics [54]. As we will describe in this chapter, whole tissue loaded over a region commensurate with the imaging field exhibits "mechanical cross talk" between its regions, not easily decoupled via commonly implemented mechanical models. Localized excitation methods aim to avoid such effects by interrogating only a local region, without loading the rest, whereas computational methods aim to decouple this influence with more sophisticated models, usually requiring superior measurement sensitivity. Thus, as detailed in the later sections, although much of optical elastographic research has been in the improvement of the imaging systems, in the future, understanding these interconnections and improving our capacity to extract unambiguous mechanical properties at high resolution will be a major direction of the field.

Probing with optics, at resolutions higher than those of the other forms of elastography, brings with it a range of distinctive features. The shallow aspect ratio (1−2 mm depth vs. 10−15 mm transverse fields of view) of optical techniques makes them particularly prone to boundary effects, such as friction in contact loading methods. In particular, it causes a higher proportion of the field of view to be influenced by the surface topography. Furthermore, the higher resolution makes optical elastography more sensitive to fine-scale tissue heterogeneities, such as the presence of microvasculature, lymphatic ducts, and other features [9,52], that invalidate the common assumption of a tissue as a continuous solid [4], which would otherwise be averaged and perhaps lost in coarser resolution MR- and US-based elastography.

Comparing closely to US elastography, the shear wave speed in tissue is typically in the range ∼1−10 m/s; thus, optical methods, with typically five times smaller field of view than US elastography, are required to have an acquisition speed five times higher to achieve the same elastic wave front sampling density. Furthermore, in most approaches, the capacity to resolve and distinguish tissue mechanical properties is proportional to one or more of the wavelength of the acoustic waves, the bandwidth of an impulsive load, and the ability to sufficiently confine loading in space and in time. The relative increase in the resolution scale of optical methods over US elastography necessitates higher frequency or higher bandwidth loading (to achieve the same relative performance), which poses a challenge because of the high attenuation or dispersion in both transmitting the excitation to tissue and propagating the mechanical wave over an appreciable field of view. Interestingly, in the early applications of ARF loading in US elastography, the load was considered to be localized [55], but such a load was considered as a uniform load over the field of view of optical elastography [56].

We detail these methods in Section 2.4, and these broader effects of scale are discussed throughout this chapter and, indeed, can be seen to influence many of the research directions and applications.

2. Optical coherence elastography

2.1 Optical coherence tomography

OCE is based on the underlying imaging modality of optical coherence tomography (OCT). OCT is able to capture volume images of tissue microstructures with high resolution, ∼1−10 μm [57], by weakly focusing near-infrared light into tissues. Owing to the presence of refractive index gradients in the tissue microstructure, a tiny portion of the incident light is backscattered from the tissue microstructure, and this is often conceptualized as reflecting *echoes* of light. The magnitude and time-of-flight information extracted from these echoes is used to reconstruct a depth profile of the tissue structure. Detection of changes in the time-of-flight of light in OCT corresponding to micrometer-scale resolution requires sampling on the order of a few femtoseconds, which makes direct detection impossible [58]. Instead,

OCT employs low-coherence interferometry [57] to determine the time of flight. Low temporal coherence is achieved by employing a source with a broad optical bandwidth (3 dB bandwidth typically in the range 50–200 nm; for center wavelengths, typically in the range 800–1350 nm), which can be either contemporaneous from a partially coherent or incoherent optical source or sequential by scanning a narrow optical bandwidth over a broad range of frequencies. The light from the source is split into two paths: a reference path and a sample path. The light returned from the sample, by backscattering or reflection, is combined (interfered) with the light returned from a known reference point (typically provided by a mirror) (Fig. 9.1A). The superposition of waves of the interfered light conveys depth-resolved information on the backscattering properties of the tissue. The wider the optical bandwidth, the better the spatial resolution, where the product of the two quantities is a constant. OCT is sensitive to both the absorption and scattering of light; however, in tissues, over the wavelength range probed by OCT, absorption is generally insignificant. OCT is also sensitive to dispersion, i.e., the different wavelengths of light possessing different phase velocities. When dispersion differs between the sample and reference arms, a broadening of the coherence function causes axial spatial resolution to be degraded. Dispersion in OCT systems can typically be compensated for physically (with the introduction of additional optical components) and numerically (in postprocessing). Dispersion in tissues is generally negligible and only becomes significant in systems with sub-2 to sub-3 μm depth resolution [59].

OCT is also sensitive to the polarization of light, which is affected by the source, the optics, and the sample. Similar to dispersion, it is the differences in the polarization state of light returning from both branches of the interferometer that affect the signal strength being maximized when they are matched. Additionally, some samples, particularly fibrous or other tissues that have regular organized structures on the molecular scale up to that accessed by OCT, may cause light to propagate

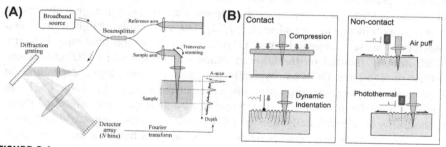

FIGURE 9.1

Optical coherence tomography (OCT). (A) A Fourier-domain OCT system. (B) Some of the common contact and noncontact methods of mechanical loading.

(A) Adapted from T. R. Hillman, Microstructural Information beyond the Resolution Limit: Studies in Two Coherent, Wide-Field Biomedical Imaging Systems, Ph.D. thesis, The University of Western Australia (2007).

more slowly in one polarization state than another: a phenomenon known as birefringence. Current realizations of OCT commonly incorporate polarization-diversity detection [60,61] to make OCT immune to variations in the polarization state, and this can also form the basis of polarization-sensitive OCT, which contrasts features in tissues through the measurement of birefringence [62,63].

Using low-coherence interferometry for tissue measurement dates back to at least the 1980s [65], for example, in measuring the length of the eyes in vivo. OCT was first demonstrated by Huang et al. in 1991 [66] in forming a cross-sectional image by laterally scanning the optical beam used to perform low-coherence interferometry in a raster pattern across the tissue surface. Beam scanning forms the basis of most OCT systems, which borrows its nomenclature from ultrasonography. Axial depth scans, or A-scans, are captured sequentially, whereas scanning across the fast x axis forms cross-sectional images, or B-scans. Scanning across the slow y axis permits multiple B-scans to be assembled into 3D image volumes. The transverse spatial resolution in such systems is set by the focused beam diameter. As the signal from all depths is usually captured at the same time, to maintain a relatively high lateral resolution the beam diameter should not vary too much with depth over $1-2$ mm, which means weak focusing (numerical aperture [NA] in the order of 0.05) is typically used. An alternative configuration that supports strong focusing and, therefore, higher lateral resolution is full-field OCT [67]; this technique uses a two-dimensional (2D) image sensor to directly capture a *'en face'* (x,y) image at a particular depth, which can then be scanned to form a volumetric image. Furthermore, in line-field OCT, line illumination and parallel detection with a 2D sensor offers rapid contemporaneous detection of a single B-scan [68,69]. OCT is closely related to holography, and various holoscopic approaches are being pursued [70,71].

OCT light sources are typically centered close to 800 or 1300 nm, coinciding with the *optical diagnostic window* ($\sim 650-1350$ nm), where light has the maximum penetration depth in tissues. Penetration in tissues is not only dependent on but also determined by the trade-off of lower scattering versus higher absorption by water, with increasing wavelength [57,72]. The availability of light sources and detectors at these wavelengths was originally driven by the photonics industry, which initially focused on the 800-nm window, then moved to 1300 nm, before settling on 1550 nm. OCT can typically image at $1-2$ mm depth, depending on the optical scattering and absorption properties of the sample. Laterally, OCT can readily achieve fields of view of $10-15$ mm in a single acquisition, which is limited by the ability to telecentrically scan the OCT beam across the sample within a given optical aperture; however, greater fields of view can be achieved using larger apertures, or by stitching together multiple OCT images [73-75]. Ultimately, the point-by-point sampling of most OCT systems sets limits on the acquisition scale, both spatial and temporal. The axial (depth) resolution of OCT is dictated by the wavelength and coherence length of the light source, which is inversely proportional to its bandwidth. With commonly used bandwidths of $50-200$ nm, OCT resolution is commonly $\sim 5-20\ \mu m$ [58]; however, resolutions of $1-3\ \mu m$ have been demonstrated [76,77]. Notably, OCT light sources are nonionizing and noninvasive and tissues do not require added labeling to form contrast, making OCT a prime candidate for in vivo imaging.

These ideal qualities have made OCT to be directed into a myriad of applications, with by far its greatest uptake being in ophthalmology, where the low scattering of the tissues of the eye in the visual pathway permits greater penetration and the ability to form volumetric images of the anterior and posterior ocular structures [78]. Despite the limited penetration depth of OCT, catheter and endoscope-based probes have enabled its commercial application in cardiology and gastroenterology [79,80]. OCT has also been employed for superficial tissue imaging, for instance, in dermatology [81] and in cancer imaging [82].

Elastographic imaging with OCT (OCE) began with the work of Schmitt in 1998 [38], who utilized OCT depth information to extract tissue strain in response to quasi-static compression of the whole sample surface. Schmitt used digital image correlation to extract tissue displacement in 2D. The work was limited by the low sensitivity and slow imaging speed of the OCT systems available at the time, the so-called time-domain systems [57], in which the reference mirror was required to be mechanically scanned to form each depth A-scan. More importantly, the resolution available from digital image correlation of speckle was limited, and this is further discussed in Section 2.2. The development of Fourier-domain OCT [83] (Fig. 9.1A) reinvigorated the field, providing access to much faster acquisition speeds (a volume in <1 s) without sacrificing sensitivity and, importantly for OCE, easy access to the phase of the interfered OCT signal, which provided access for the first time to displacements on the nanoscale. Although it is possible to capture phase information with time-domain OCT, mechanical scanning with sufficiently small subwavelength jitter is technically not an easy feat. When available, the phase information can be used to directly infer the displacement or velocity of subresolution tissue scatterers, which was instrumental in the early development of OCT flow measurement methods [84,85], as well as in the recent elastographic methods, which we detail in the next section.

The adoption of Fourier-domain OCT was followed by numerous demonstrations of elastography in applications such as vascular biomechanics [16,86,87], tissue engineering [88], and cancer detection [89]. Although these demonstrations have been instrumental in laying the foundations of OCE, they were mostly hindered by the limited acquisition speed, OCT system sensitivity, and computing power available at the time. Only in the past 5 years, with the maturation of the technology, have many new OCE techniques and applications begun to show promise [6,7,52,90], for instance, in cancer assessment [9,10] and eye disease treatment monitoring and diagnosis [91−93]. The main principles of OCE are much like those in other elastography techniques: a tissue is subjected to a load, it deforms based on its intrinsic mechanical properties, the deformation (typically local displacement) is imaged, and a mechanical model is used to reconstruct a mechanical property (e.g., Young's modulus) or parameter (e.g., strain). Various common contact and noncontact loading methods are illustrated in Fig. 9.1B. Tissue loading schemes can be broadly classified into two groups: quasi-static and dynamic (harmonic and transient), as we describe in this section. First, we will discuss the methods for measuring displacement in OCE, and subsequently, we will consider mechanical models and future prospects.

2.2 Measuring displacement in optical coherence elastography

OCE techniques are naturally sensitive to nanometer-scale displacement essentially because measurable fractions of the optical wavelength are on the same scale. A cross-correlation approach was employed in many of the early demonstrations of OCE [38,87,94], tracking the displacement between B-scans of features visible in OCT or, more commonly, of the speckle within features. Speckle is a granular and mottled texture that arises in coherent imaging methods (such as OCT or ultra-sonography), in which the detected signal is a summation of multiple constructively or destructively interfering wavefields [95]. Speckle is temporally invariant in static tissues, yet a displacement of subresolution scatterers leads to a corresponding displacement in the OCT speckle pattern, as demonstrated in Fig. 9.2A. Using

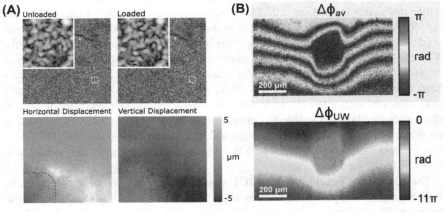

FIGURE 9.2

Measurement of displacement. (A) The evolution of speckle in a tissue-simulating phantom (with a stiff feature outlined by a dashed line in the bottom left) under compression: (top) optical coherence tomographic images with a 2-mm field of view, insets 50 μm and (bottom) digital image correlation used to track speckle displacement. (B) Phase-sensitive measurement of axial displacement of a tissue-simulating phantom with a stiff inclusion: (top) phase difference demonstrating the phenomenon of phase wrapping and (bottom) unwrapping algorithm applied to reconstruct true phase shift.

(B) Adapted from B. F. Kennedy, R. A. McLaughlin, K. M. Kennedy, L. Chin, A. Curatolo, A. Tien, B. Latham, C. M. Saunders, D. D. Sampson, Optical coherence micro-elastography: mechanical-contrast imaging of tissue microstructure, Biomed. Opt. Express 5 (7) (2014) 2113–2124.

digital image correlation on consecutive cross-sectional or volume OCT images [77,96], it is possible to track the displacement of the speckle—a technique termed *speckle tracking*. Speckle tracking can be performed in one dimension, 2D, or 3D but has most commonly been performed in 2D. Speckle techniques are limited in a number of ways. First, the spatial resolution of measured displacements is dictated by the correlation window size, which is typically no less than an area covering four to five speckles, some 20–30 pixels. Speckle tracking techniques are also limited in the range of accessible displacements. At the high end, they are limited by the phenomenon of speckle decorrelation [96,97]. As the subresolution scatterers are deformed and reorganized during tissue loading, new realizations of speckle are generated, with patterns no longer being correlated to those of the initial displacement field, which reduces the effectiveness of image correlation. While this is the subject of much research [96], speckle tracking typically breaks down above ~1% difference in tissue strain. Under uniaxial compression, for instance, over a typical 500-μm OCT depth field of view, 1% tissue strain roughly corresponds to a maximum displacement of 5 μm when the spatial resolution is around 10 μm. This maximum scales with the spatial resolution of the OCT system. Minimum displacement is set by the speckle size and the need to sample it adequately. Speckle size is determined by the coherence envelope of the light and focal spot size and is on the same scale as the OCT spatial resolution [98]. Nyquist sampling corresponds to 0.5× OCT sampled pixel size; however, strategies have been identified to enable subpixel tracking [96]. The advantage of speckle tracking, as we discuss later, is its potential to measure the 3D components of displacement, which is required, for instance, to map the full strain tensor describing tissue deformation [99,100]. Thus improving mechanical sensitivity and spatial resolution of speckle tracking remains an area of opportunity in OCE.

The introduction of Fourier-domain OCT provided easy access to the complex form (intensity and phase) of the detected depth-resolved signal. By tracking the evolution of phase in OCT images, it is possible to estimate tissue displacement in the axial dimension on a scale set by the optical wavelength, instead of by the coherence envelope—a method termed *phase-sensitive detection* [44]. Phase-sensitive detection owes its origins to Doppler imaging in OCT [84,85], in which the velocity of moving constituents in the sample produce a shift in the frequency of backscattered light. Similarly, the displacement of tissue subresolution scatterers can be accessed from the difference in phase, $\Delta\phi$, between consecutive OCT acquisitions. The axial (depth) component of displacement, u_z, is given as $u_z = \Delta\phi\lambda_0/4\pi n$, where λ_0 is the mean wavelength and n is the sample's refractive index. Interrogating the phase yields a very high sensitivity to displacement (<1 nm [101]) and resolution matching that of OCT. This is due to modern OCT systems typically reaching a phase stability of 1.5–3 mrad [91,102,103]; however, this value is typically measured as the standard deviation of phase over a large number of consecutive scans, in a location of high OCT signal, thus representing a best case in practice. Phase and displacement sensitivity is proportional to the signal-to-noise ratio (SNR) of the OCT signal [104]; therefore, in tissues, where the OCT SNR is poorer than that of an ideal test sample, the maximum displacement sensitivity can be more than 300 mrad [102]. The measured phase difference is modulo 2π, which limits the unambiguously measurable displacement to half the OCT wavelength (Fig. 9.2B). Phase unwrapping strategies have been successfully employed to extend the range of measurable displacement [101]; however, similar to speckle tracking, displacement and strain of the sample induce phase decorrelation [97], limiting the maximum measurable displacement between consecutive scans to typically $3-4$ µm. Another limitation of phase-sensitive detection is that it is typically sensitive only to displacement coaxial to the imaging beam. 3D displacements can be accessed by using multiple off-axis beams [105] or via spectral broadening when using high NAs [106].

Phase-sensitive detection has been the predominant method used for OCE in the past 5 years because of its high sensitivity, dynamic range, and resolution. However, there is continuing interest in speckle tracking, particularly in realizations where phase is not easily accessible, such as in full-field OCE [77]. Recent developments in processing speed, accuracy, and subpixel tracking [107–109] are improving the prospects of speckle tracking methods, which to date have not demonstrated performance on a par with phase-sensitive detection. Recently, a noniterative method of speckle tracking was proposed, evaluating displacements directly from a collection of cross-correlation coefficients [109], or combined with phase-sensitive detection [110]. This method was demonstrated on monitoring laser photocoagulation; however, it holds strong potential for fast 3D displacement measurement in OCE. The distinct advantages in estimating the complete displacement vector field and greater tolerance to higher displacement will likely be important in emerging computational methods and in vivo applications, which are further discussed in this chapter.

2.3 Quasi-static optical coherence elastographic methods

Quasi-static OCE methods, at their core, rely on the assumption that a tissue is loaded sufficiently slowly such that the effects of inertia are insignificant (e.g., no mechanical wave propagation or measurable motion is present) [54]. The most common quasi-static method is step compression loading. The tissue is compressed by an external load and OCT images are acquired in the compressed and uncompressed state. These images are used to derive the local displacement of tissue structures. Many of the early demonstrations of OCE used compression because of the combined simplicity of loading, measurement, and mechanical model [52], and compression loading still remains prominent. Other quasi-static methods [54], such as quasi-static cyclic methods (e.g., slow sinusoidal compression), have been demonstrated in OCE [111], but they have been less prominent.

Compression is often applied coaxially to the imaging beam, such that the tissue principally deforms along the z axis. Owing to the limits imposed on displacement measurement in both speckle tracking and phase-sensitive detection, the tissue is typically first preloaded to ensure an even and complete contact with the loading mechanism and is allowed to relax in order to minimize the effects of viscoelasticity.

A controlled microscale step load is then applied during image acquisition. The pre-load strain ($\sim$5%–20%) is typically much greater than the strain applied during imaging ($<$0.1% for phase-sensitive detection). The microscale step loading is often achieved with a piezoelectric actuator [112] or translation stage [77]. To satisfy the quasi-static assumption, step compression is performed at a frequency below $\sim$10 Hz [4]. The axial strain (the gradient of displacement) in the tissue can be estimated locally from the change in axial displacement with axial depth, $\varepsilon_z = \nabla_z u \approx \Delta u_z / \Delta z$ [113]. The process of strain estimation can be optimized by combining it with displacement estimation from phase via vectorial [114] or hybrid forms [115]. The estimated strain is directly mapped into a cross-sectional or volume image—the strain elastogram.

Strain elastograms are formed from relative mechanical contrast, and there is some evidence of their capability to produce unique textures, patterns, and hallmarks correlated with tissue structure, morphology, and disease state [9,52,75]. These hallmarks have been exploited to classify tissues by the heterogeneity in strain [116]. Compression OCE has been demonstrated in a number of tissues, including in human breast cancer [9,75] (Breakout Box 9.1) and in murine models of muscular dystrophy [117]. Strain elastograms, however, do not provide an absolute knowledge of the tissue's intrinsic mechanical properties, which is often seen as a prominent limitation of this technique. Under the assumption that the stress field inside the tissue is uniform, strain is linearly proportional to the inverse of Young's modulus [118]. This assumption requires the sample to be mechanically uniform and flat, which is rarely the case in tissues. However, whether or not strain elastograms provide sufficient information is as yet unknown and will undoubtedly depend on the particular application.

A step toward the accurate quantification of mechanical properties with compression OCE has been reported [119]. A compliant, uniform, and transparent reference layer has been used to estimate stress imparted onto a sample [120]. Retaining the assumption of stress uniformity, the estimated surface stress was projected into the sample and, in conjunction with the strain, used to reconstruct local Young's modulus using Hooke's law [119], i.e., $E = \sigma_z / \varepsilon_z$. The method is accurate for simple tissue-mimicking phantoms and has shown improved contrast in delineating features in breast cancer samples (Fig. 9.3A). It is difficult to gauge the absolute accuracy of the method in real tissues because the accuracy of the assumption of stress uniformity varies according to the sample [121]; however, it is possible to improve it by carefully preparing the tissue to have a flat surface and by employing localized tissue loading [118]. The use of a reference layer to estimate surface stress has further opened up compression to computational inverse methods [122], which are discussed later in this section.

The simplicity of loading and computation in compression OCE has enabled fast imaging over large fields of view, which is particularly beneficial for clinical imaging. OCE volumes of $5 \times 5 \times 2$ mm (x,y,z), have been acquired in 5 s [123], and separately fields of view of 50 mm [75] and near-video-rate processing [124] have been demonstrated. Strain resolution is reported to be equivalent to the OCT system

Box 9.1 Oncology

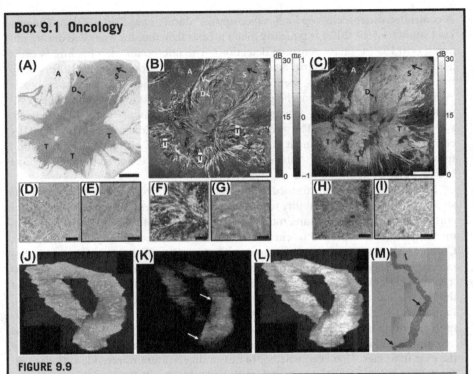

FIGURE 9.9

Oncology. (A−I) Invasive ductal carcinoma: (A, D, E) histology; (B, G, F) *en face* strain elastogram with adipose tissue masked, overlaid on optical coherence tomography (OCT); (C, H, I) OCT structural image. Insets magnify (D, F, H) a tumor region marked by the blue (gray in print version) arrow and (E, G, I) a benign region marked by the black arrow. *A*, adipose; *D*, duct; *S*, mature stroma; *T*, tumor; *V*, blood vessel. Scale bar is 3 mm in (A−C) and 0.5 mm in (D−I). (J−M) Malignant prostate biopsy: (J) OCT structural image, (K) elasticity, (L) fusion of (J,K), and (M) histology. (L, M) The arrows indicate the beginning and end of malignancy (scales unspecified; biopsies are ∼5−20 mm in length; elasticity ranges ∼0−1 MPa).

(A−I) Adapted from B. F. Kennedy, R. A. McLaughlin, K. M. Kennedy, L. Chin, P. Wijesinghe, A. Curatolo, A. Tien, M. Ronald, B. Latham, C. M. Saunders, D. D. Sampson, Investigation of optical coherence micro-elastography as a method to visualize cancers in human breast tissue, Cancer Res. 75 (16) (2015) (OFI-OFIO).

(J−M) Adapted from C. Ii, G. Guan, Y. Ling, Y.-T. Hsu, S. Song, J. T. I. Huang, S. Lang, R. K. Wang, Z. Huang, G. Nabi, Detection and characterisation of biopsy tissue using quantitative optical coherence elastography (OCE) in men with suspected prostate cancer, Cancer Lett. 357 (1) (2015) 121−128.

It is well known that cancer pathologies significantly alter the tissue's mechanical properties [1,225,226]. Thus given its prominence among diseases, it is not surprising that it has been the subject of much effort in elastography [4,52]. US elastography has been well established through clinical studies and in a commercial capacity; however, optical elastography is still emerging and largely at the demonstration and pilot study stage. The immediate challenge in applying optical elastography to

Box 9.1 Oncology—cont'd

cancer imaging lies in the limit of optical penetration (1–2 mm), making it difficult to see the whole tumor, or to observe the basement membrane penetration in epithelial tumors. However, in some applications such as in the detection of tumor in excised tissue margins, such a penetration depth may be sufficient. Indeed, strain imaging with compression OCE has shown a capacity to delineate the presence of malignant tumors at the surface by a unique heterogeneous mechanical signature (Figs. 9.9A–I) in 58 freshly excised breast cancer samples [9], in lymph nodes [227], and, recently, extended to image strain [75] and elasticity [228] in whole lumpectomies. The technique offers promise for rapid intraoperative assessment of excised tissue margins, with the goal of reducing the currently high rates of surgical re-excision. In another promising study, ARF OCE was used to characterize 120 prostate biopsy samples (Fig. 9.9J–M), showing an ability to discriminate between malignant and nonmalignant tumors [10]. Probe-based methods (Section 2.5) offer imaging beyond superficial tissues, and it is evident that optical elastography is suited to rapid intraoperative or biopsy imaging.

resolution in the lateral directions; however, owing to the necessarily finite range required for displacement gradient estimation [113], axial resolution is typically degraded to ~50–100 μm. In full-field OCE, resolution depends on the availability of strong speckle or feature contrast and appears to be in the order of ~30–50 μm,

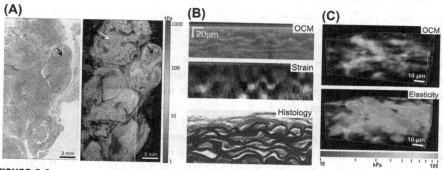

FIGURE 9.3

Quasi-static methods. (A) Compression optical coherence elastography (OCE) with a compliant layer of an ex vivo fibroadenoma: (left) hematoxylin-eosin histology and (right) stiffness elastogram, with black arrows showing a region of dense fibrous tissue and white arrows showing a region comprising small fibrous nodules. (B,C) Ultrahigh-resolution compression OCE images of ex vivo mouse aorta, demonstrating delineation of the mechanics of individual elastin sheaths and smooth muscle cell layers, validated by comparison to representative histology. *OCM*, optical coherence microscopy.

(A) Adapted from K. M. Kennedy, L. Chin, R. A. McLaughlin, B. Latham, C. M. Saunders, D. D. Sampson, B. F. Kennedy, Quantitative micro-elastography: imaging of tissue elasticity using compression optical coherence elastography, Sci. Rep. 5 (2015) 15538; (B) Strain imaging, B-scan (x,z), adapted from A. Curatolo, M. Villiger, D. Lorenser, P. Wijesinghe, A. Fritz, B. F. Kennedy, D. D. Sampson, Ultrahigh-resolution optical coherence elastography, Opt. Lett. 41 (1) (2016) 21–24 (C) Elasticity imaging, adapted from P Wijesinghe, N. J. Johansen, A. Curatolo, D. D. Sampson, R. Ganss, B. F. Kennedy, Ultrahigh-resolution optical coherence elas-tography images cellular-scale stiffness of mouse aorta, Biophys. J. 113 (11) (2017) 2540–2551.

isotropic [77]. Recently, strain imaging was demonstrated using optical coherence microscopy (OCM) [19] (Fig. 9.3B) and quantification was achieved with an isotropic resolution of 15 μm [14] (Fig. 9.3C), showing the capacity to distinguish features as small as individual elastin sheaths interleaved with smooth muscle cells and opening up prospects for cellular-scale mechanical measurement. A further step toward the use of OCE in cell mechanics is the recent report of a method to measure 3D displacement fields for traction force OCE [125]. Additionally, compression OCE was used alongside AFM to characterize the stiffness of hydrogels used in assessing stem cell migration and mechanotransduction [20] and in characterizing the mechanical properties of cancer spheroids [126].

2.4 Dynamic optical coherence elastographic methods

Dynamic methods rely on the effects of inertia and the time-dependent behavior of tissues to probe the mechanical properties [54]. Dynamic behavior typically manifests from acoustic or higher frequency loading of the tissue, which can be internal or external and localized or applied over a wide field [7]. Many of the dynamic OCE methods currently being investigated are based on measurement of the transverse acoustic wave propagation velocity, an approach that was first investigated in US elastography [127]. The precise nature of the wave propagation and its dependence on mechanical properties is dependent on the distance from the surface of the tissue and the associated tissue structure. Acoustic waves propagating on or near the surface of a tissue can be described as surface acoustic waves; propagation in the bulk of thick tissue can be described as shear waves and propagation in layered tissues can be described as Lamb waves. In each case, the relationship is slightly different, but the advantage in measuring wave propagation is that, under the assumptions of linear elasticity, incompressibility, and local mechanical homogeneity, wave speed can be directly related to shear modulus. For instance, shear wave speed, c_s, is related to shear modulus, μ, as $c_s^2 = \mu/\rho$, where ρ is the density. Shear wave speed in soft tissues is typically in the range 1−10 m/s and can propagate for 0.520 mm [90]. For a more comprehensive treatment of acoustic wave models, the reader is directed to Refs. [7,90].

Early demonstrations of OCE have measured surface acoustic waves [128−132]; however, current techniques predominantly focus on subsurface shear wave imaging [43,91,133], and recently, Rayleigh-Lamb wave imaging [131,134,135]. To excite acoustic waves, these techniques impart a transient (pulsed) load, either externally or internally, to tissues. A great variety of contact or noncontact methods have been demonstrated [90]. In early demonstrations, indenters have been used to impart short localized pulses to the tissue surface [42,130,136]. Although contact indentation loading is perhaps the most trivial to employ, it poses challenges in delicate tissues, such as the cornea, where noncontact methods are preferable. Following the developments in US elastography, ARF has also been used to provide a transient load by focusing an US acoustic wave into the tissue [137−139] (Fig. 9.4A). A force is generated depending on the acoustic scattering and absorption properties of the

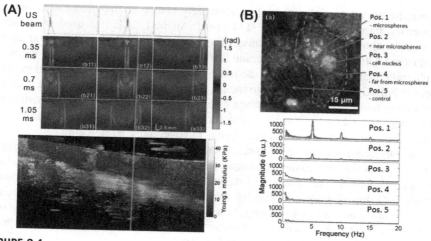

FIGURE 9.4

Dynamic methods. (A) (top) Shear waves generated by an impulsive acoustic radiation force at varying incident angles focused internally in a tissue-simulating phantom, and (bottom) stiffness of porcine retina estimated from local shear wave velocity. (B) (top) Magnetic microspheres engulfed by a single macrophage, outlined in green (white in print version), and (bottom) displacement frequency spectrum showing the local response to magnetic field modulation. *US*, ultrasound.

(A) Adapted from [138]. (B) Adapted from V Crecea, B. W. Graf, T. Kim, G. Popescu, S. A. Boppart, High resolution phase-sensitive magnetomotive optical coherence microscopy for tracking magnetic microbeads and cellular mechanics, IEEE J. Sel. Top. Quantum Electron. 20 (2) (2014) 25–31.

tissue and shear waves are generated, typically, by the impulse created by a focused burst of US. ARF loading has been predominantly a contact method [8], requiring a material to provide acoustic matching between a tissue and the transducer. In US elastography, such contact is required for detection; however, in OCE, its potential for noncontact measurement has only recently been exploited. Air-coupled ARF has been demonstrated using the acoustic reflection, as opposed to subsurface scattering, to generate displacement [140]. This loading method was utilized for noncontact shear wave elasticity imaging of the eye, termed acoustic micro-tapping [92]. Shear waves have also been generated using short air-puff bursts [91,141], The noncontact nature of air-puff, and recently ARF, loading is particularly well suited for applications in ophthalmology (Breakout Box 9.2). Similar to air-puff tonometry, early demonstrations imparted force over a wide surface area of the eye and captured the deflection or applanation of the cornea [142,143]. More recent developments have sought to spatially localize loading to a <1-mm point required for the generation of a clear acoustic wave front for wave-based elastography [91,141]. Air-puff loading, however, has a relatively slow response time, producing an impulse with a reported minimum width of 0.8 ms [8]. Despite this limitation, air-puff OCE has demonstrated substantial potential in accurately characterizing tissue mechanical properties, particularly in the eye [144–146]. Air-puff OCE was used to classify glomerulonephritis, notably combining elasticity with optical properties parameterized from OCE via supervised machine learning [147]. Another noncontact method is photothermal loading, in which the force generated from thermal absorption of a focused pulsed laser beam is used to induce shear waves [148]. By its nature, care is required to minimize the tissue damage caused by thermal absorption. Furthermore, photothermal loading beyond the tissue surface is made challenging by light scattering in tissues, which sees intensity decay exponentially with depth.

An interesting alternative for dynamic methods is to incorporate exogenous magnetic particles into the tissue. In magnetomotive OCE, these embedded magnetic nanoparticles are excited using an external magnetic field [150]. The requirement for magnetic labeling prevents magnetomotive OCE from being employed for a general case of tissue elasticity imaging; however, selective tagging of magnetic nanoparticles, akin to fluorescence microscopy, may provide additional functional contrast, attractive for cellular-scale applications.

Commonly in these dynamic methods the velocity of acoustic waves is measured by estimating the acoustic wave number from a set of local displacement measurements across a spatial fitting range or, alternatively, by measuring the time of flight of a leading acoustic wave front [90]. This typically degrades the mechanical resolution to the order of an acoustic wavelength ($\sim 100-500$ μm). In tissues that are heterogeneous over this length scale, waves change velocity and direction, as well as disperse, reducing the accuracy of this simple approach. Computational methods, discussed in Section 2.6, are likely to be important for wave techniques in heterogeneous samples. Furthermore, wave techniques require high temporal sampling rates to evaluate local wave speed, leading to potentially long acquisition times [52]. Attenuation and dispersion of acoustic waves is rapid in tissues [7], which limits the field of view or, alternatively, requires loading to be independently repeated at various locations within the field of view. However, newly developed swept-source OCT systems are enabling much more rapid acquisition [151]; recently, ultrafast OCE was deqmonstrated with megahertz OCT A-line rates for shear wave elasticity imaging [103,152]. Alternatively line-field holographic approaches can be used to capture the state of wave propagation in space in one shot [153,154].

Wave speed imaging, however, is not the only path to mechanical information in dynamic OCE. Harmonic loading with a continuous wave or pulse in the kilohertz range can be localized or applied over a wide field. In some instances, harmonic excitation leads to the formation of standing waves when acoustic waves are reflected from distinct boundaries and structures, producing modal patterns at resonant frequencies. These techniques are likely to be suited to largely homogeneous tissues that possess well-defined known boundaries, such as the cornea. Indeed, modal vibration patterns on the surface of the cornea have been observed [155]. Similarly, separately generated shear waves can be interfered in tissues, producing crawling waves [156]. Crawling waves can propagate much more slowly and, thus, relax the scan speed requirements of the imaging system.

Harmonic loading in the acoustic frequency range has been demonstrated with mechanical actuators [89] and via ARF [41], in which continuous loading generates vibration in tissues, with the amplitude varying according to local tissue absorption and acoustic impedance mismatch. The amplitude of vibration can be used to generate images of mechanical contrast [41,89,157,158]. Harmonic excitation of magnetic nanoparticles [149,159] can be used to target and probe the local mechanical response of cells and tissues (Fig. 9.4B). Lorentz forces generated in tissues have also been demonstrated for dynamic excitation [160]. Given a strong acoustic mismatch at the surface, tissues can also be strained using on-off modulated ARF [161]. Strain can be captured from the gradient of the local vibration amplitude, and mechanical modulus can be extracted by comparing the strain in a tissue to the strain in a known reference material [10]. This technique has been used to characterize prostate cancer biopsies (Breakout Box 9.1).

By sweeping the loading frequency, a frequency-dependent mechanical response of a tissue can be extracted [162]. Measured local displacement is decomposed across a frequency spectrum through Fourier analysis. Unique frequency content can be used as mechanical contrast. Furthermore, it can provide knowledge of the tissue's viscoelastic properties: the shear and loss moduli [54]. Viscoelasticity has also been measured from the frequency-dependence and dispersion of propagating mechanical waves [134,135,163,164] and step loading [165,166]. A more comprehensive treatment of viscoelastic characterization is given by Mulligan et al. [7]. Intrinsic tissue dynamics (i.e., loading generated by the tissue itself), such as blood flow [167], Brownian motion [168], and osmomechanical stress [169], have been also explored.

Despite the diversity of dynamic methods in OCE, they offer common advantages and disadvantages over quasi-static OCE. Dynamic methods often provide easier access to quantification of mechanical moduli, i.e., they do not require knowledge of the local tissue stress. Furthermore, they permit noncontact loading, which is beneficial in many in vivo applications, or applications on delicate tissues. Dynamic methods, however, add a new dimension (time or frequency) to the measurement, which increases acquisition time, limits the field of view, and represents a challenge for rapid clinical imaging.

2.5 Probe-based optical coherence elastography

Translation of OCE techniques toward clinical and biological applications faces challenges in many instances of the limited optical penetration depth in tissues. This limitation has seen most OCE applications directed toward easily accessible samples, such as skin [158,170] and cell cultures [149]; excised tissues, such as breast [9] and prostate [10]; or otherwise transparent tissues, such as the eye [90]. In vivo minimally invasive imaging of deep tissues with optical elastography still remains an elusive, but attractive prospect. The challenge in achieving this goal lies in developing application-specific probes. Such probes must deliver mechanical loading deep into tissues and miniaturize and relay the optical imaging system to the same or nearby location.

A general advantage of optics is that the sample imaging beam can be readily relayed through a flexible and small-footprint optical fiber. Thus one promising approach is the miniaturization of OCE into a fiber-based catheter probe, which would be attractive for applications in intravascular, gastrointestinal, and airway imaging. Catheterized OCT probes have already been demonstrated for diagnosis and therapy [171] and employed for commercial use in the esophagus and cardiac arteries. Such probes are very similar to commercial intravascular US probes in dimensions and scanning approaches. Still, a challenge remains in locally loading tissues, and an attractive approach would be to use the natural dynamics and fluctuations of tissue, such as breathing or blood flow. Externally applied pressure has been employed in measuring in vivo airway compliance in humans from the resulting changes in airway dimensions seen in OCT [172]. Using luminal blood pressure for intravascular OCE has also been proposed [16] (Fig. 9.5A). Other proposed methods include ARF loading using a probe-based US transducer [173]. Catheter-based OCE is still, however, in the very early stages of development and clear mechanical images have yet to be reported [174]. Alternative to fiber-based methods, rigid endoscopic probes can be used. OCM-based elastography has been demonstrated through a rigid gradient-index endoscope [175].

Another approach to deep tissue imaging is needle-based OCE. Structural imaging through a needle has already been demonstrated, aimed toward biopsylike measurement, surgical guidance, and monitoring [176]. Forward-facing needle probes have been demonstrated in capturing tissue deformation in one dimension [177,178]. In this approach, loading was applied by the blunt tip of a needle and displacement was captured using phase-sensitive detection. Distinct displacement slopes were used to identify layer boundaries in tissues, including a pig airway wall and breast cancer samples (Fig. 9.5B). In a related approach, quantification was achieved by additionally measuring stress using an integrated short Fabry-Pérot cavity on the end of the fiber probe [179].

Achieving the high sensitivity of phase-sensitive detection is very challenging in probe-based methods, in which small ambient vibrations and fluctuations in the sample or the scanning probe introduce motion artifacts on the scale of the displacement probed by OCE. Together with the relatively low values of maximum measurable displacement (discussed in Section 2.2), avoiding artifacts from ambient fluctuations is a major challenge in developing effective probe-based OCE for in vivo applications. Differential vibrations in the interferometer can be overcome, in part, with common-path detection, in which the reference is placed in proximity to, or in contact with, the sample and shares a common optical path. Common-path configurations have been used for quasi-static [101] and dynamic measurements [180]. Sample or hand-generated motion can be overcome, in part, by faster acquisition, such that between consecutive measurements, extrinsic motion is minimal. In vivo OCE has been demonstrated on skin [181] and in cornea [130]. Faster OCE systems [103,152] strengthen prospects for probe-based and in vivo applications. Strategies for motion correction have been successful in closely related probe-based OCT [182] and, similarly, we will likely see them feature in emerging applications of probe-based OCE.

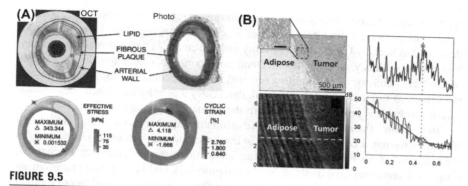

FIGURE 9.5

(A) Intravascular optical coherence elastography (OCT) of a lipid-rich coronary plaque. The segmented structure used to predict stress and strain distribution from blood flow pressure. (B) Needle OCE of ex vivo breast cancer samples, showing the capacity of OCT ID forward-looking signal and axial displacement to distinguish adipose and tumor.

(A) Adapted from A. H. Chau, R. C. Chan, M. Shishkov, B. MacNeill, N. Iftimiia, G. J. Tearney, R. D. Kamm, B. E. Bouma, M. R. Kaazempur-Mofrad, Mechanical analysis of atherosclerotic plaques based on optical coherence tomography, Ann. Biomed. Eng. 32 (11) (2004) 1494–1503. (B) Adapted from K. M. Kennedy, R. A. McLaughlin, B. F. Kennedy, A. Tien, B. Latham, C. M. Saunders, D. D. Sampson, Needle optical coherence elastography for the measurement of microscale mechanical contrast deep within human breast tissues. J. Biomed. Opt. 18 (12) (2013) 121510.

The attractions of probe-based OCE methods are various. In cardiovascular OCT, the reliable characterization of atherosclerotic plaques is still lacking. In other areas, the benefits of mechanical contrast are less clear, but there is a wealth of potential applications in cancer, for example. Although probe-based OCE is yet to demonstrate practical advances in deep tissue mechanical imaging, early demonstrations have shown promise. Likely, we will see considerable research in probe-based OCE in the coming years.

2.6 Computational inverse methods in optical coherence elastography

To be rigorously quantitative, OCE methods invariably require combination with techniques that solve the *inverse elasticity problem*, i.e., where an intrinsic mechanical property is estimated from measured data. Absolute quantification of mechanical properties is desirable, as it allows for system- and method-independent, intersample, and longitudinal comparisons. Attempts to date to extract quantitative data rely, in some fashion, on a series of often stringent assumptions on the nature of the mechanical behavior in the imaged sample. For instance, they typically assume that the tissue is linear elastic and homogeneous and undergoes infinitesimal strain in order to simplify the model of mechanical behavior to the point where it can be readily inverted, thus providing a clear relation between what is measured (e.g., displacement) and an intrinsic mechanical property (e.g., Young's modulus) [54,183]. These relations, often termed *first-order approximations*, are quick to compute and robust to measurement noise—important qualities in clinical imaging [53], However, as we described in Section 1.2, incompressibility and continuity of tissues creates a mechanical interdependency, whereby the mechanical response in a local area depends on the mechanical behavior of surrounding regions; in heterogeneous tissues, first-order approximations fail to account for this and, thus, carry a penalty in the accuracy of estimated mechanical properties and lead to the formation of image artifacts [7,52].

To minimize the assumptions made in estimating mechanical moduli (particularly the pervasive but necessary assumption of the structural and mechanical uniformity in tissues), and in an effort to work toward an accurate quantification of mechanical properties, computational methods have been proposed. Unlike first-order approximations, the majority of computational inverse methods cannot be solved through direct inversion and a solution has to be iterated toward, keeping strict to a series of equilibrium equations that define the mechanical behavior at all points in the imaged sample. Computational inverse methods have been studied in US and MR elastography and comprehensively reviewed by Barbone and Oberai [184]. The use of computational methods in OCE to date has been limited, however. Such methods were employed in the early attempts at application of OCE in analyzing atherosclerotic plaques [16,185,186] (Fig. 9.5A); however, in these examples, either synthetic data was used or the computational problem was significantly constrained based on a priori structural information. These investigations did not proceed beyond the proof-of-principle stage. A more general approach to solve the inverse elasticity problem in compression OCE was presented [122]. The method was made computationally feasible by its use of adjoint equations, thereby significantly reducing computation time. Fig. 9.6 shows a 2D reconstruction of the spatially resolved shear modulus in a tissue-simulating phantom, demonstrating the ability to characterize elasticity, notably without assuming a uniform and uniaxial stress distribution. This method has also been demonstrated in volume reconstructions [187].

Solving the equivalent inverse problem for the dynamic case has not been attempted in OCE to date. Such a reconstruction has been found to be tractable when all components of the displacement field are known and, consequently, has been primarily addressed in MR elastography [53]. However, there have been substantial developments in improving wave-based OCE methods toward more accurate estimation of shear modulus and, further, to gain access to viscoelastic properties [8]. These developments have focused primarily on the eye, whose geometry and fluid boundaries pose a challenging but constrained problem. For instance, using finite-element methods (FEMs), the curvature and the thickness of the cornea was found to directly impact the velocity of elastic waves [188]. Similarly, boundary fluid pressure, such as that imparted by the intraocular pressure, reduces group velocity [189]. Strong internal reflectors (internal mechanical boundaries) further affect the accuracy of recovered elasticity and vary with the nature of the imparted dynamic load [190]. Toward the accurate estimation of elasticity, Han et al. [144] compared a number of direct inversion models and FEM, demonstrating that the Rayleigh-Lamb frequency equation is the most accurate direct method. FEM, which

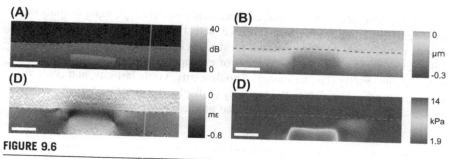

FIGURE 9.6

Tissue-simulating phantom in compression optical coherence elastography with a compliant layer: (A) optical coherence tomographic structural image (with layer masked); (B) measured axial displacement; (C) calculated axial strain, where mε is the millistrain; and (D) shear modulus computed by solving the inverse elasticity problem (scale bar, 500 μm).

Adapted from L. Dong, E Wijesinghe, J. T. Dantuono, D. D. Sampson, P. R. T. Munro, B. F. Kennedy, A. A. Oberai, Quantitative compression optical coherence elastography as an inverse elasticity problem, IEEE J. Sel. Top. Quantum Electron. 22 (3) (2016) 277–287.

was the most accurate overall, was performed by manually choosing the mechanical properties that resulted in the closest match in velocities to the experiment—the first step toward iterative inverse reconstruction. Consequently, the faster Rayleigh-Lamb methods were used to quantify the viscoelastic properties of tissues [134] and, specifically, of the cornea [135].

We anticipate that the use of computational methods to accurately determine mechanical properties will become important in OCE, as has been the case in MR and US elastography [184]. Furthermore, the added generality (reduced number of assumptions) of the computational inverse methods is likely to facilitate the development of probe-based methods in OCE by enabling the solution of a wider range of mechanical models and, thus, accommodating a wider range of tissue loading methods. However, there remain challenges in progressing these methods. Computational methods are often ill-posed and ill-conditioned [184], and thus they are prone to measurement noise, which is often present in OCT images. Whether or not sufficient accuracy can be obtained from computational methods given the typically noisy input displacement data remains to be seen. Furthermore, OCE routinely generates volumetric images, which adds a significant computational overhead relative to cross-sectional images, as are typical in US elastography. The shallow field of view and proximity to the tissue surface makes OCE measurement particularly sensitive to boundary effects, which so far have mostly been left untreated in the development of these methods [184]. The shallow imaging depth and the high noise levels at deeper depths caused by the strong attenuation (typically $1-10 \text{ mm}^{-1}$), in particular, make it challenging to prescribe the deep boundary condition and may add errors to elasticity values recovered in the field of view. OCE would undoubtedly benefit from the capture of vectorial displacement (vs. the single axial displacement component available from phase-sensitive detection) with sufficient sensitivity. Computational methods, however, will most likely not displace simpler approaches to elasticity imaging. Within the context of any application, a trade-off exists between their generality (i.e., treatment of nonlinearity, viscoelasticity, and compressibility) and the requirement for the accuracy of displacement measurement (which is coupled to the acquisition field of view and time).

3. Brillouin microscopy

Brillouin microscopy utilizes the phenomenon of Brillouin scattering to capture images of tissue mechanical properties. Brillouin scattering is a form of inelastic scattering arising from the interaction of light with high-frequency (gigahertz) acoustic waves (phonons) [191]. Acoustic phonons are generated by the thermodynamic fluctuations inherent to all materials. The frequency of light may be downshifted through excitation (Stokes shifted) or upshifted through relaxation (anti-Stokes shifted) of acoustic phonons that must be phase-matched with the light in order to conserve momentum [192] (Fig. 9.7B). The Brillouin frequency shift (upward and downward), v_B is given by, $v_B = (2n/\lambda)V\sin(\theta/2)$, where n is the sample's refractive index, λ is the optical wavelength, θ is the angle between the incident and scattered light, and V is the acoustic phonon velocity [191]. Phonon velocity is dependent on the local mechanical properties of tissues and, thus, the Brillouin shift can be used to estimate tissue elasticity, as we detail later in this section.

Light scattering via acoustic phonons was first predicted by Brillouin in 1922 [193] and was experimentally verified by Gross in 1930 [194]. The first quantification of the mechanical properties using Brillouin scattering was made in the lens and the cornea of the eye in 1980 [195]. Brillouin scattering was predominantly employed in single-point spectroscopic techniques for material property characterization, rheology, environmental sensing, and structural monitoring [191]. In biology, Brillouin scattering has been used to reveal physical properties, such as viscoelasticity, tensile and compressive strains, temperature, and acoustic velocity, and has been applied to muscle, bone, and eye characterization [196]. Although Brillouin spectroscopy has been extensively used, until recently it has remained a single-point measurement technique because of the typically long acquisition time per spectrum. The main measurement challenge is separating the Brillouin scattered light from the orders-of-magnitude stronger elastically scattered (Rayleigh-scattered) light [191], which is only separated from it in optical frequency by a few to a few tens of gigahertz. Thus the detection of the Brillouin shift requires narrow linewidth lasers and spectrometers with sub-gigahertz spectral resolution.

The capacity of Brillouin spectroscopy to form images (Brillouin microscopy) was first demonstrated by Scarcelli and Yun in 2008 [191] (Fig. 9.7A), enabled by their utilization of a *virtually imaged phased array* (VIPA) spectrometer. Since then, Brillouin microscopy has been applied to mechanical imaging of single cells [18], eyes

Box 9.2 Ophthalmology

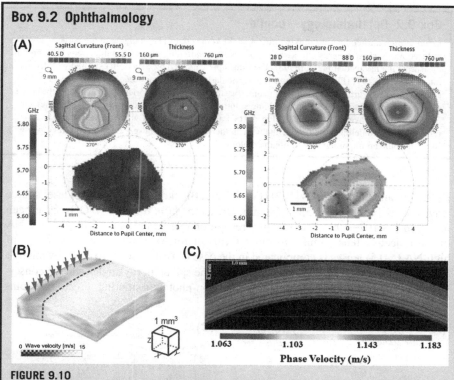

FIGURE 9.10

(A) In vivo Brillouin microscopy of (left) normal and (right) keratoconus eye; top insets are curvature (*D*, diopter) and pachymetry map. (B) Corneal wave group velocity in response to noncontact acoustic radiation force loading. (C) Corneal Lamb wave phase velocity in response to air-puff loading.

(A) Adapted from G. Scarcelli, S. Besner, R. Pineda, P. Kalout, S. H. Yun, In vivo biomechanical mapping of normal and keratoconus corneas, JAMA Ophthalmology 133 (4) (2015) 480–482. (B) Adapted from L. Ambrozinski, S. Song, S. J. Yoon, I. Pelivanov, D. Li, L. Gao, T. T. Shen, R. K. Wang, M. O'Donnell, Acoustic micro-tapping for non-contact 4D imaging of tissue elasticity, Sci. Rep. 6. (C) Adapted from S. Wang, K. V Larin, Noncontact depth-resolved micro-scale optical coherence elastography of the cornea, Biomed. Opt. Express 5 (11) (2014) 3807–3821.

The biomechanics of the eye can be used to assess and monitor several diseases, such as glaucoma and keratoconus [230,231]. Optical elastography provides an attractive means of for monitoring corneal and lens mechanics, and possibly intraocular pressure, with potential to be used for clinical diagnosis of a range of conditions throughout the eye. Indeed, this has been one of the main applications of optical elastographic research in the past decade [6–8]. An enabling feature in imaging the eye is to apply a load via a noncontact and minimally invasive method, preferred over contact methods due to patient tolerance and the delicate nature of the eye. Dynamic OCE methods, with air puff and, recently, ARF loading [92] (Fig. 9.10B), are particularly promising in this regard and have already been demonstrated

Continued

Box 9.2 Ophthalmology—cont'd

in monitoring ultraviolet cross-linking treatment for keratoconus [11] (bulging of the cornea caused by reduced mechanical integrity), in assessing age-related stiffening of the crystalline lens [12], and in measuring the intraocular pressure [93]. The same applications have also been explored with Brillouin microscopy [13,199] (Fig. 9.10A), which is particularly encouraging because it avoids the need for external loading. Furthermore, the clear geometric structure and boundaries of the eye, as in the cornea, can simplify mechanical modeling, enabling solutions through complex wave propagation, such as Rayleigh-Lamb waves [135,144,229] (Fig. 9.10C) and through modal vibration patterns [155]. With refinements in modeling and loading, and with an increase in imaging speed, optical elastography will likely become an important tool in ophthalmology.

[198,199], and blood vessels [197] (Fig. 9.7 and Breakout Box 9.2). With laser source wavelengths in the range 532–780 nm, Brillouin shifts of ~6–15 GHz have been observed [18,197–199]. The imaging resolution can be tailored by varying the NA of the objective lens (similar to other forms of microscopy); however, the use of high-NA (>0.5) lenses is complicated in the case of Brillouin microscopy by angular spectral broadening [192], which results from the spread in ray angles and the conservation of momentum implied by the phonon-photon scattering process. Depth-

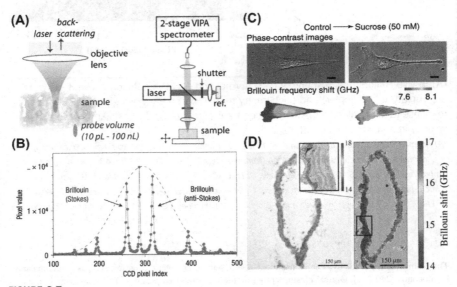

FIGURE 9.7

Brillouin microscopy. (A) Principle and schematic. (B) Brillouin spectrum of distilled water. (C) Brillouin shift images (coregistered with phase microscopy) of a cell before and after hyperosmotic shock (scale bar, 10 μm). (D) Mouse carotid artery (inset performed at higher resolution). *CCD*, charge-coupled device; *VIPA*, virtually imaged phase array.

sectioning in Brillouin microscopy is not necessarily required (for example, when imaging a thin layer of cells); however, it can be achieved through use of a confocal optical configuration [191]. Brillouin microscopy with a resolution of $\sim 0.5 \times 0.5 \times 2 \ \mu m^3$, a spectral extinction of 70 dB, and spectral resolution of 600 MHz has been demonstrated [18]. Achieving such impressive specifications in in vivo applications of Brillouin microscopy, however, is still challenged by the relatively long acquisition times (20 min per volume scan) [199]. Until recently, Brillouin microscopy has exploited spontaneous Brillouin scattering, arising from the random thermal fluctuations. Just as with the laser, however, such spontaneous emission can be stimulated to produce enhanced signal through gain. Recently, stimulated Brillouin scattering was demonstrated for the first time, providing the prospect of dramatically decreasing the required acquisition time [196,200]. Dark-field Brillouin microscopy was demonstrated to reduce the detection of elastically scattered light [201].

The Brillouin shift provides an innate link to acoustic wave velocity and thus a window into the sample's mechanical properties. As such, Brillouin microscopy, to an extent, is similar to dynamic elastographic methods, but with two major distinctions. First, acoustic waves probed by Brillouin microscopy are high-frequency pressure waves, whose velocity is set by the bulk modulus, as we detail later. Second, as phase-matching is required for photons and phonons to interact, the phonon wavelength ($\lambda/2n$) required is typically 100−250 nm (4−10 GHz) for visible light [191]. These short-wavelength acoustic phonons in viscoelastic tissues are extremely short lived, and thus they only propagate for a few to tenths of micrometers. Therefore local phonon behavior is unlikely to be influenced by the boundaries and mechanical properties of the surrounding material on the micrometer scale.

The Brillouin modulus, M', can be computed as $M' = pV^2$, where ρ is the tissue density. In backward (epi-) detection, typically employed for Brillouin microscopy [18], the Brillouin modulus is commonly regarded as equivalent to the high-frequency longitudinal modulus, i.e., the ratio of axial stress over axial strain, given the transverse strain is negligible. Furthermore, the peak Brillouin frequency shift occurs at $\theta = \pi$ [191]; thus the longitudinal modulus of tissues can be estimated using $M' = \rho \lambda^2 v_B^2 / (4n^2)$. Furthermore, the longitudinal modulus is related to bulk modulus, K, and shear modulus, G, as $M = K + 4G/3$. As the bulk modulus in tissues is high (tissue is nearly incompressible), the longitudinal modulus is often in the gigapascal range, much larger than the kilopascal range of shear and Young's moduli. It is possible, in theory, to probe the shear wave speed directly through the orthogonal measurement of scattered versus incident light [191,192]; however, on a cellular length scale, shear waves are likely to be suppressed by the biphasic (solid and fluid) composition of cells (fluid is unable to shear).

It is important to note that the longitudinal modulus derives from solid mechanics, particularly the theory of linear elasticity. Furthermore, the assumption of isotropic and homogeneous materials is inherent in its formulation. Although these assumptions may be adequate for the low-frequency dynamic elastographic methods (Section 2.4), it is difficult to assert their relevance to gigahertz acoustic phonon behavior, where nonlinear elasticity and viscoelasticity is dominant. In addition, the longitudinal modulus is dominated by the bulk modulus, which is much more sensitive to the compressibility of tissue than to its stiffness. Thus the correct biophysical interpretation of high-frequency longitudinal modulus, and its relation to commonly used moduli, would appear problematic. There is empirical evidence of the strong correlation between the Brillouin modulus and Young's modulus [18]; however, the exact mapping is different for different materials. Whilst the exact relationship may not be understood, it might not be the case that it reduces the utility of Brillouin microscopy to a biologist. Such clarity is not an absolute requirement: the Brillouin modulus could become a new biophysical representation of cell and tissue mechanics. Indeed, new biophysical models beyond linear elasticity are already being explored [202].

Despite its young age, Brillouin microscopy has rapidly gained traction since its inception in 2008. It is distinguished from most other elastographic methods, such as those presented in Section 2, in that external tissue loading is not required. Intrinsic tissue loading (via acoustic phonons) is an attractive concept, enabling Brillouin microscopy to be noncontact and minimally invasive. Brillouin microscopy will likely grow as a useful tool for diagnostic in vivo imaging and cell mechanics research and may yet play a larger role in clinical applications, as improvements in instrumentation and moves toward endoscopic implementation continue.

4. Other techniques

Over the past 5 years, OCE and Brillouin microscopy have demonstrated the greatest advances in optical elastography; however, a number of other optical techniques have been demonstrated for tissue elasticity imaging, which we briefly survey in this section.

Laser speckle imaging (LSI) was introduced in the 1980s and has become a convenient and extensively used tool for blood flow imaging, particularly in neuroscience [17]. In LSI the tissue is illuminated with a coherent beam of light and the backscattered speckle-modulated image is captured by a camera. The time evolution of the imaged speckle pattern, commonly quantified by a correlation coefficient, relates proportionally to the rate of movement of tissue microstructure. Although being particularly suited to flow measurement, it has also been used for the characterization of rheologic parameters [203], e.g., viscoelasticity. LSI has been demonstrated for imaging vasculature [204,205] and skin [206]. Fig. 9.8B shows a map of a time constant of speckle decorrelation measured in a necrotic-core fibroatheroma, where the objective is to assess differences in the mechanics of the cap of the fibroatheroma and thereby the propensity to rupture [204]. LSI is readily incorporated into intravascular probes, directed toward in vivo vascular monitoring [207], but the requirement for subsecond data acquisition necessitated by the short-duration flushing of blood needed to see the vessel wall presents major challenges because the temporal evolution of speckles is typically on the millisecond scale.

Photoacoustic tomography (PAT) has rapidly developed over the past 10 years [208,209]. By combining optical excitation with US detection, PAT provides optical

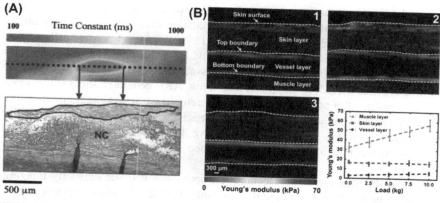

FIGURE 9.8

Alternative optical elastography techniques. (A) Time constant of laser speckle fluctuations mapped across the surface of a necrotic-core (NC) fibroatheroma (top), with the corresponding cross-sectional histologic examination at the dotted line (bottom). (B) Photoacoustic elastography of skin: movement of structure under increasing compressive load (panels 1–3) is used to estimate Young's modulus (bottom right).

contrast, while overcoming the 1–2 mm limit in the optical penetration depth. PAT exploits the photoacoustic effect—a pulse of focused light is absorbed by the tissue, generating acoustic waves, which are then detected by a single US transducer or an array of US transducers. The same mechanism has been employed in OCE (Section 2.4) in reverse, instead of the acoustic detection of acoustic waves, optical detection of the local displacement caused by the waves is employed. Optical absorption contrast provided by PAT has been primarily employed for the detection and characterization of vasculature, hemodynamics, and biomarkers [209]. By contrast, PAT elastography is very recent; however, it has already been demonstrated to measure viscoelastic [210] and elastic contrast [211,212], as well as Young's modulus in vivo [213] (Fig. 9.8A). In contrast to many other optical elastography techniques, PAT images do not possess speckle; displacement tracking in PAT elastography, which uses digital image correlation, is, thus, more challenging because it must rely on the sufficient intrinsic textural contrast in tissues. However, strategies to artificially generate texture are already being explored [214]. PAT elastography may become a useful tool for providing mechanical information, complementing the functional information PAT provides, such as blood oxygen concentration [212].

Similar to PAT, ultrasound-modulated optical tomography (UOT) aims to reach beyond 1–2 mm optical penetration depth while retaining high resolution and sensitivity to optical contrast [215]. In UOT, coherent light passes through a section of tissue and is highly scattered. Orthogonal to the propagation of light, US is focused to a spot, modulating the refractive index of tissues and, thus, the phase of light waves passing through it. The wave front emerging from tissue comprises a superposition of light waves scattered by the tissue (i.e., producing speckle), and thus local modulation of phase leads to a modulation of the wave front light intensity; the local optical properties can be inferred by extracting the acoustically modulated component of the recorded light intensity. US-generated ARF can further be used to generate propagating shear waves in tissues by spatiotemporally modulating its refractive index. Trading off spatial resolution, ARF-generated shear waves have been demonstrated to boost the UOT signal strength, while enabling simultaneous detection and discrimination of optical and mechanical contrast by observing the time evolution of the recorded optical wave front. Mechanical imaging with UOT, however, has only been demonstrated on tissue-simulating phantoms [215–217].

Beyond the aforementioned techniques, optics-based tissue elasticity imaging has been demonstrated using digital holography [218,219], diffuse wave spectroscopy [220], digital photography [221,222], and shearography [223], among others [21].

5. Outlook

Optical elastography has seen remarkable progress over the past 5 years or so. Demonstrations of microscale mechanical contrast in breast and prostate cancer [9,10] and rapid advances in in vivo assessment of eye diseases [11–13] have demonstrated the potential of optical elastography as a medical imaging tool. The further advancement of high-resolution techniques [14,18,19] will drive the translational efforts of optical elastography toward the cellular scale. The various techniques offer different advantages, such as noncontact imaging possible with dynamic OCE and intrinsic to Brillouin microscopy, or rapid wide-field assessment of compression OCE, and overall provide a comprehensive suite of tools with many possible applications in medical translation and biological research.

Despite the recent progress, however, a great journey remains for the field. In many instances, the estimated mechanical properties of tissues present open questions, both in how they relate to the form and function of tissues and diseases and in whether they are, indeed, properly describing the true behavior of complex biological tissues. The future will likely lie with the hand-in-hand progression of the medical translation and biological research fields of endeavor—phenomenologically observing whether the mechanical images provide sufficient sensitivity and specificity to distinguish disease on the one hand, and translating research in cellular scale mechanics to volumetric multicellular scale structures and tissues on the other. This requires the proof-of-concept demonstrations described here to be translated into the hands of clinicians, medical researchers, and biologists [6–8]. Multiscale elastography, combining OCE with other, more established, elastographic methods, holds promise toward this goal and has already been demonstrated by combining Brillouin microscopy with AFM to relate cellular density to retinal mechanics [224].

Future progress hinges on continued technological advancement, including the continuation of the improvement in optoelectronics and photonics technology to enhance miniaturization, sensitivity, acquisition speed, and image quality. The emergence of computational methods, enabled by dramatic increases in computation speed and data volume, will likely not only improve accuracy but also enable more complex tissue loading schemes. The development of probe-based techniques should further expand the applications of optical elastography to intravascular and airway imaging, among others. These developments should bring us much closer to a comprehensive understanding of tissue biomechanics across the full range of scales, from the cell to the organ.

References

[1] G.Y.H. Lee, C.T. Lim, Biomechanics approaches to studying human diseases, Trends Biotechnol. 25 (3) (2007) 111—118.

[2] D. Discher, C. Dong, J.J. Fredberg, F. Guilak, D. Ingber, P. Janmey, R.D. Kamm, G.W. Schmid-Schonbein, S. Weinbaum, Biomechanics: cell research and applications for the next decade, Ann. Biomed. Eng. 37 (5) (2009) 847—859.

[3] A.A. Appel, M.A. Anastasio, J.C. Larson, E.M. Brey, Imaging challenges in biomaterials and tissue engineering, Biomaterials 34 (28) (2013) 6615—6630.

[4] K.J. Parker, M.M. Doyley, D.J. Rubens, Imaging the elastic properties of tissue: the 20 year perspective, Phys. Med. Biol. 56 (1) (2011) R1—R29.

[5] Y.C. Fung, Biomechanics: Mechanical Properties of Living Tissues, second ed., Springer-Verlag, New York, 1993.

[6] B.F. Kennedy, P. Wijesinghe, D.D. Sampson, The emergence of optical elastography in biomedicine, Nat. Photonics 11 (4) (2017) 215—221.

[7] J.A. Mulligan, G.R. Untracht, S.N. Chandrasekaran, C.N. Brown, S.G. Adie, Emerging approaches for high-resolution imaging of tissue biomechanics with optical coherence elastography, IEEE J. Sel. Top. Quantum Electron. 22 (3) (2016) 1—20.

[8] K.V. Larin, D.D. Sampson, Optical coherence elastography - OCT at work in tissue biomechanics, Biomed. Opt. Express 8 (2) (2017) 1172—1202.

[9] B.F. Kennedy, R.A. McLaughlin, K.M. Kennedy, L. Chin, P. Wijesinghe, A. Curatolo, A. Tien, M. Ronald, B. Latham, C.M. Saunders, D.D. Sampson, Investigation of optical coherence microelastography as a method to visualize cancers in human breast tissue, Cancer Res. 75 (16) (2015) 3236—3245.

[10] C. Li, G. Guan, Y. Ling, Y.-T. Hsu, S. Song, J.T.I. Huang, S. Lang, R.K. Wang, Z. Huang, G. Nabi, Detection and characterisation of biopsy tissue using quantitative optical coherence elastography (OCE) in men with suspected prostate cancer, Cancer Lett. 357 (1) (2015) 121—128.

[11] M.D. Twa, I. Li, S. Vantipalli, M. Singh, S. Aglyamov, S. Emelianov, K.V. Larin, Spatial characterization of corneal biomechanical properties with optical coherence elastography after UV cross-linking, Biomed. Opt. Express 5 (5) (2014) 1419—1427.

[12] C. Wu, Z. Han, S. Wang, T. Li, M. Singh, C.-H. Liu, S. Aglyamov, S. Emelianov, F. Manns, K.V. Larin, Assessing age-related changes in the biomechanical properties of rabbit lens using a coaligned ultrasound and optical coherence elastography

systemage-related changes in biomechanics of lens, Investig. Ophthalmol. Vis. Sci. 56 (2) (2015) 1292−1300.

[13] G. Scarcelli, E. Kim, S.H. Yun, In vivo measurement of age-related stiffening in the crystalline lens by Brillouin optical microscopy, Biophys. J. 101 (6) (2011) 1539−1545.

[14] P. Wijesinghe, N.J. Johansen, A. Curatolo, D.D. Sampson, R. Ganss, B.F. Kennedy, Ultrahigh-resolution optical coherence elastography images cellular-scale stiffness of mouse aorta, Biophys. J. 113 (11) (2017) 2540−2551.

[15] A.L. Oldenburg, G. Wu, D. Spivak, F. Tsui, A.S. Wolberg, T.H. Fischer, Imaging and elastometry of blood clots using magnetomotive optical coherence tomography and labeled platelets, IEEE J. Sel. Top. Quantum Electron. 18 (3) (2012) 1100−1109.

[16] A.H. Chau, R.C. Chan, M. Shishkov, B. MacNeill, N. Iftimiia, G.J. Tearney, R.D. Kamm, B.E. Bouma, M.R. Kaazempur-Mofrad, Mechanical analysis of atherosclerotic plaques based on optical coherence tomography, Ann. Biomed. Eng. 32 (11) (2004) 1494−1503.

[17] D.A. Boas, A.K. Dunn, Laser speckle contrast imaging in biomedical optics, J. Biomed. Opt. 15 (1) (2010), 011109.

[18] G. Scarcelli, W.J. Polacheck, H.T. Nia, K. Patel, A.J. Grodzinsky, R.D. Kamm, S.H. Yun, Noncontact three-dimensional mapping of intracellular hydromechanical properties by Brillouin microscopy, Nat. Methods 12 (12) (2015) 1132−1134.

[19] A. Curatolo, M. Villiger, D. Lorenser, P. Wijesinghe, A. Fritz, B.F. Kennedy, D.D. Sampson, Ultrahigh-resolution optical coherence elastography, Opt. Lett. 41 (1) (2016) 21−24.

[20] W.J. Hadden, J.L. Young, A.W. Holle, M.L. McFetridge, D.Y. Kim, P. Wijesinghe, H. Taylor-Weiner, J.H. Wen, A.R. Lee, K. Bieback, B-N. Vo, D.D. Sampson, B.F. Kennedy, J.P. Spatz, A.J. Engler, Y.S. Choi, Stem cell migration and mechanotransduction on linear stiffness gradient hydrogels, PNAS, 2017, 201618239.

[21] J.F. Orr, J.C. Shelton, Optical Measurement Methods in Biomechanics, Springer, 1997.

[22] A.E. Pelling, M.A. Horton, An historical perspective on cell mechanics, Pflügers Archiv. 456 (1) (2008) 3−12.

[23] H.W. Turnbull, The Correspondence of Isaac Newton, vol. 2, Cambridge University Press, 1960, pp. 1676−1687.

[24] F.N. Egerton, A history of the ecological sciences, Part 19: Leeuwenhoek's microscopic natural history, Bull. Ecol. Soc. Am. 87 (1) (2006) 47−58.

[25] W. Seifriz, Methods of research on the physical properties of protoplasm, Plant Physiol. 12 (1) (1937) 99−116.

[26] L. V Heilbrunn, The viscosity of protoplasm, Q. Rev. Biol. 2 (2) (1927) 230−248.

[27] J. Gray, The mechanism of ciliary movement. VI. Photographic and stroboscopic analysis of ciliary movement, in: Proceedings of the Royal Society of London. Series B, Containing Papers of a Biological Character, vol. 107 (751), 1930, pp. 313−332.

[28] H.E. von Gierke, H.L. Oestreicher, E.K. Franke, H.O. Parrack, W.W. von Wittern, Physics of vibrations in living tissues, J. Appl. Physiol. 4 (12) (1952) 886−900.

[29] J.M. Mitchison, M.M. Swann, The mechanical properties of the cell surface, J. Exp. Biol. 31 (3) (1954) 461−472.

[30] J. Ophir, Elastography: a quantitative method for imaging the elasticity of biological tissues, Ultrason. Imaging 13 (2) (1991) 111−134.

[31] R. Muthupillai, D.J. Lomas, P.J. Rossman, J.F. Greenleaf, A. Manduca, R.L. Ehman, Magnetic resonance elastography by direct visualization of propagating acoustic strain waves, Science 269 (1995) 1854−1857.

[32] J.B. Fowlkes, Magnetic-resonance imaging techniques for detection of elasticity variation, Med. Phys. 22 (11) (1995) 1771–1778.

[33] H. Czichos, T. Saito, L. Smith, Springer Handbook of Materials Measurement Methods, vol. 978, Springer Berlin, 2006.

[34] J.W. Goodman, Some fundamental properties of speckle, J. Opt. Soc. America 66 (11) (1976) 1145–1150.

[35] W.H. Peters, W.F. Ranson, Digital imaging techniques in experimental stress analysis, Opt. Eng. 21 (3) (1982), 213427-213427.

[36] A. Oulamara, G. Tribillon, J. Duvernoy, Biological activity measurement on botanical specimen surfaces using a temporal decorrelation effect of laser speckle, J. Mod. Opt. 36 (2) (1989) 165–179.

[37] K. Wardell, A. Jakobsson, G.E. Nilsson, Laser Doppler perfusion imaging by dynamic light scattering, IEEE Trans. Biomed. Eng. 40 (4) (1993) 309–316.

[38] J. Schmitt, OCT elastography: imaging microscopic deformation and strain of tissue, Opt. Express 3 (6) (1998) 199–211.

[39] S.J. Kirkpatrick, R.K. Wang, D.D. Duncan, M. Kulesz-Martin, K. Lee, Imaging the mechanical stiffness of skin lesions by in vivo acousto-optical elastography, Opt. Express 14 (21) (2006) 9770–9779.

[40] X. Liang, M. Orescanin, K.S. Toohey, M.F. Insana, S.A. Boppart, Acousto-motive optical coherence elastography for measuring material mechanical properties, Opt. Lett. 34 (19) (2009) 2894–2896.

[41] W. Qi, R. Chen, L. Chou, G. Liu, J. Zhang, Q. Zhou, Z. Chen, Phase-resolved acoustic radiation force optical coherence elastography, J. Biomed. Opt. 17 (11) (2012) 110505.

[42] C. Li, G. Guan, R. Reif, Z. Huang, R.K. Wang, Determining elastic properties of skin by measuring surface waves from an impulse mechanical stimulus using phase-sensitive optical coherence tomography, J. R. Soc. Interface 9 (70) (2011) 831–841.

[43] M. Razani, A. Mariampillai, C. Sun, T.W.H. Luk, V.X.D. Yang, M.C. Kolios, Feasibility of optical coherence elastography measurements of shear wave propagation in homogeneous tissue equivalent phantoms, Biomed. Opt. Express 3 (5) (2012) 972–980.

[44] R.K. Wang, S. Kirkpatrick, M. Hinds, Phase-sensitive optical coherence elastography for mapping tissue microstrains in real time, Appl. Phys. Lett. 90 (16) (2007) 164105.

[45] G. Bao, S. Suresh, Cell and molecular mechanics of biological materials, Nat. Mater. 2 (11) (2003) 715–725.

[46] A.M. Stoneham, J.H. Harding, Not too big, not too small: the appropriate scale, Nat. Mater. 2 (2) (2003) 77–83.

[47] M. Meyers, P. Chen, A. Lin, Y. Seki, Biological materials: structure and mechanical properties, Prog. Mater. Sci. 53 (1) (2007) 1–206.

[48] A. Itoh, E. Ueno, E. Tohno, H. Kamma, H. Takahashi, T. Shiina, M. Yamakawa, T. Matsumura, Breast disease: clinical application of US elastography for diagnosis, Radiology 239 (2) (2006) 341–350.

[49] L. Sandrin, B. Fourquet, J.M. Ilasquenoph, S. Yon, C. Fournier, F. Mai, C. Christidis, M. Ziol, B. Poulet, F. Kazemi, M. Beaugrand, R. Palau, Transient elastography: a new noninvasive method for assessment of hepatic fibrosis, Ultrasound Med. Biol. 29 (12) (2003) 1705–1713.

[50] J.M. Chang, W.K. Moon, N. Cho, A. Yi, H.R. Koo, W. Han, D.-Y. Noh, H.-G. Moon, S.J. Kim, Clinical application of shear wave elastography (SWE) in the diagnosis of benign and malignant breast diseases, Breast Canc. Res. Treat. 129 (1) (2011) 89–97.

[51] K. Haase, A.E. Pelling, Investigating cell mechanics with atomic force microscopy, J. R. Soc. Interface 12 (104) (2015) 20140970.

[52] B.F. Kennedy, K.M. Kennedy, D.D. Sampson, A review of optical coherence elastography: fundamentals, techniques and prospects, IEEE J. Sel. Top. Quantum Electron. 20 (2) (2014) 272–288.

[53] M.M. Doyley, Model-based elastography: a survey of approaches to the inverse elasticity problem, Phys. Med. Biol. 57 (3) (2012) R35–R73.

[54] K.J. Parker, L.S. Taylor, S. Gracewski, D.J. Rubens, A unified view of imaging the elastic properties of tissue, J. Acoust. Soc. Am. 117 (5) (2005) 2705–2712.

[55] K. Nightingale, M.S. Soo, R. Nightingale, G. Trahey, Acoustic radiation force impulse imaging: in vivo demonstration of clinical feasibility, Ultrasound Med. Biol. 28 (2) (2002) 227–235.

[56] W. Qi, R. Li, T. Ma, K. Kirk Shung, Q. Zhou, Z. Chen, Confocal acoustic radiation force optical coherence elastography using a ring ultrasonic transducer, Appl. Phys. Lett. 104 (12) (2014) 123702.

[57] W. Drexler, J.G. Fujimoto, Optical Coherence Tomography: Technology and Applications, second ed., Springer, 2015.

[58] D.D. Sampson, T.R. Hillman, Optical coherence tomography, in: G. Palumbo, R. Pratesi (Eds.), Lasers and Current Optical Techniques in Biology, vol. 4, Royal Society of Chemistry, Cambridge, UK, 2004.

[59] M. Wojtkowski, V.J. Srinivasan, T.H. Ko, J.G. Fujimoto, A. Kowalczyk, J.S. Duker, Ultrahigh-resolution, high-speed, Fourier domain optical coherence tomography and methods for dispersion compensation, Opt. Express 12 (11) (2004) 2404–2422.

[60] Y. Lim, Y.-J. Hong, L. Duan, M. Yamanari, Y. Yasuno, Passive component based multifunctional Jones matrix swept source optical coherence tomography for Doppler and polarization imaging, Opt. Lett. 37 (11) (2012) 1958–1960.

[61] B. Baumann, W. Choi, B. Potsaid, D. Huang, J.S. Duker, J.G. Fujimoto, Swept source/Fourier domain polarization sensitive optical coherence tomography with a passive polarization delay unit, Opt. Express 20 (9) (2012) 10229–10241.

[62] L. Chin, X. Yang, R.A. McLaughlin, E.B. Noble, D.D. Sampson, En face parametric imaging of tissue birefringence using polarization-sensitive optical coherence tomography, J. Biomed. Opt. 18 (6) (2013), 066005.

[63] M. Villiger, D. Lorenser, R.A. McLaughlin, B.C. Quirk, R.W. Kirk, B.E. Bouma, D.D. Sampson, Deep tissue volume imaging of birefringence through fibre-optic needle probes for the delineation of breast tumour, Sci. Rep. 6 (2016) 28771.

[64] T.R. Hillman, Microstructural Information beyond the Resolution Limit: Studies in Two Coherent, Wide-Field Biomedical Imaging Systems, Ph.D. thesis, The University of Western Australia, 2007.

[65] A.F. Fercher, K. Mengedoht, W. Werner, Eye-length measurement by interferometry with partially coherent light, Opt. Lett. 13 (3) (1988) 186–188.

[66] D. Huang, E.A. Swanson, C.E. Lin, J.S. Schuman, W.G. Stinson, W. Chang, M.R. Hee, T. Flotte, K. Gregory, C.A. Puliafito, Optical coherence tomography, Science 254 (5035) (1991) 1178–1181.

[67] A. Dubois, L. Vabre, A.-C. Boccara, E. Beaurepaire, High-resolution fullfield optical coherence tomography with a Linnik microscope, Appl. Opt. 41 (4) (2002) 805–812.

[68] T. Endo, Y. Yasuno, S. Makita, M. Itoh, T. Yatagai, Profilometry with line-field Fourier-domain interferometry, Opt. Express 13 (3) (2005) 695–701.

[69] B. Grajciar, M. Pircher, A.F. Fercher, R.A. Leitgeb, Parallel Fourier domain optical coherence tomography for in vivo measurement of the human eye, Opt. Express 13 (4) (2005) 1131–1137.

[70] D. Hillmann, C. Liihrs, T. Bonin, P. Koch, G. Huttmann, Holoscopy - holographic optical coherence tomography, Opt. Lett. 36 (13) (2011) 2390–2392.

[71] D. Hillmann, G. Franke, C. Liihrs, E. Koch, G. I luttmann, Digital holoscopy, in: Optical Coherence Tomography: Technology and Applications, 2015, pp. 839–863.

[72] T.R. Hillman, D.D. Sampson, The effect of water dispersion and absorption on axial resolution in ultrahigh-resolution optical coherence tomography, Opt. Express 13 (6) (2005) 1860–1874.

[73] T. Klein, W. Wieser, C.M. Eigenwillig, B.R. Biedermann, R. Huber, Megahertz OCT for ultrawide-field retinal imaging with a 1050nm Fourier domain mode-locked laser, Opt. Express 19 (4) (2011) 3044–3062.

[74] L. Reznicek, T. Klein, W. Wieser, M. Kernt, A. Wolf, C. Haritoglou, A. Kampik, R. Huber, A.S. Neubauer, Megahertz ultra-wide-field swept-source retina optical coherence tomography compared to current existing imaging devices, Graefe's Arch. Clin. Exp. Ophthalmol. 252 (6) (2014) 1009–1016.

[75] W.M. Allen, L. Chin, P. Wijesinghe, R.W. Kirk, B. Latham, D.D. Sampson, C.M. Saunders, B.F. Kennedy, Wide-field optical coherence micro- elastography for intraoperative assessment of human breast cancer margins, Biomed. Opt. Express 7 (10) (2016) 4139–4152.

[76] R.A. Leitgeb, M. Villiger, A.H. Bachmann, L. Steinmann, T. Lasser, Extended focus depth for Fourier domain optical coherence microscopy, Opt. Lett. 31 (16) (2006) 2450–2452.

[77] A. Nahas, M. Bauer, S. Roux, A.C. Boccara, 3D static elastography at the micrometer scale using Full Field OCT, Biomed. Opt. Express 4 (10) (2013) 2138–2149.

[78] S. Mrejen, R.F. Spaide, Optical coherence tomography: imaging of the choroid and beyond, Surv. Ophthalmol. 58 (5) (2013) 387–429.

[79] G. Ferrante, E. Presbitero, R. Whitbourn, P. Barlis, Current applications of optical coherence tomography for coronary intervention, Int. J. Cardiol. 165 (1) (2013) 7–16.

[80] T-H. Tsai, J.G. Fujimoto, H. Mashimo, Endoscopic optical coherence tomography for clinical gastroenterology, Diagnostics 4 (2) (2014) 57–93.

[81] E. Sattler, R. Kastle, J. Welzel, Optical coherence tomography in dermatology, J. Biomed. Opt. 18 (6) (2013), 061224.

[82] B.J. Vakoc, D. Fukumura, R.K. Jain, B.E. Bouma, Cancer imaging by optical coherence tomography: preclinical progress and clinical potential, Nat. Rev. Cancer 12 (5) (2012) 363–368.

[83] A.F. Fercher, C.K. Hitzenberger, G. Kamp, S.Y. El-Zaiat, Measurement of intraocular distances by backscattering spectral interferometry, Opt. Commun. 117 (1) (1995) 43–48.

[84] J.A. Izatt, M.D. Kulkarni, S. Yazdanfar, J.K. Barton, A.J. Welch, In vivo bidirectional color Doppler flow imaging of picoliter blood volumes using optical coherence tomography, Opt. Lett. 22 (18) (1997) 1439–1441.

[85] R.K. Wang, Z. Ma, S.J. Kirkpatrick, Tissue Doppler optical coherence elastography for real time strain rate and strain mapping of soft tissue, Appl. Phys. Lett. 89 (14) (2006), 144103.

[86] R.C. Chan, A.H. Chau, W.C. Karl, S. Nadkarni, A.S. Khalil, N. Iftimia, M. Shishkov, G.J. Tearney, M.R. Kaazempur-Mofrad, B.E. Bouma, OCT-based arterial

elastography: robust estimation exploiting tissue biomechanics, Opt. Express 12 (19) (2004) 4558–4572.

[87] G. van Soest, F. Mastik, N. de Jong, A.F.W. van der Steen, Robust intravascular optical coherence elastography by line correlations, Phys. Med. Biol. 52 (9) (2007) 2445–2458.

[88] H.J. Ko, W. Tan, R. Stack, S.A. Boppart, Optical coherence elastography of engineered and developing tissue, Tissue Eng. 12 (1) (2006) 63–73.

[89] X. Liang, A.L. Oldenburg, V. Crecea, E.J. Chaney, S.A. Boppart, Optical micro-scale mapping of dynamic biomechanical tissue properties, Opt. Express 16 (15) (2008) 11052–11065.

[90] S. Wang, K.V. Larin, Optical coherence elastography for tissue characterization: a review, J. Biophot. 8 (4) (2015) 279–302.

[91] S. Wang, K.V. Larin, Shear wave imaging optical coherence tomography (SWI-OCT) for ocular tissue biomechanics, Opt. Lett. 39 (1) (2014) 41–44.

[92] L. Ambrozinski, S. Song, S.J. Yoon, I. Pelivanov, D. Li, L. Gao, T.T. Shen, R.K. Wang, M. O'Donnell, Acoustic micro-tapping for non-contact 4D imaging of tissue elasticity, Sci. Rep. 6 (2016), 38967.

[93] M. Singh, Z. Han, A. Nair, A. Schill, M.D. Twa, K.V. Larin, Applanation optical coherence elastography: noncontact measurement of intraocular pressure, corneal biomechanical properties, and corneal geometry with a single instrument, J. Biomed. Opt. 22 (2) (2017), 020502.

[94] J. Rogowska, N.A. Patel, J.G. Fujimoto, M.E. Brezinski, Optical coherence tomographic elastography technique for measuring deformation and strain of atherosclerotic tissues, Heart 90 (5) (2004) 556–562.

[95] J.M. Schmitt, S.H. Xiang, K.M. Yung, Speckle in optical coherence tomography, J. Biomed. Opt. 4 (1) (1999) 95–105.

[96] V.Y. Zaitsev, A.L. Matveyev, L.A. Matveev, G.V. Gelikonov, V.M. Gelikonov, A. Vitkin, Deformation-induced speckle-pattern evolution and feasibility of correlational speckle tracking in optical coherence elastography, J. Biomed. Opt. 20 (7) (2015), 075006.

[97] L. Chin, A. Curatolo, B.F. Kennedy, B.J. Doyle, P.R.T. Munro, R.A. McLaughlin, D.D. Sampson, Analysis of image formation in optical coherence elastography using a multiphysics approach, Biomed. Opt. Express 5 (9) (2014) 2913–2930.

[98] A. Curatolo, B.F. Kennedy, D.D. Sampson, T.R. Hillman, Speckle in optical coherence tomography, in: Advanced Biophotonics, Series in Optics and Optoelectronics, Taylor & Francis, 2013, pp. 211–277.

[99] J. Fu, F. Pierron, P. Ruiz, Elastic stiffness characterization using three-dimensional full-field deformation obtained with optical coherence tomography and digital volume correlation, J. Biomed. Opt. 18 (12) (2013) 121512.

[100] J. Fu, M. Haghighi-Abayneh, F. Pierron, P.D. Ruiz, Depth-resolved full-field measurement of corneal deformation by optical coherence tomography and digital volume correlation, Exp. Mech. 56 (7) (2016) 1203–1217.

[101] B.F. Kennedy, R.A. McLaughlin, K.M. Kennedy, L. Chin, A. Curatolo, A. Tien, B. Latham, C.M. Saunders, D.D. Sampson, Optical coherence micro-elastography: mechanical-contrast imaging of tissue microstructure, Biomed. Opt. Express 5 (7) (2014) 2113–2124.

[102] W. Choi, B. Potsaid, V. Jayaraman, B. Baumann, I. Grulkowski, J.J. Liu, C.D. Lu, A.E. Cable, D. Huang, J.S. Duker, J.G. Fujimoto, Phase-sensitive swept-source optical

coherence tomography imaging of the human retina with a vertical cavity surface-emitting laser light source, Opt. Lett. 38 (3) (2013) 338–340.

[103] S. Song, W. Wei, B-Y. Hsieh, I. Pelivanov, T.T. Shen, M. O'Donnell, R.K. Wang, Strategies to improve phase-stability of ultrafast swept source optical coherence tomography for single shot imaging of transient mechanical waves at 16 kHz frame rate, Appl. Phys. Lett. 108 (19) (2016), 191104.

[104] J.W. Goodman, Statistical Optics, John Wiley & Sons, 2015.

[105] C. Blatter, B. Grajciar, L. Schmetterer, R.A. Lcitgeb, Angle independent flow assessment with bidirectional Doppler optical coherence tomography, Opt. Lett. 38 (21) (2013) 4433–4436.

[106] A. Bouwens, D. Szlag, M. Szkulmowski, T. Bolmont, M. Wojtkowski, T. Lasser, Quantitative lateral and axial flow imaging with optical coherence microscopy and tomography, Opt. Express 21 (15) (2013) 17711–17729.

[107] H. Ammari, E. Bretin, E. Millien, L. Seppecher, J.-K. Seo, Mathematical modeling in full-field optical coherence elastography, SIAM J. Appl. Math. 75 (3) (2015) 1015–1030.

[108] A. Nahas, M. Tanter, T-M. Nguyen, J-M. Chassot, M. Fink, A.C. Boccara, From supersonic shear wave imaging to full-field, optical coherence shear wave elastography, J. Biomed. Opt. 18 (12) (2013), 121514.

[109] K. Kurokawa, S. Makita, Y.-l. Hong, Y. Yasuno, In-plane and out-of-plane tissue micro-displacement measurement by correlation coefficients of optical coherence tomography, Opt. Lett. 40 (9) (2015) 2153–2156.

[110] K. Kurokawa, S. Makita, Y.-J. Hong, Y. Yasuno, Two-dimensional micro-displacement measurement for laser coagulation using optical coherence tomography, Biomed. Opt. Express 6 (1) (2015) 170–190.

[111] B.F. Kennedy, M. Wojtkowski, M. Szkulmowski, K.M. Kennedy, K. Karnowski, D.D. Sampson, Improved measurement of vibration amplitude in dynamic optical coherence elastography, Biomed. Opt. Express 3 (12) (2012) 3138–3152.

[112] B.F. Kennedy, T.R. Hillman, R.A. McLaughlin, B.C. Quirk, D.D. Sampson, In vivo dynamic optical coherence elastography using a ring actuator, Opt. Express 17 (24) (2009) 21762–21772.

[113] B.F. Kennedy, S.H. Koh, R.A. McLaughlin, K.M. Kennedy, P.R.T. Munro, D.D. Sampson, Strain estimation in phase-sensitive optical coherence elastography, Biomed. Opt. Express 3 (8) (2012) 1865–1879.

[114] V.Y. Zaitsev, A.L. Matveyev, L.A. Matveev, G.V. Gelikonov, A.A. Sovetsky, A. Vitkin, Optimized phase gradient measurements and phase-amplitude interplay in optical coherence elastography, J. Biomed. Opt. 21 (11) (2016) 116005.

[115] V.Y. Zaitsev, A.L. Matveyev, L.A. Matveev, G. V Gelikonov, E.V. Gubarkova, N.D. Gladkova, A. Vitkin, Hybrid method of strain estimation in optical coherence elastography using combined sub-wavelength phase measurements and supra-pixel displacement tracking, J. *Biophoton.* 9 (5) (2016) 499–509.

[116] L. Chin, B. Latham, C.M. Saunders, D.D. Sampson, B.F. Kennedy, Simplifying the assessment of human breast cancer by mapping a micro-scale heterogeneity index in optical coherence elastography, J. *Biophoton.* 10 (5) (2016) 690–700.

[117] L. Chin, B.F. Kennedy, K.M. Kennedy, P. Wijesinghe, G.J. Pinniger, J.R. Terrill, R.A. McLaughlin, D.D. Sampson, Three-dimensional optical coherence micro-elastography of skeletal muscle tissue, Biomed. Opt. Express 5 (9) (2014) 3090–3102.

[118] K.M. Kennedy, C. Ford, B.F. Kennedy, M.B. Bush, D.D. Sampson, Analysis of mechanical contrast in optical coherence elastography, J. Biomed. Opt. 18 (12) (2013) 121508.

[119] K.M. Kennedy, L. Chin, R.A. McLaughlin, B. Latham, C.M. Saunders, D.D. Sampson, B.F. Kennedy, Quantitative micro-elastography: imaging of tissue elasticity using compression optical coherence elastography, Sci. Rep. 5 (2015) 15538.

[120] K.M. Kennedy, S. Es'haghian, L. Chin, R.A. McLaughlin, D.D. Sampson, B.F. Kennedy, Optical palpation: optical coherence tomography-based tactile imaging using a compliant sensor, Opt. Lett. 39 (10) (2014) 3014–3017.

[121] V.Y. Zaitsev, A.L. Matveyev, L.A. Matveev, E. V Gubarkova, A.A. Sovetsky, M.A. Sirotkina, G. V Gelikonov, E.V. Zagaynova, N.D. Gladkova, A. Vitkin, Practical obstacles and their mitigation strategies in compressional optical coherence elastography of biological tissues, J. Innov. Opt. Health Sci. 10 (06) (2017) 1742006.

[122] L. Dong, E. Wijesinghe, J.T. Dantuono, D.D. Sampson, P.R.T. Munro, B.F. Kennedy, A.A. Oberai, Quantitative compression optical coherence elastography as an inverse elasticity problem, IEEE J. Sel. Top. Quantum Electron. 22 (3) (2016) 277–287.

[123] B.F. Kennedy, F.G. Malheiro, L. Chin, D.D. Sampson, Three-dimensional optical coherence elastography by phase-sensitive comparison of C-scans, J. Biomed. Opt. 19 (7) (2014), 076006.

[124] R.W. Kirk, B.F. Kennedy, D.D. Sampson, R.A. McLaughlin, Near video-rate optical coherence elastography by acceleration with a graphics processing unit, IEEE/OSA J. Lightwave Technol. 33 (16) (2015) 3481–3485.

[125] J.A. Mulligan, F. Bordeleau, C.A. Reinhart-King, S.G. Adie, Measurement of dynamic cell-induced 3D displacement fields in vitro for traction force optical coherence microscopy, Biomed. Opt. Express 8 (2) (2017) 1152–1171.

[126] D. Pokharel, P. Wijesinghe, V. Oenarto, J.F. Lu, D.D. Sampson, B.F. Kennedy, V.P. Wallace, M. Bebawy, Deciphering cell-to-cell communication in acquisition of cancer traits: extracellular membrane vesicles are regulators of tissue biomechanics, OMICS A J. Integr. Biol. 20 (8) (2016) 462–469.

[127] A. E Sarvazyan, O.V. Rudenko, S.D. Swanson, J. Fowlkes, S.Y. Emelianov, Shear wave elasticity imaging: a new ultrasonic technology of medical diagnostics, Ultrasound Med. Biol. 24 (9) (1998) 1419–1435.

[128] C. Li, G. Guan, X. Cheng, Z. Huang, R.K. Wang, Quantitative elastography provided by surface acoustic waves measured by phase-sensitive optical coherence tomography, Opt. Lett. 37 (4) (2012) 722–724.

[129] C. Ii, G. Guan, Z. Huang, M. Johnstone, R.K. Wang, Noncontact all-optical measurement of corneal elasticity, Opt. Lett. 37 (10) (2012) 1625–1627.

[130] R.K. Manapuram, S.R. Aglyamov, F.M. Monediado, M. Mashiatulla, J. Li, S.Y. Emelianov, K.V. Larin, In vivo estimation of elastic wave parameters using phase-stabilized swept source optical coherence elastography, J. Biomed. Opt. 17 (10) (2012) 15–18.

[131] S. Wang, J. Li, R.K. Manapuram, F.M. Menodiado, D.R. Ingram, M.D. Twa, A.J. Lazar, D.C. Lev, R.E. Pollock, K.V. Larin, Noncontact measurement of elasticity for the detection of soft-tissue tumors using phase-sensitive optical cohcrence tomography combined with a focused air-puff system, Opt. Lett. 37 (24) (2012) 5184–5186.

[132] X. Liang, S.A. Boppart, Biomechanical properties of in vivo human skin from dynamic optical coherence elastography, IEEE Trans. Biomed. Eng. 57 (4) (2010) 953–959.

[133] S. Song, Z. Huang, R.K. Wang, Tracking mechanical wave propagation within tissue using phase-sensitive optical coherence tomography: motion artifact and its compensation, J. Biomed. Opt. 18 (12) (2013) 121505.

[134] Z. Han, M. Singh, S.R. Aglyamov, C.-H. Liu, A. Nair, R. Raghunathan, C. Wu, J. Li, K.V. Larin, Quantifying tissue viscoelasticity using optical coherence elastography and the Rayleigh wave model, J. Biomed. Opt. 21 (9) (2016), 090504.

[135] Z. Han, J. Li, M. Singh, C. Wu, C.-H. Liu, R. Raghunathan, S.R. Aglyamov, S. Vantipalli, M.D. Twa, K.V. Larin, Optical coherence elastography assessment of corneal viscoelasticity with a modified Rayleigh-Lamb wave model, J. Mech. Behav. Biomed. Mater. 66 (2017) 87–94.

[136] T-M. Nguyen, S. Song, B. Arnal, E.Y. Wong, Z. Huang, R.K. Wang, M. O'Donnell, Shear wave pulse compression for dynamic elastography using phase-sensitive optical coherence tomography, J. Biomed. Opt. 19 (1) (2014), 016013.

[137] T-M. Nguyen, B. Arnal, S. Song, Z. Huang, R.K. Wang, M. O'Donnell, Shear wave elastography using amplitude-modulated acoustic radiation force and phase-sensitive optical coherence tomography, J. Biomed. Opt. 20 (1) (2015), 016001.

[138] S. Song, N.M. Le, Z. Huang, T. Shen, R.K. Wang, Quantitative shear-wave optical coherence elastography with a programmable phased array ultrasound as the wave source, Opt. Lett. 40 (21) (2015) 5007–5010.

[139] J. Zhu, Y. Qu, T. Ma, R. Li, Y. Du, S. Huang, K.K. Shung, Q. Zhou, Z. Chen, Imaging and characterizing shear wave and shear modulus under orthogonal acoustic radiation force excitation using OCT Doppler variance method, Opt. Lett. 40 (9) (2015) 2099–2102.

[140] L. Ambroziriski, I. Pelivanov, S. Song, S.J. Yoon, D. Li, L. Gao, T.T. Shell, R.K. Wang, M. O'Donnell, Air-coupled acoustic radiation force for non- contact generation of broadband mechanical waves in soft media, Appl. Phys. Lett. 109 (4) (2016) 043701.

[141] S. Wang, K.V. Larin, J. Li, S. Vantipalli, R.K. Manapuram, S. Aglyamov, S. Emelianov, M.D. Twa, A focused air-pulse system for optical- coherence-tomography-based measurements of tissue elasticity, Laser Phys. Lett. 10 (7) (2013) 075605.

[142] D. Alonso-Caneiro, K. Karnowski, B.F. Kaluzny, A. Kowalczyk, M. Wojtkowski, Assessment of corneal dynamics with high-speed swept source optical coherence to-mography combined with an air puff system, Opt. Express 19 (15) (2011) 14188–14199.

[143] C. Dorronsoro, D. Pascual, P. Perez-Merino, S. BQing, S. Marcos, Dynamic OCT measurement of corneal deformation by an air puff in normal and cross-linked corneas, Biomed. Opt. Express 3 (3) (2012) 473–487.

[144] Z. Han, J. Li, M. Singh, C. Wu, C.-h. Liu, S. Wang, R. Idugboe, R. Raghunathan, N. Sudheendran, S.R. Aglyamov, M.D. Twa, K.V. Larin, Quantitative methods for reconstructing tissue biomechanical properties in optical coherence elastography: a comparison study, Phys. Med. Biol. 60 (9) (2015) 3531–3547.

[145] M. Singh, J. Li, S. Vantipalli, S. Wang, Z. Han, A. Nair, S.R. Aglyamov, M.D. Twa, K.V. Larin, Noncontact elastic wave imaging optical coherence elastography for evaluating changes in corneal elasticity due to crosslinking, IEEE J. Sel. Top. Quantum Electron. 22 (3) (2016) 266–276.

[146] M. Singh, J. Li, S. Vantipalli, Z. Han, K.V. Larin, M.D. Twa, Optical coherence elastography for evaluating customized riboflavin/UV-A corneal collagen crosslinking, J. Biomed. Opt. 22 (9) (2017) 091504.

[147] C.-H. Liu, Y. Du, M. Singh, C. Wu, Z. Han, J. Li, A. Chang, C. Mohan, K.V. Larin, Classifying murine glomerulonephritis using optical coherence tomography and optical coherence elastography, J. Biophotonics 9 (8) (2016) 781−791.

[148] C. Li, G. Guan, F. Zhang, S. Song, R.K. Wang, Z. Huang, G. Nabi, Quantitative elasticity measurement of urinary bladder wall using laser-induced surface acoustic waves, Biomed. Opt. Express 5 (12) (2014) 4313−4328.

[149] V. Crecea, B.W. Graf, T. Kim, G. Popescu, S.A. Boppart, High resolution phase-sensitive magnetomotive optical coherence microscopy for tracking magnetic microbeads and cellular mechanics, IEEE J. Sel. Top. Quantum Electron. 20 (2) (2014) 25−31.

[150] A. Ahmad, J. Kim, N.A. Sobh, N.D. Shemonski, S.A. Boppart, Magnetomotive optical coherence elastography using magnetic particles to induce mechanical waves, Biomed. Opt. Express 5 (7) (2014) 2349−2361.

[151] W. Wieser, B.R. Biedermann, T. Klein, C.M. Eigenwillig, R. Huber, Multi- megahertz OCT: high quality 3D imaging at 20 million A-scans and 4.5 GVoxels per second, Opt. Express 18 (14) (2010) 14685−14704.

[152] M. Singh, C. Wu, C.-H. Liu, J. Li, A. Schill, A. Nair, K.V. Larin, Phase-sensitive optical coherence elastography at 1.5 million A-Lines per second, Opt. Lett. 40 (11) (2015) 2588−2591.

[153] C.-H. Liu, A. Schill, M. Singh, C. Wu and K.V. Larin, Line-field low coherence holography for ultra-fast assessment of tissue biomechanical properties, in: Optical Coherence Tomography and Coherence Domain Optical Methods in Biomedicine XXI. p. 100531Z

[154] C.-H. Liu, A. Schill, R. Raghunathan, C. Wu, M. Singh, Z. Han, A. Nair, K.V. Larin, Ultra-fast line-field low coherence holographic elastography using spatial phase shifting, Biomed. Opt. Express 8 (2) (2017) 993−1004.

[155] B.I. Akca, E.W. Chang, S. Kling, A. Ramier, G. Scarcelli, S. Marcos, S.H. Yun, Observation of sound-induced corneal vibrational modes by optical coherence tomography, Biomed. Opt. Express 6 (9) (2015) 3313−3319.

[156] E. Meemon, F. Yao, Y.-J. Chu, F. Zvietcovich, K.J. Parker, J. P. Rolland, Crawling wave optical coherence elastography, Opt. Lett. 41 (5) (2016) 847−850.

[157] S.G. Adie, B.F. Kennedy, J. J. Armstrong, S.A. Alexandrov, D.D. Sampson, Audio frequency in vivo optical coherence elastography, Phys. Med. Biol. 54 (10) (2009), 3129-3139.

[158] B.F. Kennedy, X. Liang, S.G. Adie, D.K. Gerstmann, B.C. Quirk, S.A. Boppart, D.D. Sampson, In vivo three-dimensional optical coherence elastography, Opt. Express 19 (7) (2011) 6623−6634.

[159] A.L. Oldenburg, S.A. Boppart, Resonant acoustic spectroscopy of soft tissues using embedded magnetomotive nanotransducers and optical coherence tomography, Phys. Med. Biol. 55 (4) (2010) 1189−1201.

[160] C. Wu, M. Singh, Z. Han, R. Raghunathan, C.-H. Liu, J. Li, A. Schill, K.V. Larin, Lorentz force optical coherence elastography, J. Biomed. Opt. 21 (9) (2016), 090502.

[161] G. Guan, C. Li, Y. Ling, Y. Yang, J.B. Vorstius, R.P. Keatch, R.K. Wang, Z. Huang, Quantitative evaluation of degenerated tendon model using combined optical coherence elastography and acoustic radiation force method, J. Biomed. Opt. 18 (11) (2013), 111417.

[162] A. Ahmad, P.C. Huang, N.A. Sobh, P. Pande, J. Kim, S.A. Boppart, Mechanical contrast in spectroscopic magnetomotive optical coherence elastography, Phys. Med. Biol. 60 (17) (2015) 6655–6668.

[163] S. E Kearney, A. Khan, Z. Dai, T.J. Royston, Dynamic viscoelastic models of human skin using optical elastography, Phys. Med. Biol. 60 (17) (2015) 6975–6990.

[164] Z. Han, S.R. Aglyamov, J. Li, M. Singh, S. Wang, S. Vantipalli, C. Wu, C.-h. Liu, M.D. Twa, K.V. Larin, Quantitative assessment of corneal viscoelasticity using optical coherence elastography and a modified Rayleigh-Lamb equation, J. Biomed. Opt. 20 (2) (2015), 020501.

[165] V. Crecea, A. Ahmad, S.A. Boppart, Magnetomotive optical coherence elastography for microrheology of biological tissues, J. Biomed. Opt. 18 (12) (2013) 121504.

[166] P. Wijesinghe, R.A. McLaughlin, D.D. Sampson, B.F. Kennedy, Parametric imaging of viscoelasticity using optical coherence elastography, Phys. Med. Biol. 60 (6) (2015) 2293–2307.

[167] H. Spahr, D. Hillmann, C. Ilain, C. Pfaffle, H. Sudkamp, G. Franke, G. Hüttmann, Imaging pulse wave propagation in human retinal vessels using full-field swept-source optical coherence tomography, Opt. Lett. 40 (20) (2015) 4771–4774.

[168] C. Apelian, F. Harms, O. Thouvenin, A.C. Boccara, Dynamic full field optical coherence tomography: subcellular metabolic contrast revealed in tissues by interferometric signals temporal analysis, Biomed. Opt. Express (4) (2016) 1511–1524.

[169] C.-E. Leroux, J. Palmier, A.C. Boccara, G. Cappello, S. Monnier, Elastography of multicellular aggregates submitted to osmo-mechanical stress, New J. Phys. 17 (7) (2015), 073035-073035.

[170] S. Es'haghian, P. Gong, K.M. Kennedy, P. Wijesinghe, D.D. Sampson, R.A. McLaughlin, B.F. Kennedy, In vivo optical elastography: stress and strain imaging of human skin lesions, Proc. SPIE 9327 (2015), 93270C-8.

[171] F. Prati, G. Guagliumi, G.S. Mintz, M. Costa, E. Regar, T. Akasaka, P. Barlis, G.J. Tearney, I.-K. Jang, E. Arbustini, others, Expert review document part 2: methodology, terminology and clinical applications of optical coherence tomography for the assessment of interventional procedures, Eur. Heart J. 33 (20) (2012) 2513–2520.

[172] J. P Williamson, R.A. McLaughlin, W.J. Noffsinger, A.L. James, V.A. Baker, A. Curatolo, J.J. Armstrong, A. Regli, K.L. Shepherd, G.B. Marks, D.D. Sampson, Elastic properties of the central airways in obstructive lung diseases measured using anatomical optical coherence tomography, Am. J. Respir. Crit. Care Med. 183 (5) (2011) 612–619.

[173] Y. Qu, T. Ma, Y. He, M. Yu, I. Zhu, Y. Miao, C. Dai, R. Patel, K.K. Shung, Q. Zhou, Z. Chen, Miniature probe for mapping mechanical properties of vascular lesions using acoustic radiation force optical coherence elastography, Sci. Rep. 7 (1) (2017) 4731.

[174] B.F. Kennedy, K.M. Kennedy, A.L. Oldenburg, S.G. Adie, S.A. Boppart, D.D. Sampson, in: W. Drexler, J.G. Fujimoto (Eds.), Optical Coherence Tomography Technology and Applications, Second ed., Springer-Verlag, 2015, pp. 1007–1054.

[175] Q. Fang, A. Curatolo, P. Wijesinghe, Y.L. Yeow, J. Hamzah, R. Noble, K. Karnowski, D.D. Sampson, R. Ganss, J.K. Kim, W.M. Lee, B.F. Kennedy, Ultrahigh-resolution optical coherence elastography through a micro- endoscope: towards in vivo imaging of cellular-scale mechanics, Biomed. Opt. Express 8 (11) (2017) 5127–5138.

[176] R.A. McLaughlin, B.C. Quirk, A. Curatolo, R.W. Kirk, L. Scolaro, D. Lorenser, P.D. Robbins, B.A. Wood, C.M. Saunders, D.D. Sampson, Imaging of breast cancer

with optical coherence tomography needle probes: feasibility and initial results, IEEE J. Sel. Top. Quantum Electron. 18 (3) (2012) 1184–1191.

[177] K.M. Kennedy, R.A. McLaughlin, B.F. Kennedy, A. Tien, B. Latham, C.M. Saunders, D.D. Sampson, Needle optical coherence elastography for the measurement of microscale mechanical contrast deep within human breast tissues, J. Biomed. Opt. 18 (12) (2013), 121510.

[178] K.M. Kennedy, B.F. Kennedy, R.A. McLaughlin, D.D. Sampson, Needle optical coherence elastography for tissue boundary detection, Opt. Lett. 37 (12) (2012) 2310–2312.

[179] Y. Qiu, Y. Wang, Y. Xu, N. Chandra, J. Haorah, B. Hubbi, B.J. Pfister, X. Liu, Quantitative optical coherence elastography based on fiber-optic probe for in situ measurement of tissue mechanical properties, Biomed. Opt. Express 7 (2) (2016) 688–700.

[180] G. Lan, M. Singh, K.V. Larin, M.D. Twa, Common-path phase-sensitive optical coherence tomography provides enhanced phase stability and detection sensitivity for dynamic elastography, Biomed. Opt. Express 8 (11) (2017) 5253–5266.

[181] S. Es'haghian, K.M. Kennedy, E. Gong, Q. Li, L. Chin, P. Wijesinghe, D.D. Sampson, R.A. McLaughlin, B.F. Kennedy, In vivo volumetric quantitative micro-elastography of human skin, Biomed. Opt. Express 8 (5) (2017) 2458–2471.

[182] G.L. Monroy, J. Won, D.R. Spillman, R. Dsouza, S.A. Boppart, Clinical translation of handheld optical coherence tomography: practical considerations and recent advancements, J. Biomed. Opt. 22 (12) (2017) 121715.

[183] J.F. Greenleaf, M. Fatemi, M. Insana, Selected methods for imaging elastic properties of biological tissues, Annu. Rev. Biomed. Eng. 5 (1) (2003) 57–78.

[184] P.E. Barbone, A.A. Oberai, A review of the mathematical and computational foundations of biomechanical imaging, in: Computational Modeling in Biomechanics, Springer, 2010, pp. 375–408.

[185] A.S. Khalil, R.C. Chan, A.H. Chau, B.E. Bouma, M.R.K. Mofrad, Tissue elasticity estimation with optical coherence elastography: toward mechanical characterization of in vivo soft tissue, Ann. Biomed. Eng. 33 (11) (2005) 1631–1639.

[186] A.S. Khalil, B.E. Bouma, M.R. Kaazempur Mofrad, A combined FEM/genetic algorithm for vascular soft tissue elasticity estimation, Cardiovasc. Eng. 6 (3) (2006) 93–102.

[187] L. Dong, E. Wijesinghe, J.T. Dantuono, D.D. Sampson, P.R.T. Munro, B.F. Kennedy, A.A. Oberai, Quantitative optical coherence elastography as an inverse elasticity problem, Proc. SPIE 9710 (2016), 971011-1.

[188] Z. Han, J. Li, M. Singh, S.R. Aglyamov, C. Wu, C.-h. Liu, K.V. Larin, Analysis of the effects of curvature and thickness on elastic wave velocity in cornea-like structures by finite element modeling and optical coherence elastography, Appl. Phys. Lett. 106 (23) (2015), 233702-233702.

[189] Z. Han, J. Li, M. Singh, S. Vantipalli, S.R. Aglyamov, C. Wu, C.-Hao Liu, R. Raghunathan, M.D. Twa, K.V. Larin, Analysis of the effect of the fluid- structure interface on elastic wave velocity in cornea-like structures by OCE and FEM, Laser Phys. Lett. 13 (3) (2016) 035602.

[190] F. Zvietcovich, J. R Rolland, J. Yao, P. Meemon, K.J. Parker, Comparative study of shear wave-based elastography techniques in optical coherence tomography, J. Biomed. Opt. 22 (3) (2017) 035010.

[191] G. Scarcelli, S.H. Yun, Confocal Brillouin microscopy for three- dimensional mechanical imaging, Nat. Photonics 2 (1) (2008) 39–43.

[192] G. Antonacci, M.R. Foreman, C. Paterson, P. Torok, Spectral broadening in Brillouin imaging, Appl. Phys. Lett. 103 (22) (2013), 221105.

[193] L. Brillouin, Diffusion de la lumiere et des rayons X par un corps transparent homo gene. Influence de l'agitation thermique, Ann. Phys. 17 (21) (1922) 88–122.

[194] E. Gross, Change of wave-length of light due to elastic heat waves at scattering in liquids, Nature 126 (1930) 201–202.

[195] J.M. Vaughan, J.T. Randall, Brillouin scattering, density and elastic properties of the lens and cornea of the eye, Nature 284 (5755) (1980) 489–491.

[196] C.W. Ballmann, J.V. Thompson, A.J. Traverso, Z. Meng, M.O. Scully, V.V. Yakovlev, Stimulated Brillouin scattering microscopic imaging, Sci. Rep. 5 (2015), 18139.

[197] G. Antonacci, R.M. Pedrigi, A. Kondiboyina, V.V. Mehta, R. de Silva, C. Paterson, R. Krams, P. Tordk, Quantification of plaque stiffness by Brillouin microscopy in experimental thin cap fibroatheroma, J. R. Soc. Interface 12 (112) (2015), 20150843.

[198] G. Scarcelli, R. Pineda, S.H. Yun, Brillouin optical microscopy for corneal biomechanics, Investig. Ophthalmol. Vis. Sci. 53 (1) (2012) 185–190.

[199] G. Scarcelli, S. Besner, R. Pineda, P. Kalout, S.H. Yun, In vivo biomechanical mapping of normal and keratoconus corneas, JAMA Ophthalmol. 133 (4) (2015) 480–482.

[200] C.W. Ballmann, Z. Meng, A.J. Traverso, M.O. Scully, V.V. Yakovlev, Impulsive Brillouin microscopy, Optica 4 (1) (2017) 124–128.

[201] G. Antonacci, Dark-field Brillouin microscopy, Opt. Lett. 42 (7) (2017) 1432–1435.

[202] K.E. Kasza, A.C. Rowat, J. liu, T.E. Angelini, C.P. Brangwynne, G.H. Koen- derink, D.A. Weitz, The cell as a material, Curr. Opin. Cell Biol. 19 (1) (2007) 101–107.

[203] Z. Hajjarian, S.K. Nadkarni, Evaluating the viscoelastic properties of tissue from laser speckle fluctuations, Sci. Rep. 2 (2012), 316.

[204] S.K. Nadkarni, A. Bilenca, B.E. Bouma, G.J. Tearney, Measurement of fibrous cap thickness in atherosclerotic plaques by spatiotemporal analysis of laser speckle images, J. Biomed. Opt. 11 (2) (2006), 021006.

[205] S.K. Nadkarni, B.E. Bouma, T. Helg, R. Chan, E. Halpern, A. Chau, M.S. Minsky, J.T. Motz, S.L. Ilouser, G.J. Tearney, Characterization of atherosclerotic plaques by laser speckle imaging, Circulation 112 (6) (2005) 885–892.

[206] S.J. Kirkpatrick, D.D. Duncan, L. Fang, Low-frequency surface wave propagation and the viscoelastic behavior of porcine skin, J. Biomed. Opt. 9 (6) (2004) 1311–1319.

[207] Z. Hajjarian, J. Xi, F.A. Jaffer, G.J. Tearney, S.K. Nadkarni, Intravascular laser speckle imaging catheter for the mechanical evaluation of the arterial wall, J. Biomed. Opt. 16 (2) (2011), 026005.

[208] L.V. Wang, Prospects of photoacoustic tomography, Med. Phys. 35 (12) (2008) 5758–5767.

[209] L.V. Wang, S. Hu, Photoacoustic tomography: in vivo imaging from organelles to organs, Science 335 (6075) (2012) 1458–1462.

[210] Y. Zhao, S. Yang, C. Chen, D. Xing, Simultaneous optical absorption and viscoelasticity imaging based on photoacoustic lock-in measurement, Opt. Lett. 39 (9) (2014) 2565–2568.

[211] E. Hai, Y. Zhou, J. Liang, C. Li, L. V Wang, Photoacoustic tomography of vascular compliance in humans, J. Biomed. Opt. 20 (12) (2015), 126008.

[212] E. Hai, I. Yao, G. Li, C. Li, L. V Wang, Photoacoustic elastography, Opt. Lett. 41 (4) (2016) 725−728.

[213] E. Hai, Y. Zhou, L. Gong, L.V. Wang, Quantitative photoacoustic elastography in humans, J. Biomed. Opt. 21 (6) (2016), 066011.

[214] T. Glatz, O. Scherzer, T. Widlak, Texture generation for photoacoustic elastography, J. Math. Imaging Vis. 52 (3) (2015) 369−384.

[215] D.S. Elson, R. Li, C. Dunsby, R. Eckersley, M.-X. Tang, Ultrasound-mediated optical tomography:A review of current methods, Interface Focus 1 (4) (2011) 632−648.

[216] R. Li, D.S. Elson, C. Dunsby, R. Eckersley, M.-X. Tang, Effects of acoustic radiation force and shear waves for absorption and stiffness sensing in ultrasound modulated optical tomography, Opt. Express 19 (8) (2011) 7299−7311.

[217] K. Daoudi, A.C. Boccara, E. Bossy, Detection and discrimination of optical absorption and shear stiffness at depth in tissue-mimicking phantoms by transient optoelastography, Appl. Phys. Lett. 94 (15) (2009), 154103.

[218] K.D. Mohan, A.L. Oldenburg, Elastography of soft materials and tissues by holographic imaging of surface acoustic waves, Opt. Express 20 (17) (2012) 18887−18897.

[219] C.-H. Liu, A. Schill, C. Wu, M. Singh, K.V. Larin, Non-contact single shot elastography using line field low coherence holography, Biomed. Opt. Express 7 (8) (2016) 3021−3031.

[220] C.U. Devi, R.S.B. Chandran, R.M. Vasu, A.K. Sood, Measurement of visco-elastic properties of breast-tissue mimicking materials using diffusing wave spectroscopy, J. Biomed. Opt. 12 (3) (2007), 034035.

[221] Y. Zhang, R.T. Brodell, E.N. Mostow, C.J. Vinyard, H. Marie, In vivo skin elastography with high-definition optical videos, Skin Res. Technol. 15 (3) (2009) 271−282.

[222] N.T. Clancy, G.E. Nilsson, C.D. Anderson, M.I. Leahy, A new device for assessing changes in skin viscoelasticity using indentation and optical measurement, Skin Res. Technol. 16 (2) (2010) 210−228.

[223] D. Francis, R. R Tatam, R.M. Groves, Shearography technology and applications: a review, Meas. Sci. Technol. 21 (10) (2010), 102001.

[224] I. Weber, S-H. Yun, G. Scarcelli, K. Franze, The role of cell body density in ruminant retina mechanics assessed by atomic force and Brillouin microscopy, Phys. Biol. 14 (6) (2017) 065006.

[225] R. Friedl, S. Alexander, Cancer invasion and the microenvironment: plasticity and reciprocity, Cell 147 (5) (2011) 992−1009.

[226] P. Wellman, R.D. Howe, E. Dalton, K.A. Kern, Breast Tissue Stiffness in Compression Is Correlated to Histological Diagnosis, Harvard BioRobotics Laboratory Technical Report, 1999, pp. 1−15.

[227] K.M. Kennedy, L. Chin, P. Wijesinghe, R.A. McLaughlin, B. Latham, D.D. Sampson, C.M. Saunders, B.F. Kennedy, Investigation of optical coherence micro-elastography as a method to visualize micro-architecture in human axillary lymph nodes, BMC Cancer 16 (2016) 874.

[228] W.M. Allen, K.M. Kennedy, Q. Fang, L. Chin, A. Curatolo, L. Watts, R. Zilkens, S.L. Chin, B.F. Dessauvagie, B. Latham, C.M. Saunders, B.F. Kennedy, Wide-field quantitative micro-elastography of human breast tissue, Biomed. Opt. Express 9 (3) (2018) 1082−1096.

[229] S. Wang, K.V. Larin, Noncontact depth-resolved micro-scale optical coherence elastography of the cornea, Biomed. Opt. Express 5 (11) (2014) 3807−3821.

[230] D.P. Pinero, N. Alcon, Corneal biomechanics: a review, Clin. Exp. Optom. 98 (2) (2015) 107–116.

[231] M.R. Ford, A.S. Roy, A.M. Rollins, W.J. Dupps, Serial biomechanical comparison of edematous, normal, and collagen crosslinked human donor corneas using optical coherence elastography, J. Cataract Refract. Surg. 40 (6) (2014) 1041–1047.

Index

A

Acoustic phonons, 208
Acoustic radiation force, 9, 86—90
  imaging techniques
    applied to tissues, 90—91
    impulse imaging. *See* Impulse imaging,
      acoustic radiation force
  from ultrasonography, 86—90
Altered point spread functions, 171
Angular beam-steered data acquisitions, 174
  approach, 171—173
Angular displacement noise, 173
Artifact, 71—73
Artifact-free regions-of-interest (ROIs), 135—136
Axicon CUSE (AxCUSE), 112

B

Beam-steered data acquisition, 172—173
Bessel functions, 50
Blatz model, 74
Brain, 141—142
Brillouin microscopy, 208—212
Brillouin modulus, 211—212
Brillouin scattering, 208

C

Cardiac pulsations/natural sources, 4—5
Circumferential strain, 178
Comb-push ultrasound elastography, 111—112
Commercial ultrasonographic systems, 167
Compression elastography, 75
Computational inverse methods, 205—208
Conventional medical ultrasonographic
  imaging, 1
Coupled axial estimation, 174
Crawling waves, 52—54
Cylindrical coordinate systems, 25—26

D

Deformation, 17—18
Dimensionless perturbation parameter, 88
Direct axial estimation, 173—174
Direct inversion schemes, 156
Directional beamforming, 175
Dispersion, 40
Displacement and strain, 17—20
Displacement tracking, 93—94

Displacement vector of particle, 17—18
Dual transducers, perpendicular insonification
  using, 170
Dynamic deformation, 30—32
  of viscoelastic medium, 39—41
Dynamic optical coherence elastographic
  methods, 200—203
Dynamic problems
  motion of solid sphere under dynamic load,
    33—34
  plane wave propagation, 32—33

E

Eigenfrequencies, 46—47
Elastic approximation, 69—70
Elastic constants, stress-strain relation for, 22—24
Elasticity imaging
  imaging tissue stiffness
    compression elastography, 7
    vibration amplitude sonoelastography, 6—7
  microscopic tissue, 10
  observations from radiologist, 1—3
  quantitative tissue stiffness determination and
    imaging, 7—8
  tissue, 8—10
  tissue elasticity determination
    Oestreicher and von Gierke, 3—4
    palpation, 3
    tissue motion studies, 4—6
Elastic material, stress-strain relation for,
  22—24
Elastic modulus, 67
Elastography
  comb-push ultrasound, 111—112
  compression, 75
  harmonic, 156—157
  optical, 10, 214
  optical coherence. *See* Optical coherence
    elastography
  photoacoustic, 10
  quasi-static. *See* Quasi-static elastography
  reconstructive. *See* Reconstructive elastography
  shear wave, 30
  ultrasound. *See* Ultrasound elastography
Equilibrium equations, 24—25
Externally controllable sources, tissue stimulation
  by, 5—6

F

Fetal lung elasticity, 5
FibroScan, 7—8
Finite-element methods (FEMs), 206—207
Finite strain tensors, 18—19
Fourier-domain OCT, 193, 195—196

G

Generalized viscoelastic models, 38—39
Governing theory of elasticity imaging, 17
    cylindrical and spherical coordinate systems,
        25—26
    displacement and strain, 17—20
    dynamic deformation, 30—32
        of viscoelastic medium, 39—41
    dynamic problems
        motion of solid sphere under dynamic load,
            33—34
        plane wave propagation, 32—33
    equilibrium equations, 24—25
    forces, stress, and equilibrium equation, 21—22
    stress-strain relation for elastic material and
        elastic constants, 22—24
    uniaxial deformation, 26—27
        of tissue with spherical inclusion, 28—30
    viscoelastic models
        generalized viscoelastic models, 38—39
        viscoelastic response to harmonic excitation,
            38
        viscoelastic tissue response, 34—38

H

Harmonic elastography, 156—157
Harmonic motion imaging (HMI), 102
Harmonic tracking, 95—96
    of shear waves, 116
Helmholtz decomposition, 31
High speed tracking techniques for acoustic
        radiation force impulse, 94—95
Hooke's law, 69, 71—73, 197

I

Imaging tissue stiffness
    compression elastography, 7
    vibration amplitude sonoelastography, 6—7
Impulse imaging, acoustic radiation force, 92—97
    displacement tracking, 93—94
    harmonic motion imaging (HMI), 102
    harmonic tracking, 95—96
    high speed tracking techniques for, 94—95
    motion filters, 96—97
    preliminary applications of, 97—98

vibroacoustography, 98—101
viscoelastic response imaging and model-based
    aproaches, 102—104
Incompressibility assumption, 169—170
Instrument-enhanced palpation, 4
Interpretation artifacts, 71—73
Inverse harmonic elastography problem, 157
Iterative inversion techniques, 156

K

Kelvin-Voigt model, 37, 41

L

Lagrangian finite strain tensor, 19
Lame constants, 23
Large deformations, tracking, 66—67
Lateral displacement estimation, 173—174
Lateral estimation methods, 169
Lateral strain estimation methods, 174
Linear elastic imaging
    nonlinear elastic imaging, 73—74
    poroelastic imaging, 74—75
Linearized strain tensor, 68
Liver, 138—141
Longitudinal ultrasound pulses, 9
Longitudinal wave, 32—33

M

Magnetic resonance elastography (MRE), 129
    acquisition
        generating and delivering mechanical waves,
            131—132
        imaging waves with, 132—134
    applications
        brain, 141—142
        liver, 138—141
        organs, 144—145
        tumors, 142—144
    artifacts and quality control, 145—147
    inversions
        outputs, 136—138
        and processing, 134—136
Maxwell model, 36—37
Mechanical imaging, 9
Microscopic tissue elasticity imaging, 10
Microvessel density (MVD), 74—75
Minimization and regularization, 174—175
Model-based elastography, 174
Modulus of elasticity, 70
Motion filters, 96—97
Motion tracking algorithms, 63—64
Motion tracking performance and error, 65—66

Multimodel direct inversion (MMDI), 134—135
Multitracking line (MTL), 114
  methods, 114

N

Navier-Cauchy equations, 25
Neo-Hookean hyperelastic model, 158—160
Nonlinear elastic imaging, 73—74

O

One-dimensional assumption, 70—71
One-dimensional deformation tracking and
  estimation
  altered point spread functions and synthetic
    aperture system, 171
  angular beam-steered data acquisition approach,
    171—173
  incompressibility assumption, 169—170
  perpendicular insonification using dual trans-
    ducers, 170
  weighted interpolation and recorrelation
    approach, 170
Optical coherence elastography
  computational inverse methods in, 205—208
  dynamic optical coherence elastographic
    methods, 200—203
  measuring displacement in, 194—196
  optical coherence tomography, 190—194
  probe-based optical coherence elastography,
    203—205
  quasi-static optical coherence elastographic
    methods, 196—200
Optical coherence microscopy (OCM), 197—200
Optical coherence tomography, 190—194
Optical elastography, 10, 214
Optical elastography on microscale, 185—188,
  214—215
  brillouin microscopy, 208—212
  matter of length scale, 188—190
  optical coherence elastography
    computational inverse methods in,
      205—208
    dynamic optical coherence elastographic
      methods, 200—203
    measuring displacement in, 194—196
    optical coherence tomography, 190—194
    probe-based optical coherence elastography,
      203—205
    quasi-static optical coherence elastographic
      methods, 196—200
  techniques, 212—214
Organs, 144—145

P

Partial differential equations (PDEs), 156
Perpendicular insonification using dual
  transducers, 170
Phase-sensitive detection, 195—196
Phase unwrapping strategies, 195—196
Phase velocity, 32
Photoacoustic absorption, 10
Photoacoustic elastography, 10
Photoacoustic imaging, 10
Photoacoustic tomography (PAT), 212—213
Physically linear stress-strain relation, 22—23
Plane harmonic shear wave, 33
Plane wave compounding, 175
Plane wave propagation, 32—33
Point spread function (PSF), 66—67
Poroelastic imaging, 74—75
Poroelastography, 177—178
Probe-based optical coherence elastography,
  203—205
Pure-tone frequency modulation process, 50—51

Q

Quadratic extrapolation filter, 97
Quasi-static approximation, 67—68
Quasi-static elastography, 61—62, 68, 156—158
  deformation application and measurement
    motion tracking algorithms, 63—64
    motion tracking performance and error,
      65—66
    tracking large deformations, 66—67
  interpretation of measured deformation
    elastic approximation, 69—70
    one-dimensional assumption, 70—71
    quasi-static approximation, 67—68
    strain, 68
    strain image interpretation, 71—73
  linear elastic imaging
    nonlinear elastic imaging, 73—74
    poroelastic imaging, 74—75
Quasi-static optical coherence elastographic
  methods, 196—200

R

Radial strain, 178
Radiation force, shear wave generation by, 104
Rayleigh-Lamb wave imaging, 200—202
Reconstructive elastography, 155
  nonlinearity, 160—161
  reconstruction approaches to elastography
    harmonic elastography, 156—157
    quasi-static elastography, 156—158

Reconstructive elastography (*Continued*)
    transient elastography, 157—158
   viscoelasticity, 158—161
Relaxation time, 37
Reverberant shear wave fields, 55

S
Shear deformation, 71—73
Shear elastic modulus, 23
Shear modulus, 70, 157
   by inversion of shear wave equation, 106—107
Shear strains, 20
   estimation methods, 169
Shear stress, 22
Shear wave dispersion ultrasound vibrometry,
    112—114
Shear wave elastography techniques, 30
   shear wave generation by radiation force, 104
   shear wave imaging techniques, 105—114
   shear wave tracking methods
     harmonic tracking of shear waves, 116
     multitracking lines, 114
     single tracking line, 114—116
     three-dimensional shear wave imaging, 116
     ultrafast tracking of shear waves, 116
Shear wave equation
   shear modulus by inversion of, 106—107
Shear wave generation by radiation force, 104
Shear wave imaging techniques, 105—114
Shear wave speed by time-to-peak displacement,
    107—111
Shear wave tracking methods
   harmonic tracking of shear waves, 116
   multitracking lines, 114
   single tracking line, 114—116
   three-dimensional shear wave imaging, 116
   ultrafast tracking of shear waves, 116
Signal-to-noise ratio (SNR), 195—196
Single tracking line, 114—116
Solenoid active drivers, 131—132
Sonoelasticity, 6
Speckle tracking techniques, 194—195
Speed of sound, 31
Spherical coordinate systems, 25—26
Standard linear body model, 37—38
Strain, 68
Strain creep, 35
Strain elastograms, 197
Strain image interpretation, 71—73
Strain relaxation, 35
Stress, 21
Subsample displacement estimation, 173

Synthetic aperture system, 171
Synthetic lateral phase information, 171

T
Three-dimensional shear wave imaging, 116
Time-domain systems, 193
Time-to-peak displacement, 107—111
Tissue elasticity determination
   Oestreicher and von Gierke, 3—4
   palpation, 3
   tissue motion studies
     cardiac pulsations/natural sources, 4—5
     externally controllable sources, 5—6
     instrument-enhanced palpation, 4
Tissue stimulation
   by cardiac pulsations/natural sources, 4—5
   by externally controllable sources, 5—6
Transient elastography, 7—8
   based on arrival time estimation, 157—158
Transverse wave, 32—33
Trigonometric identities, 50
Tumors, 142—144
Two-dimensional deformation tracking and
    estimation
   of angular beam-steered data acquisitions, 174
   coupled axial and lateral displacement estimation
    using, 174
   direct axial and lateral displacement estimation,
    173—174
   minimization and regularization, 174—175
   model-based elastography, 174
   plane wave compounding and directional
    beamforming, 175
   subsample displacement estimation, 173

U
Ultrafast coherent compounding, 116
Ultrafast tracking of shear waves, 116
Ultrasonography, 86—90
Ultrasound elastography
   commercial ultrasonographic systems, 167
   lateral and shear strain estimation methods, 169,
    175—178
   one-dimensional deformation tracking and
    estimation
     altered point spread functions and synthetic
      aperture system, 171
     angular beam-steered data acquisition
      approach, 171—173
     incompressibility assumption, 169—170
     perpendicular insonification using dual trans-
      ducers, 170

weighted interpolation and recorrelation approach, 170
two-dimensional deformation tracking and estimation
  of angular beam-steered data acquisitions, 174
  coupled axial and lateral displacement estimation using, 174
  direct axial and lateral displacement estimation, 173—174
  minimization and regularization, 174—175
  model-based elastography, 174
  plane wave compounding and directional beamforming, 175
  subsample displacement estimation, 173
Ultrasound-modulated optical tomography (UOT), 213—214
Uniaxial deformation, 26—30

V

Vibration-amplitude model, 48
Vibration-amplitude sonoelastography, 54
Vibration phase gradient sonoelastography, 50—52
Vibration sonoelastography, 45
  crawling waves, 52—54
  reverberant shear wave fields, 55
  theory, 46—50

vibration phase gradient sonoelastography, 50—52
Vibroacoustography, 98—101
  preliminary applications of, 101
Viscoelastic behavior, 35
Viscoelasticity, 34—35
"Viscoelastic material", 69
Viscoelastic models
  generalized viscoelastic models, 38—39
  viscoelastic response to harmonic excitation, 38
  viscoelastic tissue response, 34—38
Viscoelastic response
  to harmonic excitation, 38
  imaging and model-based aproaches, 102—104
Viscoelastic tissue response, 34—38
VisR imaging, 102—103

W

Wave images, 135—136
Weighted interpolation and recorrelation approach, 170

Y

Young's modulus, 70, 74, 130, 155, 197, 211—212
  of elasticity, 51—52

Printed in the United States
By Bookmasters